W9-CCK-581

Angiogenesis Protocols

METHODS IN MOLECULAR MEDICINE™

John M. Walker, *Series Editor*

60. **Interleukin Protocols,** edited by *Luke A. J. O'Neill and Andrew Bowie,* 2001
59. **Molecular Pathology of the Prions,** edited by *Harry F. Baker,* 2001
58. **Metastasis Research Protocols: *Volume 2, Cell Behavior In Vitro and In Vivo,*** edited by *Susan A. Brooks and Udo Schumacher,* 2001
57. **Metastasis Research Protocols: *Volume 1, Analysis of Cells and Tissues,*** edited by *Susan A. Brooks andUdo Schumacher,* 2001
56. **Human Airway Inflammation:** *Sampling Techniques and Analytical Protocols,* edited by *Duncan F. Rogers and Louise E. Donnelly,* 2001
55. **Hematologic Malignancies:** *Methods and Protocols,* edited by *Guy B. Faguet,* 2001
54. **Mycobacterium Tuberculosis Protocols,** edited by *Tanya Parish and Neil G. Stoker, 2001*
53. **Renal Cancer:** *Methods and Protocols,* edited by *Jack H. Mydlo,* 2001
52. **Atherosclerosis:** *Experimental Methods and Protocols***,** edited by *Angela F. Drew,* 2001
51. **Angiotensin Protocols**, edited by *Donna H. Wang,* 2001
50. **Colorectal Cancer: *Methods and Protocols***, edited by *Steven M. Powell,* 2001
49. **Molecular Pathology Protocols**, edited by *Anthony A. Killeen,* 2001
48. **Antibiotic Resistance Methods and Protocols,** edited by *Stephen H. Gillespie,* 2001
47. **Vision Research Protocols,** edited by *P. Elizabeth Rakoczy,* 2001
46. **Angiogenesis Protocols,** edited by *J. Clifford Murray*, 2001
45. **Hepatocellular Carcinoma: *Methods and Protocols*, edited by** *Nagy A. Habib*, 2000
44. **Asthma: *Mechanisms and Protocols*, edited by** *K. Fan Chung and Ian Adcock*, 2001
43. **Muscular Dystrophy: *Methods and Protocols,* edited by** *Katherine B. Bushby and Louise Anderson*, 2001
42. **Vaccine Adjuvants: *Preparation Methods and Research Protocols***, edited by *Derek T. O'Hagan*, 2000
41. **Celiac Disease: *Methods and Protocols***, edited by *Michael N. Marsh,* 2000
40. **Diagnostic and Therapeutic Antibodies,** edited by *Andrew J. T. George and Catherine E. Urch*, 2000
39. **Ovarian Cancer: *Methods and Protocols***, edited by *John M. S. Bartlett*, 2000
38. **Aging Methods and Protocols,** edited by *Yvonne A. Barnett and Christopher R. Barnett*, 2000
37. **Electrochemotherapy, Electrogenetherapy, and Transdermal Drug Delivery:** *Electrically Mediated Delivery of Molecules to Cells*, edited by *Mark J. Jaroszeski, Richard Heller, and Richard Gilbert,* 2000
36. **Septic Shock Methods and Protocols**, edited by *Thomas J. Evans*, 2000

METHODS IN MOLECULAR MEDICINE™

Angiogenesis Protocols

Edited by

J. Clifford Murray

University of Nottingham and City Hospital
Nottingham, UK

Humana Press Totowa, New Jersey

999 Riverview Drive, Suite 208
Totowa, New Jersey 07512

This publication is printed on acid-free paper. ∞

ANSI Z39.48-1984 (American Standards Institute) Permanence of Paper for Printed Library Materials.

Cover design by Patricia F. Cleary.

Cover art: Photomicrograph of postconfluent monolayer of HuMMEC. *See* full caption on p. 218.

Production Editor: John Morgan

For additional copies, pricing for bulk purchases, and/or information about other Humana titles, contact Humana at the above address or at any of the following numbers: Tel: 973-256-1699; Fax: 973-256-8341; E-mail: humana@humanapr.com, or visit our Website at www.humanapress.com.

Printed in the United States of America. 10 9 8 7 6 5 4 3 2 1

Library of Congress Cataloging-in-Publication Data

Angiogenesis protocols / edited by J. Clifford Murray.
p. ; cm. -- (Methods in molecular medicine ; 46)
Includes index.
ISBN 0-89603-698-7 (alk. paper)
1. Neovascularization--Laboratory manuals. 2. Blood-vessels--Growth--Molecular aspects--Laboratory manuals. I. Murray, J. Clifford. II. Series.
[DNLM: 1. Angiogenesis Factor--physiology. 2. Neovascularization, Physiologic. QU 107 A588 2001]
QP106.6.A554 2001
612.1'3--dc21

2001016852

Preface

In the last few years, we have been deluged with information on angiogenesis. Scientists and the public at large are exposed daily to this "new" science, not just in specialist journals and texts, but in the tabloid press, where popular articles refer to angiogenic therapies as magic bullets and miracle cures for cancer, arthritis, retinopathies, heart disease, and circulatory problems. Is there no ill this approach will not cure? The fact that so much time, effort, and resource have been and continue to be dedicated to this new science is clear testament to its importance. Yet many fundamental aspects of angiogenesis remain poorly understood, in particular cues that activate the process. This fact has to some extent been masked behind a surfeit of fine detail; we can't see the wood for the trees. Most studies of angiogenesis identify single links in a long chain of events. Furthermore, each study is itself hampered by the limitations of the biological end-point chosen. For instance, though endothelial proliferation may well be necessary for angiogenesis, it is not sufficient. Therefore, measuring endothelial proliferation in response to a novel growth factor, and on the basis of this observation, stating that the factor is "angiogenic," is unsound logic. It is important that researchers in this field, and perhaps more importantly those experimenting at its periphery, recognize the limitations of their chosen biological end-points.

The appearance of *Angiogenesis Protocols*, which brings together many currently used assays of angiogenesis, is therefore timely. We believe this work represents a new and important resource for scientists, which will prove valuable not only to those already involved in, and familiar with, the complex field of angiogenesis, but also to those for whom this is new territory. Some of those individuals may be a little intimidated at the prospect of setting up "meaningful" assays. This text should help to allay those fears, providing easy access to a variety of angiogenesis assays likely to suit laboratories with differing technical expertise and material and, most important, financial resources. We have intentionally included a range of in-vitro assays where low cost, ease-of-use, and reproducibility are paramount. However, we have also recognized the need for clearly documented access to "cutting-edge" in-vivo models, such as the dorsal window chamber, that demand high levels of surgical skill as well as relatively expensive, custom-made equipment.

We hope you find *Angiogenesis Protocols* instructive and useful.

J. Clifford Murray

Contents

Preface .... *v*
Contributors .... *ix*

I Special Review Article

1 Therapeutic Inhibition of Angiogenesis
Hua-Tang Zhang and Roy Bicknell 3

II Angiogenesis Protocols In Vivo

2 Microscopic Assessment of Angiogenesis in Tumors
Stephen B. Fox 29

3 In Vivo Matrigel Migration and Angiogenesis Assays
Katherine M. Malinda 47

4 Alginate Microbead Release Assay of Angiogenesis
Clifford Y. Ko, Vivek Dixit, William W. Shaw, and Gary Gitnick 53

5 Disc Angiogenesis Assay
Anthony C. Allison and Luis-F. Fajardo 59

6 Sponge Implant Model of Angiogenesis
Silvia P. Andrade 77

7 Hollow Fiber Assay for Tumor Angiogenesis
Roger M. Phillips and Michael C. Bibby 87

8 Dorsal Skinfold Chamber Preparation in Mice:
Studying Angiogenesis by Intravital Microscopy
Axel Sckell and Michael Leunig 95

9 Angiogenesis Assays Using Chick Chorioallantoic Membrane
David C. West, W. Douglas Thompson, Paula G. Sells, and Mike F. Burbridge 107

10 Corneal Assay for Angiogenesis
Marina Ziche 131

III Angiogenesis Protocols In Vitro

11 Collagen Gel Assay for Angiogenesis:
Induction of Endothelial Cell Sprouting
Ana M. Schor, Ian Ellis, and Seth L. Schor ***145***

12 Chemotaxis and Chemokinesis in 3D Macromolecular Matrices:
Relevance to Angiogenesis
Ana M. Schor, Ian Ellis, and Seth L. Schor ***163***

13 Rat Aortic Ring: *3D Model of Angiogenesis In Vitro*
Mike F. Burbridge and David C. West ***185***

14 In Vitro Matrigel Angiogenesis Assays
M. Lourdes Ponce ***205***

IV Associated Techniques

15 Microvessel Endothelial Cells from Human Adipose Tissues:
Isolation, Identification, and Culture
Peter W. Hewett ***213***

16 Transfection and Transduction of Primary Human Endothelial Cells
Stewart G. Martin ***227***

17 Vascular Smooth Muscle Cells:
Isolation, Culture, and Characterization
Richard C. M. Siow and Jeremy D. Pearson ***237***

18 Bovine Retinal Microvascular Pericytes: *Isolation, Propagation, and Identification*
Ramesh C. Nayak and Ira M. Herman ***247***

Index ***265***

Contributors

ANTHONY C. ALLISON • *SurroMed Inc, Palo Alto, CA*
SILVIA P. ANDRADE • *Department of Physiology and Biophysics, Institute of Biological Sciences, Federal University of Minas Gerais, Belo Horizonte-MG, Brazil*
ROY BICKNELL • *Molecular Angiogenesis Laboratory, Imperial Cancer Research Fund; Institute of Molecular Medicine, University of Oxford; John Radcliffe Hospital, Oxford, UK*
MICHAEL C. BIBBY • *Cancer Research Unit, University of Bradford, Bradford, UK*
MIKE F. BURBRIDGE • *Experimental Oncology Division, Institut de Rechèrches Servier, Suresnes, France*
VIVEK DIXIT • *Division of Digestive Diseases, UCLA School of Medicine, Los Angeles, CA*
IAN ELLIS • *Oral Diseases Group, Cell and Molecular Biology Unit, The Dental School, University of Dundee, Dundee, UK*
LUIS-F. FAJARDO • *Stanford Medical School and Veterans Affairs Medical Center, Palo Alto, CA*
STEPHEN B. FOX • *Department of Anatomical Pathology, Christchurch Hospital, Christchurch, New Zealand*
GARY GITNICK • *Division of Digestive Diseases, UCLA School of Medicine, Los Angeles, CA*
IRA M. HERMAN • *Departments of Physiology, Anatomy and Cell Biology, and Ophthalmology, Tufts University School of Medicine; Tufts Center for Vision Research, Boston, MA*
PETER W. HEWETT • *Laboratory of Molecular Oncology, University of Nottingham; Cancer Research Campaign, Department of Clinical Oncology, City Hospital, Nottingham, UK*
CLIFFORD Y. KO • *Department of Surgery, UCLA School of Medicine, Los Angeles CA*
MICHAEL LEUNIG • *Department of Orthopedic Surgery, Inselspital, University of Berne, Berne, Switzerland*
KATHERINE M. MALINDA • *Craniofacial Developmental Biology and Regeneration Branch, National Institute of Dental Research, National Institutes of Health, Bethesda, MD*

STEWART G. MARTIN • *Laboratory of Molecular Oncology, University of Nottingham; Cancer Research Campaign, Department of Clinical Oncology, City Hospital, Nottingham, UK*

J. CLIFFORD MURRAY • *Laboratory of Molecular Oncology, University of Nottingham; Cancer Research Campaign, Department of Clinical Oncology, City Hospital, Nottingham, UK*

RAMESH C. NAYAK • *New England Eye Center, New England Medical Center Hospitals; Department of Physiology and Department of Ophthalmology, Tufts University School of Medicine; Tufts Center for Vision Research, Boston, MA*

JEREMY D. PEARSON • *Centre for Cardiovascular Biology and Medicine, School of Biomedical Sciences, King's College London, London, UK*

ROGER M. PHILLIPS • *Cancer Research Unit, University of Bradford, Bradford, UK*

M. LOURDES PONCE • *Craniofacial Developmental Biology and Regeneration Branch, National Institute of Dental Research, National Institutes of Health, Bethesda MD*

ANA M. SCHOR • *Oral Diseases Group, Cell and Molecular Biology Unit, The Dental School, University of Dundee, Dundee, UK*

SETH L. SCHOR • *Oral Diseases Group, Cell and Molecular Biology Unit, The Dental School, University of Dundee, Dundee, UK*

AXEL SCKELL • *Department of Orthopaedic Surgery, University of Heidelberg, Heidelberg, Germany*

PAULA G. SELLS • *Alistair Reid Venom Research Unit, Liverpool School of Tropical Medicine, Liverpool, UK*

WILLIAM W. SHAW • *Division of Plastic and Reconstructive Surgery, UCLA School of Medicine, Los Angeles, CA*

RICHARD C. M. SIOW • *Division of Cardiovascular Medicine, School of Clinical Medicine, University of Cambridge, Cambridge, UK*

W. DOUGLAS THOMPSON • *Department of Pathology, Medical School, Aberdeen Royal Infirmary, Aberdeen, UK*

DAVID C. WEST • *Department of Immunology, Faculty of Medicine, University of Liverpool, Liverpool, UK*

HUA-TANG ZHANG • *Molecular Angiogenesis Laboratory, Imperial Cancer Research Fund; Institute of Molecular Medicine, University of Oxford; John Radcliffe Hospital, Oxford, UK*

MARINA ZICHE • *Institute of Pharmacological Sciences, University of Siena, Siena, Italy*

I

Special Review Article

1

Therapeutic Inhibition of Angiogenesis

Hua-Tang Zhang and Roy Bicknell

1. Introduction

The idea of antiangiogenesis as a therapeutic strategy has been around for several decades *(1)*. Vigorously pursued as a novel anticancer strategy (reviewed in *(2–6)*, it is now widely considered to be a promising approach to the treatment of a range of pathologies of which uncontrolled vascular proliferation is a component (*see* **Table 1**). To date, therapeutic benefit has been achieved with antiangiogenic therapy in the treatment of life-threatening infantile hemangioma, pulmonary hemangiomatosis, and in the treatment of some vascular tumors *(7,8)*.

Recent advances in the field of antiangiogenesis and vascular targeting include: (1) the discovery of two novel natural endogenous angiogenic inhibitors, called angiostatin and endostatin *(9,10)*, (2) the demonstration that in three mouse xenograft models, cycled endostatin therapy showed no evidence for acquired drug resistance but induced several wave-like regression–regrowth–regression cycles followed by indefinite tumor dormancy *(11)*, (3) destruction of tumor endothelium and induction of tumor infarction by selective clotting in tumor vessels *(12)*, and (4) in vitro and in vivo selection of peptides that bind to endothelium in an organ-specific and tumor-selective fashion *(13,14)*. The latter was followed by successful targeted delivery of peptide-doxorubicin conjugates to transplantable MDA-MB-435 and MDA-MB-231 breast carcinomas *(15)*. Stimulated by these recent developments, further interest and investment in these areas are expected from both the scientific community and the pharmaceutical industry. In this chapter we will review the current status of antiangiogenic therapy

From: *Methods in Molecular Medicine, Vol. 46: Angiogenesis Protocols*
Edited by: J. C. Murray © Humana Press Inc., Totowa, NJ

Table 1
Pathologies Likely to Benefit from Therapeutic Intervention in Angiogenesis

Excess angiogenesis	Insufficient angiogenesis
Arthritis	Angiology
Inflammatory,	Vascular malformation
Rheumatoid,	Hemifacial micromia
Kaposi's sarcoma	Bone fracture nonunion
Leukemia, lymphoma, and myeloma	Chronic wounds
Macular degeneration	Ischemia/infarction
Paget's disease	Cerebral
Psoriasis	Intestinal
Retinopathy (and its vascular complications)	Myocardial
Proliferative	Peripheral
Of prematurity	Pyrogenic granuloma
Solid carcinomas	Ulcer
Primary	Duodenal
Secondary (metastasis)	Gastric
Vascular tumors	
Hemangioma	
Capillary	
Juvenile (infantile)	
Hemangiomatosis	
Hemagioblastoma	
Other benign vascular proliferations	

2. Angiogenesis and Antiangiogenesis

2.1. An Angiogenic Switch Is a Critical Progression Point in a Range of Pathologies

Following the recognition that tumor growth is angiogenesis dependent, evidence has accumulated that angiogenesis is a component of many pathologies. In fact, as summarized by Pepper *(16)*, virtually every subspecialty in medicine and surgery deals in one way or another with physiological and pathological conditions associated with angiogenesis. It follows that control of angiogenesis offers hope in the treatment of many illnesses and that an effective antiangiogenic therapy may have wide-spectrum applicability. Pathologies that may benefit from either anti- or pro-angiogenic therapy have been collected in **Table 1** and readers are referred to other chapters for molecular details of physiological and pathological angiogenesis.

2.2. Vascular-Derived Tumors

Blood vessels are known to be the site of origin of venous malformations, hemartomas, and some neoplasms. The latter range from a benign, tumor-like growth of vessels (hemangiomas) through intermediate malignancies (hemangioendotheliomas) to fully malignant tumors (angiosarcomas). Although the pathogenesis of such vascular diseases is largely unknown, it has been hypothesized that uncontrolled vessel growth and a malformed vascular structure may arise as a result of deregulated angiogenesis. Thus, it was no surprise that the first successful clinical application of antiangiogenic therapy was in the treatment of life-threatening infantile hemangiomas with IFN-α *(7,8)*.

Kaposi's sarcoma (KS) is a multifocal tumor characterized histologically by a prominent microvasculature and the presence of bundles of spindle-shaped cells. The spindle cells are generally considered to be the tumor cells of the KS lesion and are thought by some to be of either endothelial or vascular smooth muscle origin, in that they express various markers of vascular endothelial cells and newly-formed vessels *(17)*. Evidence accumulated over the last few years has left little doubt that human herpesvirus (HHV8) is the infectious cause of KS and that infection by HHV8 induces an angiogenic switch *(18,19)*. Several angiogenic growth factors and cytokines such as basic fibroblast growth factor (bFGF), vascular endothelial growth factor (VEGF), platelet-derived growth factor (PDGF), and hepatocyte growth factor (HGF) have been shown to play a role in the initiation and progression of KS based on the stimulation of KS spindle cell proliferation and recruitment of microvessels into KS lesions *(17,18)*. Angiostatic compounds such as PNU 153429 *(20)* and antisense oligonucleotides directed against bFGF mRNA *(21)* have been shown to inhibit the growth of spindle cells and KS lesions in nude mice.

2.3. Leukemia Has Recently Been Reported to Be an Angiogenesis-Driven Malignancy

Recent studies have shown that angiogenesis may also play a role in leukemia and other "liquid" tumors. Angiogenesis was quantitated as an increase in overall microvessel density and in vascular hot spots in bone marrow biopsies from leukemic compared to those from healthy children *(22)*. Remarkably, clumps of leukemic cells were shown to be laden with microvessels, forming grape-like associates *(22)*. Elevated bFGF was also detected in the urine of leukemic patients *(22,23)*, and in another study VEGF and the VEGF receptors (KDR and Flt-1) were shown to be expressed in leukemic cells *(24)*. It has been hypothesized that leukemic cells are nurtured by angiogenesis before breaking off into the general circulation *(22)*. These observations broaden the potential applicability of antiangiogenic therapies in oncology.

2.4. Antiangiogenesis Is a Component of Conventional AntiCancer Therapies

Antiangiogenesis is now a recognized component of the antitumor activity of many conventional cytotoxics in clinical use. Several clinically effective chemotherapeutic compounds, such as bleomycin, cyclophosphamide, doxorubicin, and nitrosoureas, have angiostatic activity that may, at least in part, account for their efficacy as antitumor agents. There is now a growing interest in the exploration of cytostatic drugs as angiogenic inhibitors.

Further evidence for this "unexpected" antiangiogenic activity of conventional as well as experimental cancer therapies comes from a growing number of studies *(25)*. For example, both chemotherapy and radiotherapy have been shown to cause ultrastructural damage to endothelial cells, and radiotherapy often elicits vascular occlusion by thrombosis within capillaries. There is accumulating evidence that damage to blood vessels precedes or accompanies tumor regression after radiation therapy, hyperthermia, photodynamic therapy, or administration of a variety of biological response modifiers such as interferon, tumor necrosis factor, interleukins, or endotoxin (reviewed in **ref *4***).

3. The Biological Basis of Therapeutic Antiangiogenesis

Because endothelial cells (ECs) play a central role in the angiogenic process, they naturally constitute a primary target for therapeutic antiangiogenesis. The phenotypic properties of ECs at the site of angiogenesis serve as determinants for the selectivity in antiangiogenic and vascular targeting strategies.

3.1. Endothelial Cells Are Highly Accessible to Therapeutic Agents

ECs form a continuum throughout the circulation and all ECs are in direct contact with the blood. This means that they are readily accessed by therapeutic drugs delivered intravenously. When the vessels in diseased tissue are the target, the difficulty in getting a therapeutic drug to penetrate sufficiently into, for example, solid tumors is not a big problem.

3.2. Therapeutic Window I: Angiogenic ECs Proliferating and Vulnerable to Therapeutic Attack

The vascular network of capillaries constitutes a vast interface between the blood and the tissues. Indeed, when considered as a systematically disseminated organ, this apparently simple endothelial membranous lining of the inside of blood vessels is composed of more than 10^{12} ECs covering a surface area of more than 1000 m^2, and weighing almost 1 kg ***(26)***. It follows from this that specific targeting of tumor endothelium is highly desirable.

Fortunately, ECs undergoing angiogenesis exhibit phenotypic differences compared to mature endothelium. One such difference is in the proliferative index. Thymidine labelling has shown that as few as 0.01–1.0% (with a median of 0.2%) of ECs in healthy mouse tissues are cycling at any given time, whereas 1–32% (with a median of 9%) of ECs in xenografted tumors are proliferating ***(27–29)***. In human tumors, the ratio of proliferating to quiescent ECs is much lower (0.8–5.3%, mean 2.2%) but still significantly higher than that in the surrounding normal tissues ***(30)***. This difference in proliferative index offers a therapeutic window for antiangiogenesis, which has been termed antiproliferating endothelial therapy (APET) ***(31)***.

It has recently been demonstrated that the quiescent ECs lie in a protected state in that they express "protective genes," including genes that prevent the up-regulation of proinflammatory genes and certain genes that prevent the cells from undergoing apoptosis ***(32)***. Angiogenic ECs display other distinct differences to mature endothelium in, for example, basement membrane (BM) composition and permeability. These differences make the ECs in new vessels vulnerable to therapeutic attack compared to mature, quiescent, and protected ECs in the established vasculature.

3.3. Therapeutic Window II: Angiogenic EC Phenotypes Are Expressed in an Organ-Specific Fashion

Endothelial organ heterogeneity is well documented ***(33,34)***. Recently, the success of peptides selected from phage display libraries in vivo in the vascular targeting of anticancer drugs ***(15)*** has excited general interest in organ targeting via the endothelium. When targeted to the tumor vasculature, doxorubicin-peptide conjugates showed greater antitumor activity and were less toxic to other organs than a comparable dose of doxorubicin alone when administered to tumor-bearing mice ***(15)*** . It is foreseeable that novel organ-specific EC markers will be identified and thus make it feasible to target organ-specific ECs.

Numerous molecular changes occur in the EC on activation. These include, amongst others, elevated expression of (1) surface molecules, (2) intercellular receptor protein kinases, (3) integrins, (4) nonintegrin adhesion molecules, and (5) proteases. For example, the $\alpha v\beta 3$ and $\alpha v\beta 5$ integrins have been shown to mediate crucial interactions between ECs and the extracellular matrix (ECM) during angiogenesis ***(35–37)***. Blocking αv binding to vitronectin by either antibodies or peptide antagonists induced EC apoptosis and inhibited neovascularisation in tumors and retinopathy ***(38–40)***.

3.4. Destruction of the Tumor Vasculature Can Lead to the Death of Many Tumor Cells

A recent estimate showed that 1 g of mouse RIF-1 fibrosarcoma contained 1.1–1.9 × 10^7 ECs quantitated by angiotensin converting enzyme (ACE)-positive staining ***(41)***. This represents 4–7% of the total cell population. A gram of tumor has been estimated to contain ~10^8–10^9 tumor cells, and thus the ratio of EC to tumor cell is in the range of 1/10 to 1/100. This means that EC-targeted therapies have a built-in amplification in that a relatively small insult to the vascular endothelium induces vascular failure and extensive ischemic necrosis results. Experimental evidence supports this notion. For example, necrosis was induced in advanced-stage melanomas by inhibition of bFGF-induced angiogenesis, following injection of antisense bFGF or FGF receptor (FGFR)-1 cDNA ***(42)***. Selective occlusion of the tumor vasculature by induction of blood-clotting-induced tumor infarction ***(12)*** and eradication of large solid tumors ***(43)***. Selective clotting has been triggered by means of bispecific antibodies binding to tissue factor ***(12)*** and by specific antibodies conjugated with toxin ***(43)***. In both cases, ECs in the tumor vessels were induced to express MHC class II antigens in response to IFN-γ secreted by transfected neuroblastoma cells. The immunotoxins or bispecific antibodies were then targeted against the major histocompatibility complex (MHC) class II that was only present on the tumor vasculature.

3.5 Genetically Stable ECs Do Not Develop Drug Resistance

Drug resistance has not been observed in trials of antiangiogenic therapy, even after repeated dosing ***(3,11)***. A generally accepted explanation for this absence of resistance is that ECs, like other genetically stable diploid cells, are less prone to develop mechanisms for acquired drug resistance than the genetically unstable and thus heterogeneously populated tumors. Together with other above-mentioned advantages, the absence of drug resistance makes antiangiogenic therapy particularly attractive.

4. Strategies and Targets for Anti-Angiogenic Therapies

Agents that suppress or block angiogenesis often do so by interfering with an essential critical step in the process (**Fig. 1**). Six broad categories of therapeutic strategies for antiangiogenesis have been delineated, namely:

1. Inhibition of EC activation.
2. Inhibition of EC proliferation.
3. Inhibition of EC migration.
4. Disruption of the organization of a 3D structure (formation of capillary tubules and loops).

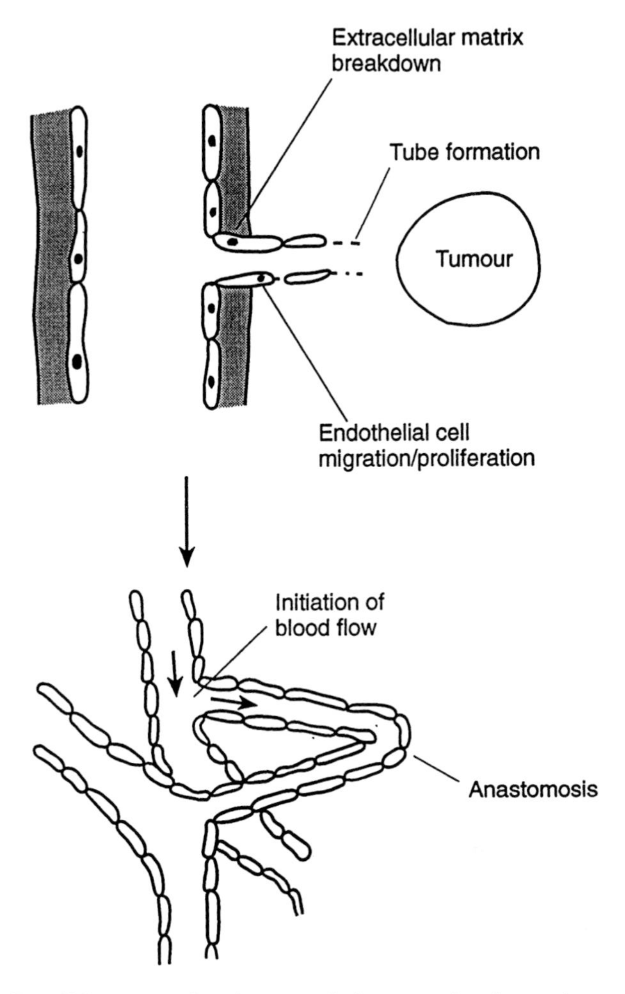

Fig. 1. Possible targets for therapeutic intervention in angiogenesis.

5. Interference with the biosynthesis and remodelling of BM and ECM (e.g., by inhibition of proteases secreted by the ECs).
6. Induction of EC apoptosis and direct killing of ECs.

There may be more than one way to block each individual step in angiogenesis and a single agent or therapy may affect more than one step. For instance, inhibition of EC activation can be achieved by removing angiogenic signals, by inhibiting cellular production of angiogenic factors or their release from ECM or a combination of both. Protease inhibitors may interfere with the release of angiogenic factors sequestered in the ECM, with EC migration and with ECM biosynthesis and remodelling. Further examples of antiangiogenic strategies are given in **Table 2**.

At the molecular level, an elaborate array of molecular events can be exploited as potential targets to block angiogenesis. These molecular targets include: (1) secreted proteins such as motility-stimulatory cytokines, (2) growth factors and their receptors, (3) adhesion molecules and receptors, (4) regulatory proteins and pathways inside the cell, and (5) ECM-degrading proteases and their inhibitors. As our knowledge of molecular angiogenesis grows, so too does the number of potential molecular targets increase. Nevertheless, all these strategic approaches and discrete molecular targets may not necessarily be equally effective. Further effort should be made to find out which particular cellular and molecular event(s) or combination(s) results in the most effective suppression of angiogenesis.

5. Inhibitors of Angiogenesis

5.1. The Search for Angiogenic Inhibitors

Substantial effort has been made in the past thirty years to identify, purify, and synthesize angiogenic inhibitory molecules. However, as Folkman recently summarized *(44)*, most of the known antiangiogenics were discovered by serendipity, for example, protamine, platelet factor-4, angiostatic steroids, and numerous microbial (fungal and bacterial) derivatives such as fumagillin. Initially, antiangiogenics were identified in extracts from naturally avascular tissues such as cartilage and the vitreous humour of the eye (reviewed in **refs.** ***45,46***. Relatively little research has been directed at exploiting avascular epithelium as a source of potentially useful angiogenesis inhibitors, although the idea was validated by the identification of protease inhibitory activity in bladder epithelium *(47)*.

Three rational strategies to identify natural, endogenous inhibitors have recently been described. Angiogenic inhibitory activity was found in a baby hamster kidney line BHK21/cl13 whose tumor suppressor gene activity was inactivated on transformation to anchorage independent growth *(48)*. Subsequently, the inhibitory activity was found to be a fragment of thrombospondin (TSP-1) *(49,50)* and its expression regulated by the wild type *p53* tumor suppressor gene *(51)*. A second strategy, logically following from that just described, was to analyze angiogenic inhibitory activity after transfection of a highly angiogenic tumor line with a tumor suppressor gene *(52)*. Evidence is accumulating that tumor suppressor genes play a role in the genetic switch that either turns off angiogenesis or keeps it in check **(Table 3)**.

Recently, a novel strategy has led to the discovery of two potent endogenous antiangiogenics, called angiostatin and endostatin *(10,53)*. The strategy

Table 2
Strategies of AntiAngiogenic Therapy

Strategic approaches	Examples
Inhibiting cellular production of angiogenic factors	IFN-α and -β down-regulate bFGF expression Antisense oligos to bFGF Ribozyme to PTN
Reducing release of angiogenic factors sequestrated in ECM	Neutralizing heparin or interfering with heparin-binding moieties: Suramin Platelet factor-4, small molecule heparin sequence mimics
Disrupting ligand/receptor interactions	Antagonists of angiogenic factors, neutralizing antibodies to VEGF and bFGF, VEGF-R chimeric protein
Disrupting intracellular signal transduction pathways	Protein kinase inhibitors: Eponemycin, epoxomicin, staurosporine, erbstatin Insensitization of ECs to angiogenic factors: Phorbol esters Inhibition of GTP-binding proteins: Octreotides
Inhibiting EC proliferation and/or migration	TGF-β, interferons, suramin, AGM 1470, paclitaxel, angiostatin, endostatin
Inhibiting cell adhesion and capillary morphogenesis	RGD-containing peptides, antibody, against integrins, E-selectin, vitronectin, and fibronectin receptors
Interfering with BM degradation, ECM biosynthesis, and remodelling	Metalloprotease inhibitors, cartilage-derived inhibitors, minocycline, angiostatic steroids, plasminogen activator inhibitor
Inhibiting recruitment of inflammatory cells that release angiogenic cytokines	Cyclosporin, cortisone acetate plus heparin
Interfering with co-factors of angiogenesis	D-penicillamine
Inducing other angiogenesis inhibitors	Induction of IL-12 by interferon-γ
Inducing EC apoptosis	Antibody against αvβ3 integrin, retinoids
Direct killing of proliferating ECs	CM101, bFGF-toxin conjugate, immunotoxins

Table 3
Tumor Suppressor Genes and Other Candidates for an Angiogenic Off-Switch

Tumor suppressor genes	Mechanisms
Much studied genes	
p53	Induction of TSP-1 and VEGF synthesis
RB	Induction of TSP-1 synthesis
VHL	Inhibition of VEGF synthesis
Some other reported candidates	
Locus Loh2 on mouse chromosome 16	Induction of an unknown angiogenic inhibitory activity
A gene in baby hamster kidney line BHK21/cl13	Induction of TSP-1 synthesis
A gene on human chromosome 10q	Induction of TSP-1 synthesis

involved the purification of angiogenic inhibitory activity from the serum and urine of experimental animals bearing transplantable tumors that suppressed the growth of micrometastases in distant organs. Peptide sequencing showed that angiostatin and endostatin are proteolytic fragments of plasminogen and collagen type XVIII, respectively. More importantly, the discovery of angiostatin and endostatin has revealed a novel regulatory mechanism of angiogenesis. That is, potent negative regulators of angiogenesis are "stored" encrypted within larger proteins that are devoid of antiangiogenic activity. Prototypes of cryptic antiangiogenics, a 29-kDa fragment of fibronectin ***(54)*** and a 16-kDa fragment of prolactin ***(55)***, had been described several years earlier but these findings were not immediately followed up. The repertoire of cryptic fragments that are antiangiogenics is expanding (**Table 4**; for further discussion *see* **ref.** ***56***).

Angiostatin has been shown to be released from plasminogen by the action of elastases such as metalloelastase (MME) released by tumor-infiltrating macrophages ***(57)*** and serine proteases from several human prostate carcinoma lines ***(58)***. Further efforts are needed to elucidate the mechanisms by which angiogenic inhibitory fragments are released from the parent molecules. Screening of cancer cell lines for circulating angiogenic inhibitory activity in, for example, a mouse corneal neovascularisation model ***(59)***, helps to show what cell types normally serve to release cryptic inhibitors and under what circumstances.

5.2. Overview of Angiogenesis Inhibitors

Several hundred antiangiogenics have now been identified, some of which are listed in **Tables 5–7**. Most antiangiogenics can be grouped into the follow-

Table 4
Physiological Angiogenic Inhibitors Encrypted Within Larger Molecules Lacking Activity

Precursor proteins	Cleavage products	Main action(s) on ECs (inhibition of)
Collagen type IV	MMP-2-generated fragments	Adhesion
Collagen type XVIII	20-kDa C-terminal fragment (endostatin)	Migration and proliferation
EGF	Synthetic fragment of amino acid 33–42	Inhibition of EGF- and laminin- dependent EC migration
Fibronectin	29-kDa fragment	Proliferation
HGF	Spliced form containing the first two domains	Unknown
High-molecular-weight kininogen	His-rich fragment	Adhesion
Osteopontin	Thrombin-generated fragment containing RGD	
Plasminogen	38-kDa fragment (angiostatin)	Migration and proliferation
Platelet factor-4[a]	N-terminal processed form	Proliferation
Prolactin	16-kDa N-terminal fragment	Migration and Proliferation
SPARC[b]	Fragment(s) containing KGHK	Stimulation of proliferation Enhancement of tube formation[b]

[a]The parent protein has angiogenesis inhibitory activity but the cleavage products are more potent.
[b]The parent protein is a reversible G1-phase inhibitor and exhibits, if any, a negative effect on angiogenesis while the proteolytic product stimulates angiogenesis in vivo.

ing broad categories: (1) physiologically occurring inhibitors (**Table 5**), many of which appear to play a role in suppressing unwanted angiogenesis and may also render healthy tissues resistant to tumor invasion, (2) natural products (**Table 6**); (3) microbial derivatives (**Table 6**), and (4) synthetic or semisynthetic molecules (**Table 7**). Among the wide variety of antiangiogenics, some are in clinical trials (**Table 8**) while others are drugs that are already in use in the clinic because of other activities, e.g., bleomycin, D-penicillinamine, methotrexate, paclitaxel, and (*inter alia*) doxorubicin.

There is some perception that endogenous angiogenic inhibitors have greater specificity and potency than other antiangiogenics. Indeed, it has been demon-

Table 5
Examples of Physiological, Endogenous Angiogenesis Inhibitors

Inhibitors	Main targets/mechanisms
Angiostatin	Inhibition of EC proliferation and migration
Endostatin	Inhibition of EC proliferation and migration
Glioblastoma-derived angiogenesis inhibitory factor	Inhibition of EC migration
Heparinases	Blockage of cytokine binding to ECs
Interferons	Reduction of release of antigenic factors, inhibition of EC proliferation and migration
Platelet factor-4	Inhibition of EC proliferation and tube formation, down-regulation of FGF receptors, inhibition of VEGF activity
Proliferin-related protein	Inhibition of EC proliferation and migration
Protease inhibitors TIMPs PAIs CDI (cartilage-derived inhibitors)	Inhibition of ECM-degrading proteases, inhibition of EC proliferation and migration, interference with vessel remodelling
Thrombospondin (TSP-1)	Modulation of EC adhesion, inhibition of collagenase, inhibition of EC proliferation and migration

strated that angiostatin not only retarded tumor growth in mice but also induced dramatic shrinkage of large tumors (e.g., 400 mg, about 2% of the body weight of the mouse) to a microscopic size ***(60)***. More dramatically, endostatin induced complete regression and the indefinite dormancy of three fast growing tumor types in mice ***(11)***. Similar results have not been seen with other antiangiogenics or even cytotoxic chemotherapeutic drugs. These unprecedented observations may well be a hopeful prelude to the use of endogenous antiangiogenics in the treatment of human primary tumors and metastases.

6. Delivery of Antiangiogenic Agents

Delivery is an important consideration. To this end, numerous options have been proposed and the scattered reports in the literature can be grouped according to the nature of the agents and vectors used.

1. Direct administration: Low-molecular-weight drugs such as suramin analogs ***(61)*** and AGM-1470 ***(62)*** can normally be delivered free of specially designed vehicles or targeting partners. Other antiangiogenic agents that have been directly administered are molecules that possess an intrinsic targeting property such as synthetic peptides, neutralizing antibodies, growth factor/receptor antagonists,

Table 6
Some Natural Product Antiangiogenics

(Anti-)microbial agents	Chemotherapeutic compounds	Extracts from natural materials
Fumagillin	Paclitaxel	Gensenoside (ginseng)
Erbstatin	Bleomycin	Isoliquiritin (Licorice root)
Staurosporine	Magnosalin	Genistein (soy beans)
Tecogalan		Squalamine (shark tissue)
CM101		

and integrin antagonists. Anti-VEGF ***(63)*** and anti-VEGFR2 ***(64)*** antibodies and integrin αvβ3 antagonists ***(39)*** have been shown to inhibit angiogenesis and thus suppress tumor growth and/or tumor invasion and metastasis.

2. Antibody-directed therapy: Specific antibodies can be conjugated with various therapeutic agents such as toxins (i.e., immunotoxins), radioisotopes, or enzymes. Genetically engineered antibody fragments such as Fab and single chain antibody variable domains (scFv) ***(65)*** and bispecific antibodies ***(12)*** also serve this purpose.
3. Chimeric protein therapy: Here, a fragment of an angiogenic factor or its receptor is engineered as the targeting partner and fused with a toxin or other therapeutic agent. The fragment of growth factor or receptor serves at the same time as an antagonist to interrupt ligand-receptor activating interactions. For example, chimeric toxins (aFGF-PE) composed of aFGF fused to a mutant form of *Pseudomonas* exotoxin) ***(66)*** and chimeric proteins of VEGF receptor linked to IgG heavy chain ***(67)*** have been shown to inhibit angiogenesis in vitro and in vivo, respectively.
4. Peptide guided therapy: As demonstrated by Arap et al. ***(15)***, anticancer drugs can be delivered as conjugates with peptides that home to tumor vessels. Specific antiangiogenic agents can also be delivered in this way and the pattern of peptides used for targeted delivery will soon expand to those that bind to ECs in an organ-specific manner.
5. Gene transfer-based therapy: Examples include ribozyme targeting of pleiotrophin (PTN) ***(68)***, antisense targeting of bFGF and bFGFR-1 ***(42)*** dominant-negative VEGF receptor (Flk-1) mutant ***(69)*** and vascular targeting of gene therapy using EC-specific promoters ***(70)***, which have been employed to achieve an antiangiogenic effect or vascular targeted therapy. Interested readers are referred to (**refs. *6,71–74***) for further discussion.
6. Cell transfer-based therapy: Proliferating ECs and putative EC progenitors have been shown to incorporate into sites of active angiogenesis ***(75–77)***, suggesting that ECs can be transduced and then seeded. This may be useful for the delivery of anti- (or pro-) angiogenic agents to sites of pathologies. Antiangiogenesis

Table 7
Miscellaneous Antiangiogenics

General categories/ Factors	Description	Main mechanisms
Cytokine/chemokines		
IL-12	75-kDa glycoprotein heterodimer with an indirect anti-tumor activity	Induction of IP-10
gro-β	Growth-related oncogene β protein	Inhibition of EC proliferation
IP-10	Interferon-inducible protein 10	Inhibition of EC growth and differentiation
(Anti-)microbial products and analogs		
Heparinases I and III (but not II)	Purified from *Flavobacterium* heparinum	Depletion of heparan sulfate receptors
FR-111142	Fumagillin analog	Inhibition of collagenase
WF16775 A2	Antibiotic from soil fungus *C. erysiphoides*	Cytostatic to ECs
Minocycline	Semisynthetic tetracycline	Inhibition of collagenase and EC proliferation
Herbimycin A	Benzoquinoid ansamycin	Inhibition of EC proliferation
Peptides and polypeptides		
RGD-containing peptides		Inhibition of EC migration and tube formation
CDPGYIGSY-NH2	Synthetic laminin peptide	Inhibition of EC migration
Vitreal inhibitor	5.7-kDa glycoprotein	Inhibition of metaloproteinases and EC proliferation
RNasin	51-kDa RNase inhibitor	Inhibition of angiogenin-induced angiogenesis
Hyaluronic acid	High molecular weight glycosaminoglycan	Inhibition of EC proliferation
Polysaccharides		
SP-PG	Sulphated polysaccharide-peptidoglycan complex	Inhibition of EC proliferation
DS-4152	As above	Inhibition of bFGF binding to ECs, inhibition of EC proliferation
SCM-chitin III	Sulphate polysaccharide	Inhibition of type IV collagenase and of EC migration

Table 7 *(continued)*

General categories/ Factors	Description	Main mechanisms
Polycationic and polyanionic compounds		
Suramin	1.4-kDa polyanionic compound	Binding to various growth factors, cell membrane, and intracellular enzymes
Protamine	4.3-kDa arginine-rich cationic protein	Blockage of heparin-binding and heparin-induced EC migration, inhibition of EC proliferation
Distamycin A analog	Sulphonic derivatives of diastmycin	Blocking bFGF and PDGF-b binding to their receptors
Vitamin derivatives		
Retinoids	Natural or synthetic	Induction of cell differentiation
Vitamin D3 analog	Vitamin D3 metabolites	Induction of cell differentiation
Steroids		
Glucocorticoids (cortisone)	Antiinflammatory, immunosuppressive agents	Inhibition of IL-1α production and action, inhibition of collageous protein synthesis
Methyl-prednisolone and tetrahydrocortisol	Steroids without glucocorticoid activity, with a distinct structure on the D ring	Dissolution of BM, inhibition of EC proliferation and migration
Synthetic		
Medroxyprogestrone facetate	Progestogen analog	Inhibition of collagenase and plasminogen activator
D-penicillamine	Antirheumatic agent	Chelation of copper, inhibition of EC proliferation, collagenase and collagen synthesis, and plasminogen activation
Gold salts	Antirheumatic agents	Inhibition of EC mitosis and production of angiogenic factors by macrophages

Table 8
Examples of Antiangiogenic Agents in Clinical Trials

Agents	Main mechanism	Disease
Batimastat/BB-2516	Inhibition of metalloproteinases	Women with malignant ascites
CAI	Blocking of Ca^{++} channels	Various types of cancers
CM101	Polyexotoxin produced by Group B *Streptococcus*	Various solid tumors, Kaposi's sarcoma
CT-2584	Regulation of tumor phospholipase-D (PLD) enzyme activity and phosphatidic acid production	Solid tumors
IL-12	Induction of the antiangiogenic protein IP-10	Renal carcinoma
Marimastat	Inhibition of metalloproteinases	Advanced solid tumors (pancreatic, lung, colorectal, ovarian, prostate, and brain cancer)
Platelet factor-4	Inhibition of EC proliferation and tube formation, down-regulation of FGF receptors, inhibition of VEGF activity	Kaposi's sarcoma, renal and colon cancer, melanoma
Suramin	Binding to various growth factors, cell membrane, and intracellular enzymes	Prostate cancer
Tecogalan sodium, D-gluco-D-glalactan sulfate/DS4152	Inhibition of EC growth and migration	Kaposi's sarcoma and replased cancer of breast, lung, head-neck and soft tissues
Thalidomide	?	Brain, breast, and prostate cancer
Razoxane	?	Kaposi's sarcoma, psoriasis, and Crohn's disease
TNP470/AGM1470	Inhibition of EC growth and migration	Breast, renal, cervical cancer, Kaposi's sarcoma, glioblastoma multiforme

might also be delivered via cells of other lineages. For example, lymphokine-activated killer (LAK) cells and, in particular, tumor-infiltrating lymphocytes (TILs) *(78)* may be useful in homing antiangiogenic agents to the tumor vasculature.

More sophisticated approaches use combinations of the above methods to achieve highly specific and efficient delivery. A good example of this is antibody-directed enzyme prodrug therapy (ADEPT) *(79)*, where antibody-enzyme conjugates are combined with direct administration of a prodrug.

7. Potential Drawbacks of Antiangiogenic Therapy

Despite the fact that antiangiogenic therapy is in advanced clinical trials, there remain reservations concerning its utility. There is the possibility that antiangiogenic strategies may interfere with normal physiological angiogenesis, impairing wound healing, stunting the growth of young children, or inducing amenorrhea in women. However, arguments in favor of antiangiogenic therapies are well documented *(80)*. Potential adverse effects due to inhibition of physiological angiogenesis may be less prevalent than once feared. First, toxicity of antiangiogenesis in young patients may well be negligible in that normal angiogenesis appears to level off after infancy and EC turnover is thought to be extremely low in young infants, children, and adolescent males. Secondly, for women receiving therapy, pregnancy may be undesirable and amenorrhea may actually be advantageous. It has even been speculated that antiangiogenesis might offer a novel means of contraception in healthy women. Thirdly, reservations regarding impairment of wound healing apply equally to conventional chemotherapy and radiotherapy. In situations where wound healing is required, antiangiogenic therapy could be interrupted or alternatively angiogenic stimulation applied locally.

8. Future Perspectives

The therapeutic inhibition of angiogenesis is clearly an exciting area of research with potential for improving the care of patients with numerous disorders. However, several issues require further evaluation.

8.1. Optimal Targets and Delivery Systems

Unanswered concerns about clinical importance are stimulating the search for EC-specific and organ-specific "angiogenic" markers and delivery strategies to obtain more precise selectivity and efficiency. Novel natural and recombinant polypeptides, nonpolypeptide pharmaceuticals, and gene therapy approaches are expected to be described for this purpose. Technically, the difficulty in the culture and transfection of primary ECs and the lack of useful EC lines has hampered research in this area.

8.2. Monitoring the Efficacy of Antiangiogenic Therapy

Quantitative assessment of angiogenesis may be made by obtaining the tumor microvessel density (MVD). High vascular density has been shown to be a significant prognostic factor for many cancers. It will be possible in future to monitor the levels in patients of both positive and negative regulators of angiogenesis and establish an "angiogenic profile" for angiogenesis-associated diseases. Until now, most, if not all, published studies have focused on positive regulators such as bFGF and VEGF in the serum, plasma, or urine of patients. Other angiogenic factors detectable in patients include angiogenin, EGF, aFGF, HGF, TGF-α, TGF-β, TNF-α and IL-2, -6, -7, -8, and -10. It is important in future to also assess the level of natural, endogenous inhibitors of angiogenesis. Angiogenic profiles will not only increase our understanding of angiogenesis-associated diseases and permit design of more selective antiangiogenesis therapies and tailoring of therapies to the needs of individuals, but it will also provide parameters for diagnosis, prognosis, monitoring response to therapies, and follow-up.

Equally important is the quantitation of endothelial specific molecules released into the circulation as a result of EC damage after antiangiogenic treatment. In this regard, soluble E-selectin has been shown to be a useful marker of therapeutic efficiency in cancer patients treated with CM101 (Group B *Streptococcus* polyexotoxin) ***(81)***.

Other EC products selectively expressed in different vascular environments include endothelial-specific molecule (ESM-1) ***(82)*** and ptx3, a new member of the pentraxin family ***(83)***. These provide hope that specific molecules may be targeted for therapy and assessed for diagnosis and follow-up.

8.3. Synergy of Antiangiogenesis With Other Therapies

A frequently raised objection is that therapeutic antiangiogenesis might compromise the success of other therapeutic treatments. For example, chemotherapeutic drugs may no longer be effectively delivered to tumors because of the reduced number of tumor vessels, and the therapeutic efficacy of radiotherapy could be reduced due to increased tumor hypoxia. Paradoxically, there is good evidence that antiangiogenesis potentiates chemotherapy and radiotherapy (reviewed in **refs. *3,80,84***). An explanation of these observations is related to the reduction in interstitial pressure and the so-called "re-oxygenation" phenomenon that follows the "first wave" of death of sensitive cancer cells (unpacking) ***(85)***. Clearly, no one would envisage that therapeutic inhibition of angiogenesis will make conventional therapies obsolete, rather that it will augment them. It will certainly extend treatment options for patients who are not promising candidates for conventional therapies.

9. Conclusions

Clearly, antiangiogenesis now ranks prominently amongst novel and promising strategies to fight cancer as well as other pathologies. Clinical application of antiangiogenic strategies now looks an increasingly realistic prospect, given the plethora of agents with antiangiogenic properties. In particular, a number of carefully designed clinical trials are underway and it is hoped that answers to some of the questions raised in this chapter will soon be forthcoming. It is envisaged that the benefits of antiangiogenic approaches will soon be realized in other clinical settings aside from oncology.

References

1. Folkman, J. (1971) Tumor angiogenesis: therapeutic implications. *N. Engl. J. Med.* **285,** 1182–1186.
2. Bicknell, R. and Harris, A. L. (1992) Anticancer strategies involving the vasculature: vascular targeting and the inhibition of angiogenesis. *Semin. Cancer Biol.* **3,** 399–407.
3. Folkman, J. (1995) Clinical applications of research on angiogenesis [see comments]. *N. Engl. J. Med.* **333,** 1757–1763.
4. Baillie, C. T., Winslet, M. C., and Bradley, N. J. (1995) Tumor vasculature—a potential therapeutic target. *Anti Cancer Drugs* **6,** 438–442.
5. Battegay, E. J. (1995) Angiogenesis: mechanistic insights, neovascular diseases, and therapeutic prospects. *J. Mol. Med.* **73,** 333–346.
6. Bicknell, R., Lewis, C. E., and Ferrara, N., eds. (1997) *Tumor angiognesis.* Oxford University Press, New York, NY.
7. White, C. W., Sondheimer, H. M., Crouch, E. C., Wilson, H., and Fan, L. L. (1989) Treatment of pulmonary hemangiomatosis with recombinant interferon alfa-2a. *N. Engl. J. Med.* **320,** 1197–1200.
8. Ezekowitz, R. A., Mulliken, J. B., and Folkman, J. (1992) Interferon alfa-2a therapy for life-threatening hemangiomas of infancy [see comments] [published errata appear in *N. Engl. J. Med.* (1994) **330,** 300 and (1995) **333,** 595–596. *N. Engl. J. Med.* **326,** 1456–1463.
9. O'Reilly, M. S., Holmgren, L., Shing, Y., Chen, C., Rosenthal, R. A., Moses, M., et al. (1994) Angiostatin: a novel angiogenesis inhibitor that mediates the suppression of metastases by a Lewis lung carcinoma [see comments]. *Cell* **79,** 315–328.
10. O'Reilly, M. S., Boehm, T., Shing, Y., Fukai, N., Vasios, G., Lane, W. S., et al. (1997) Endostatin: an endogenous inhibitor of angiogenesis and tumor growth. *Cell* **88,** 277–285.
11. Boehm, T., Folkman, J., Browder, T., and O'Reilly, M. (1997) Antiangiogenic therapy of experimental cancer does not induce acquired drug resistance. *Nature* **390,** 404–407.

12. Huang, X., Molema, G., King, S., Watkins, L., Edgington, T. S., and Thorpe, P. E. (1997) Tumor infarction in mice by antibody-directed targeting of tissue factor to tumor vasculature [see comments]. *Science* **275,** 547–550.
13. Barry, M. A., Dower, W. J., and Johnston, S. A. (1996) Toward cell-targeting gene therapy vectors: selection of cell-binding peptides from random peptide-presenting phage libraries. *Nat. Med.* **2,** 299–305.
14. Pasqualini, R. and Ruoslahti, E. (1996) Organ targeting in vivo using phage display peptide libraries. *Nature* **380,** 364–366.
15. Arap, W., Pasqualini, R., and Rouslati, E. (1998) Cancer treatment by targeted drug delivery to tumor vasculature in a mouse model. *Science* **279,** 377–380.
16. Pepper, M. S. (1997) Manipulating angiogenesis. From basic science to the bedside. *Arterioscler. Thromb. Vasc. Biol.* **17,** 605–619.
17. Sturzl, M., Brandstetter, H., and Roth, W. (1992) Kaposi's sarcoma: a review of gene expression and ultrastructure of KS spindle cells in vivo. *AIDS Res Human Retroviruses* **8,** 1753–1763.
18. Davis, M., Sturzl, M., Blasig, C., Guo, H., Reitz, M., Opalenik, S., and Browning, P. (1997) Expression of human Herpesvirus 8-encoded cyclin D in Kaposi's sarcoma spindle cells. *J. Natl. Canc. Inst.* **89,** 1868–1874.
19. Bais, C., Santomasso, B., Coso, O., Arvanitakis, L., Raaka, E., Gutkind, J. S., et al. (1998) G-protein-coupled receptor of Kaposi's sarcoma-associated herpesvirus is a viral oncogene and angiogenesis activator. *Nature* **391,** 86–89.
20. Sola, F., Gualandris, A., Belleri, M., Giuliani, R., Coltrini, D., Bastaki, M., et al. (1997) Endothelial cells overexpressing basic fbroblast growth factor (FGF-2) induce vascular tumors in immunodeficient mice. *Angiogenesis* **1,** 102–116.
21. Ensoli, B., Markham, P., Kao, V., Barillari, G., Fiorelli, V., Gendelman, R., et al. (1994) Block of AIDS-Kaposi's sarcoma (KS) cell growth, angiogenesis, and lesion formation in nude mice by antisense oligonucleotide targeting basic fibroblast growth factor. A novel strategy for the therapy of KS. *J. Clin. Invest.* **94,** 1736–1746.
22. Perez, A. A., Sallan, S. E., Tedrow, U., Connors, S., Allred, E., and Folkman, J. (1997) Spectrum of tumor angiogenesis in the bone marrow of children with acute lymphoblastic leukemia. *Am J. Pathol.* **150,** 815–821.
23. Nguyen, M., Watanabe, H., Budson, A. E., Richie, J. P., Hayes, D. F., and Folkman, J. (1994) Elevated levels of an angiogenic peptide, basic fibroblast growth factor, in the urine of patients with a wide spectrum of cancers [see comments]. *J. Natl. Cancer Inst.* **86,** 356–361.
24. Fiedler, W., Graeven, U., Ergun, S., Verago, S., Kilic, N., Stockschlader, M., and Hossfeld, D. K. (1997) Vascular endothelial growth factor, a possible paracrine growth factor in human acute myeloid leukemia. *Blood* **89,** 1870–1875.
25. Watts, M. E., Woodcock, M., Arnold, S., and Chaplin, D. J. (1997) Effects of novel and conventional anti-cancer agents on human endothelial permeability: influence of tumor secreted factors. *Anticancer Res.* **17,** 71–75.
26. Jaffe, E. A. (1987) Cell biology of endothelial cells. *Hum. Pathol.* **18,** 234–239.

27. Denekamp, J. and Hobson, B. (1982) Endothelial-cell proliferation in experimental tumors. *Br. J. Cancer* **46,** 711–720.
28. Denekamp, J. (1984) Vascular endothelium as the vulnerable element in tumors. *Acta Radiol. Oncol.* **23,** 217–225.
29. Denekamp, J. (1986) Cell kinetics and radiation biology. *Int. J. Radiat. Biol.* **49,** 357–380.
30. Fox, S. B., Gatter, K. C., Bicknell, R., Going, J. J., Stanton, P., Cooke, T. G., and Harris, A. L. (1993) Relationship of endothelial cell proliferation to tumor vascularity in human breast cancer. *Cancer Res.* **53,** 4161–4163.
31. Denekamp, J. (1982) Endothelial cell proliferation as a novel approach to targeting tumor therapy. *Br. J. Cancer* **45,** 136–139.
32. Bach, F. H., Ferran, C., Hechenleitner, P., Mark, W., Koyamada, N., Miyatake, T., et al. (1997) Accommodation of vascularized xenografts: expression of "protective genes" by donor endothelial cells in a host Th2 cytokine environment. *Nat. Med.* **3,** 196–204.
33. McCarthy, S. A., Kuzu, I., Gatter, K. C., and Bicknell, R. (1991) Heterogeneity of the endothelial cell and its role in organ preference of tumor metastasis. *Trends Pharmacol. Sci.* **12,** 462–467.
34. Bicknell, ed. (1996) *Endothelial cell culture*. Cambridge University Press, Cambridge, UK.
35. Brooks, P. C., Clark, R. A., and Cheresh, D. A. (1994) Requirement of vascular integrin alpha v beta 3 for angiogenesis. *Science* **264,** 569–571.
36. Friedlander, M., Theesfeld, C. L., Sugita, M., Fruttiger, M., Thomas, M. A., Chang, S., and Cheresh, D. A. (1996) Involvement of integrins alpha v beta 3 and alpha v beta 5 in ocular neovascular diseases. *Proc. Natl. Acad. Sci. (US)* **93,** 9764–9769.
37. Brooks, P. C., Stromblad, S., Klemke, R., Visscher, D., Sarkar, F. H., and Cheresh, D. A. (1995) Antiintegrin alpha v beta 3 blocks human breast cancer growth and angiogenesis in human skin [see comments]. *J. Clin. Invest.* **96,** 1815–1822.
38. Friedlander, M., Brooks, P. C., Shaffer, R. W., Kincaid, C. M., Varner, J. A., and Cheresh, D. A. (1995) Definition of two angiogenic pathways by distinct alpha v integrins. *Science* **270,** 1500–1502.
39. Brooks, P. C., Montgomery, A. M., Rosenfeld, M., Reisfeld, R. A., Hu, T., Klier, G., and Cheresh, D. A. (1994) Integrin alpha v beta 3 antagonists promote tumor regression by inducing apoptosis of angiogenic blood vessels. *Cell* **79,** 1157–1164.
40. Hammes, H. P., Brownlee, M., Jonczyk, A., Sutter, A., and Preissner, K. T. (1996) Subcutaneous injection of a cyclic peptide antagonist of vitronectin receptor-type integrins inhibits retinal neovascularization. *Nat. Med.* **2,** 529–533.
41. Modzelewski, R. A., Davies, P., Watkins, S. C., Auerbach, R., Chang, M.-J., and Johnson, C. S. (1994) Isolation and identification of fresh tumor-derived endothelial cells from a murine RIF-1 fibrosarcoma. *Cancer Res.* **54,** 336–339.
42. Wang, Y. and Becker, D. (1997) Antisense targeting of basic fibroblast growth factor and fibroblast growth factor receptor-1 in human melanomas blocks intratumoral angiogenesis and tumor growth. *Nat. Med.* **3,** 887–893.

43. Burrows, F. J. and Thorpe, P. E. (1993) Eradication of large solid tumors in mice with an immunotoxin directed against tumor vasculature. *Proc. Natl. Acad. Sci. (US)* **90,** 8996–9000.
44. Folkman, J. (1995) Angiogenesis inhibitors generated by tumors. *Mol. Med.* **1,** 120–122.
45. Klagsbrun, M. and D'Amore, P. A. (1991) Regulators of angiogensis. *Annu. Rev. Physiol.* **53,** 217–239.
46. Auerbach, W. and Auerbach, R. (1994) Angiogenesis inhibition: a review. *Pharmacol. & Ther.* **63,** 265–311.
47. Waxler, B., Kuettner, K. E., and Pauli, B. U. (1982) The resistance of epithelia to vascularization: proteinase and endothelial cell growth inhibitory activities in urinary bladder epithelium. *Tissue Cell* **14,** 657–667.
48. Rastinejad, F., Polverini, P. J., and Bouck, N. P. (1989) Regulation of the activity of a new inhibitor of angiogenesis by a cancer suppressor gene. *Cell* **56,** 345–355.
49. Good, D. J., Polverini, P. J., Rastinejad, F., Le, B. M., Lemons, R. S., Frazier, W. A., and Bouck, N. P. (1990) A tumor suppressor-dependent inhibitor of angiogenesis is immunologically and functionally indistinguishable from a fragment of thrombospondin. *Proc. Natl. Acad. Sci. (US)* **87,** 6624–6628.
50. Tolsma, S. S., Volpert, O. V., Good, D. J., Frazier, W. A., Polverini, P. J., and Bouck, N. (1993) Peptides derived from two separate domains of the matrix protein thrombospondin-1 have anti-angiogenic activity. *J. Cell Biol.* **122,** 497–511.
51. Dameron, K. M., Volpert, O. V., Tainsky, M. A., and Bouck, N. (1994) Control of angiogenesis in fibroblasts by p53 regulation of thrombospondin-1. *Science* **265,** 1582–1584.
52. Van Mier, E. G., Polverini, P. J., Chazin, V. R., Su Huang, H., de Tribolet, N., and Cavenee, W. K. (1994) Release of an inhibitor of angiogenesis upon induction of wild type p53 expression in glioblastoma cells. *Nat. Gen.* **8,** 171–176.
53. O'Reilly, M. S., Holmgren, L., Shing, Y., Chen, C., Rosenthal, R. A., Cao, Y., et al. (1994) Angiostatin: a circulating endothelial cell inhibitor that suppresses angiogenesis and tumor growth. *Cold Spr. Harb. Symp. Quant. Biol.* **59,** 471–482.
54. Homandberg, G. A., Williams, J. E., Grant, D., Schumacher, B., and Eisenstein, R. (1985) Heparin-binding fragments of fibronectin are potent inhibitors of endothelial cell growth. *Am. J. Pathol.* **120,** 327–332.
55. Ferrara, N., Clapp, C., and Weiner, R. (1991) The 16K fragment of prolactin specifically inhibits basal or fibroblast growth factor stimulated growth of capillary endothelial cells. *Endocrinology* **129,** 896–900.
56. Hanahan, D. and Folkman, J. (1996) Patterns and emerging mechanisms of the angiogenic switch during tumorigenesis. *Cell* **86,** 353–364.
57. Dong, Z., Kumar, R., Yang, X., and Fidler, I. J. (1997) Macrophage-derived metalloelastase is responsible for the generation of angiostatin in Lewis lung carcinoma. *Cell* **88,** 801–810.
58. Gately, S., Twardowski, P., Stack, M. S., Patrick, M., Boggio, L., Cundiff, D. L., et al. (1996) Human prostate carcinoma cells express enzymatic activity that con-

verts human plasminogen to the angiogenesis inhibitor, angiostatin. *Cancer Res.* **56,** 4887–4890.
59. Chen, C., Parangi, S., Tolentino, M. J., and Folkman, J. (1995) A strategy to discover circulating angiogenesis inhibitors generated by human tumors. *Cancer Res.* **55,** 4230–4233.
60. O'Reilly, M. S., Holmgren, L., Chen, C., and Folkman, J. (1996) Angiostatin induces and sustains dormancy of human primary tumors in mice. *Nat. Med.* **2,** 689–692.
61. Braddock, P. S., Hu, D. E., Fan, T. P., Stratford, I. J., Harris, A. L., and Bicknell, R. (1994) A structure-activity analysis of antagonism of the growth factor and angiogenic activity of basic fibroblast growth factor by suramin and related polyanions. *Br. J. Cancer* **69,** 890–898.
62. Ingber, D., Fujita, T., Kishimoto, S., Sudo, K., Kanamaru, T., Brem, H., and Folkman, J. (1990) Synthetic analogs of fumagillin that inhibit angiogenesis and suppress tumor growth. *Nature* **348,** 555–557.
63. Kim, K. J., Li, B., Winer, J., Armanini, M., Gillett, N., Phillips, H. S., and Ferrara, N. (1993) Inhibition of vascular endothelial growth factor-induced angiogenesis suppresses tumor growth in vivo. *Nature* **362,** 841–844.
64. Skobe, M., Rockwell, P., Goldstein, N., Vosseler, S., and Fusenig, N. E. (1997) Halting angiogenesis suppresses carcinoma cell invasion. *Nat. Med.* **3,** 1222–1227.
65. Lorimer, I. A., Keppler, H. A., Beers, R. A., Pegram, C. N., Bigner, D. D., and Pastan, I. (1996) Recombinant immunotoxins specific for a mutant epidermal growth factor receptor: targeting with a single chain antibody variable domain isolated by phage display. *Proc. Natl. Acad. Sci. (US)* **93,** 14,815–14,820.
66. Merwin, J. R., Lynch, M. J., Madri, J. A., Pastan, I., and Siegall, C. B. (1992) Acidic fibroblast growth factor-Pseudomonas exotoxin chimeric protein elicits antiangiogenic effects on endothelial cells. *Cancer Res.* **52,** 4995–5001.
67. Aiello, L. P., Pierce, E. A., Foley, E. D., Takagi, H., Chen, H., Riddle, L., et al. (1995) Suppression of retinal neovascularization in vivo by inhibition of vascular endothelial growth factor (VEGF) using soluble VEGF-receptor chimeric proteins. *Proc. Natl. Acad. Sci. (US)* **92,** 10,457–10,461.
68. Czubayko, F., Schulte, A. M., Berchem, G. J., and Wellstein, A. (1996) Melanoma angiogenesis and metastasis modulated by ribozyme targeting of the secreted growth factor pleiotrophin. *Proc. Natl. Acad. Sci. (US)* **93,** 14,753–14,758.
69. Millauer, B., Shawver, L. K., Plate, K. H., Risau, W., and Ullrich, A. (1994) Glioblastoma growth inhibited in vivo by a dominant-negative Flk-1 mutant. *Nature* **367,** 576–579.
70. Jagger, R., Chan, H. Y., Harris, A. l., and Bicknell, R. (1997) Endothelial cell-specific expression of tumor necrosis factor-alpha from the KDR or E-selectin promoters following retroviral delivery. *Hum. Gene Ther.* **8,** 2239–2247.
71. Bicknell, R. (1994) Vascular targeting and the inhibition of angiogenesis. *Ann. Oncol.* **4,** 45–50.
72. Fan, T. P., Jaggar, R., and Bicknell, R. (1995) Controlling the vasculature: angiogenesis, anti-angiogenesis and vascular targeting of gene therapy. *Trends Pharmacol. Sci.* **16,** 57–66.

73. Harris, A. L., Fox, S., Bicknell, R., Leek, R., Relf, M., LeJeune, S., and Kaklamanis, L. (1994) Gene therapy through signal transduction pathways and angiogenic growth factors as therapeutic targets in breast cancer. *Cancer* **74(Supp.3),** 1021–1025.
74. Kong, H. and R. G., C. (1998) Gene therapy strategies for tumor antiangiogenesis. *J. Natl. Cancer Inst.* **90,** 273–286.
75. Lal, B., Indurti, R. R., Couraud, P. O., Goldstein, G. W., and Laterra, J. (1994) Endothelial cell implantation and survival within experimental gliomas. *Proc. Natl. Acad. Sci. (US)* **91,** 9695–9699.
76. Ojeifo, J. O., Forough, R., Paik, S., Maciag, T., and Zwiebel, J. A. (1995) Angiogenesis-directed implantation of genetically modified endothelial cells in mice. *Cancer Res.* **55,** 2240–2244.
77. Asahara, T., Murohara, T., Sullivan, A., Silver, M., van der Zee, R., Li, T., et al. (1997) Isolation of putative progenitor endothelial cells for angiogenesis. *Science* **275,** 964–967.
78. Adams, D. H., Yannelli, J. R., Newman, W., Lawley, T., Ades, E., Rosenberg, S. A., and Shaw, S. (1997) Adhesion of tumor-infiltrating lymphocytes to endothelium: a phenotypic and functional analysis. *Br. J. Cancer* **75,** 1421–1431.
79. Deonarain, M. P. and Epenetos, A. A. (1994) Targeting enzymes for cancer therapy: old enzymes in new roles. *Br. J. Cancer* **70,** 786–794.
80. Schweigerer, L. (1995) Antiangiogenesis as a novel therapeutic concept in pediatric oncology. *J. Mol. Med.* **73,** 497–508.
81. Wamil, B. D., Thurman, B. D., and Sundell, H. W. (1997) Soluble E-selectin in cancer patients as a marker of the therapeutic efficiacy of CM101. *J. Cancer Res. Clin. Oncol.* **123,** 173–179.
82. Lassalle, P., Molet, S., Janin, A., Heyden, J. V., Tavernier, J., Fiers, W., et al. (1996) ESM-1 is a novel human endothelial cell-specific molecule expressed in lung and regulated by cytokines. *J. Biol. Chem.* **271,** 20,458–20,464.
83. Introna, M., Alles, V. V., Castellano, M., Picardi, G., De Gioia, L., Bottazzai, B., et al. (1996) Cloning of mouse ptx3, a new member of the pentraxin gene family expressed at extrahepatic sites. *Blood* **87,** 1862–1872.
84. Denekamp, J. (1993) Review article: angiogenesis, neovascular proliferation and vascular pathophysiology as targets for cancer therapy. *Br. J. Radiol.* **66,** 181–196.
85. Folkman, J. (1995) Angiogenesis in cancer, vascular, rheumatoid and other disease. *Nat. Med.* **1,** 27–31.

II

Angiogenesis Protocols In Vivo

2

Microscopic Assessment of Angiogenesis in Tumors

Stephen B. Fox

1. Introduction

Although it has been recognized for many centuries that neoplastic tissue is more vascular than its normal counterpart, it is only since Folkmans' hypothesis on antiangiogenesis *(1)* that a more quantitative method for measuring angiogenesis in tissue sections has been pursued. Folkman and colleagues recognized that quantitation of the tumor vasculature might play an important a role in predicting tumor behavior and patient management. They therefore developed a **m**icroscopic **a**ngiogenesis **g**rading **s**ystem, designated the "MAGS" score, calculated by measuring vessel number, endothelial cell hyperplasia, and cytology in tinctorially stained tissue sections *(2)*. It was hoped that this would be an objective method for quantifying tumor angiogenesis, one that would yield important information on the relationship to other clinicopathological tumor characteristics and help in the testing of antiangiogenic therapies. However, although it was possible to classify tumors into endothelial "poor" or "rich," the technical limitations of sample selection, inter- and intra-observer variation, and conceptual biological problems were such that the technique could not be easily applied. Interest in grading tumor angiogenesis was rekindled in the 1980s with the advent of nonspecific endothelial markers *(3–5)*, but only in the last five to ten years, with the advent of more specific endothelial markers, have quantitation studies on tissues have been performed.

Most studies have employed a method based on that developed by Weidner et al. *(6)*, in which blood vessels are immunohistochemically highlighted and the number of microvessels quantified in the most vascular areas (so called "hot spots") of the tumor. These studies have confirmed that tumors are more vascular than their normal tissue counterparts and have shown that microvessel

From: *Methods in Molecular Medicine, Vol. 46: Angiogenesis Protocols*
Edited by: J. C. Murray © Humana Press Inc., Totowa, NJ

density is a powerful prognostic tool in many human tumor types [reviewed in *(7)*]. Nevertheless, due to limitations in capillary identification and quantitation, not all investigators have been able to confirm a relationship between tumor vascularity and prognosis ***(8–15)***. This chapter will briefly discuss the considerations in quantifying tumor angiogenesis in tissue sections, give the current optimal protocol for assessment, and then outline other candidate techniques with potential for the future.

Since endothelium is highly heterogeneous ***(16)***, the choice of antibody profoundly influences the number of microvessels available for assessment. Many such as those directed against vimentin ***(17)***, lectin ***(4,18)***, alkaline phosphatase ***(3)***, and type IV collagen ***(11,19,20)*** suffer from low specificity and are present on many nonendothelial elements. Others including antibodies to factor VIII related antigen, the marker used in most studies ***(6,8,9,21–25)***, which identifies only a proportion of capillaries and also detects lymphatic endothelium. The most specific and sensitive endothelial marker currently available is CD31, which is present on most capillaries and is a reliable epitope for immunostaining in routinely handled formalin-fixed paraffin-embedded tissues ***(26)***; a good alternative antigen is CD34, although this antigen is also expressed by some stromal cells ***(25)***.

Although most quantitative vessel studies have been performed on tumors, the technique can also be applied to normal tissues. The hot spot method might not be as relevant as vascularity will vary between particular tissue compartments (e.g., acini and stroma in breast, mucosa and submucosa in intestine, etc.). Nevertheless, in tumors, once the vasculature has been immunohistochemically highlighted, it is scanned at low magnification (×40–100) to identify angiogenic hot spots ***(6,22)***. The number of vessels is then quantified at high magnification (×200–400) in these regions. These areas of high vascularity are chosen on the likelihood that they will be biologically important. Tumors naturally have a limited number of hot spots and they would be diluted if too many were counted. Thus, although the number of hot spots assessed varies from one to five ***(6,8–10,22,27,28)***, most studies have examined three from a single representative tissue block (*see* **Note 1**). Nevertheless, both the magnification used and its corresponding tumor field area will determine the vessel number derived from each hot spot. A high magnification, which will identify more microvessels by virtue of increased resolution ***(31)***, used over a too small an area will always give a high vessel index whereas a low magnification over too large an area will dilute out the hot spot. It is thus recommended that three regions are examined using a microscope magnification of between ×200 and ×400 (*see* **Note 2**), which corresponds to field areas of 0.74–0.15 mm^2 depending on the microscope type ***(32,33)***.

Although less subjective than identifying angiogenic hot spots ***(33)***, the process of counting vessels has also resulted in significant variation in published series. This has been recently emphasized in the study of Axelsson et al. ***(15)*** where the authors, after an initial training period with Weidner, who defined the criteria as to what constituted individual microvessels (*see* **Subheading 3.**), did not observe a correlation between microvessel density and patient survival. Even experienced observers occasionally disagree as to what constitutes a microvessel. To overcome these problems, after selection of each hot spot, a 25-dot Chalkley microscope eyepiece graticule ***(34)*** has been used to quantify tumor angiogenesis (*see* **Subheading 3.**). This method is not only objective, because no decision is required as to whether adjacent stained structures are separate, but rapid (2–3 min per section), reproducible, and gives independent prognostic information in breast ***(32,35)*** and bladder ***(36)*** cancers. Thus, it is currently the preferred method contained in a recent multicenter discussion paper ***(33)***.

The final consideration in quantifying angiogenesis is the differences in the value used for stratification into different study groups. This alone will result in different conclusions being drawn from the same data set. Studies have used the highest, the mean, the median ***(31)***, tertiles ***(32)***, mean count in node-negative patients with recurrence ***(23)***, or variable cut-offs given as a function of tumor area ***(8,22)*** or microscope magnification ***(28)***. The median and tertile groups do not make assumptions about the relationship between tumor vascularity and other variables, including survival, and is therefore useful clinically. However, there is some loss of information making it optimal to use continuous data where possible. Arbitrary cut-points should be avoided.

2. Materials

1. Silane-coated microscope slides.
2. Dry incubator/oven at 37°C.
3. Citroclear (HD Supplies) or xylene.
4. Graded alcohols (100%, 90%, and 70% ethanol).
5. 5% H_2O_2 in methanol.
6. Phosphate-buffered saline (PBS).
7. Tris-buffered saline (TBS).
8. Protease type XXIV (Sigma).
9. Monoclonal antibody against human PECAM/CD31 (JC70a, Dako, UK).
10. Alkaline phosphatase antialkaline phosphatase (APAAP) kit (Dako, UK).
11. Streptavidin-biotin complex (StreptABC) kit (Dako, UK).
12. Levamisole (Sigma).
13. Aqueous mountant.

3. Methods

3.1. Staining procedure

It should be emphasized that time must be devoted to optimizing the immunohistochemical staining procedure, since quality staining with little background greatly facilitates assessment. Many histopathology laboratories are well versed in immunohistochemistry, necessitating only minor adjustments to the preferred protocol outlined below.

1. Cut 4 µm formalin-fixed paraffin-embedded sections (*see* **Note 3**) of the representative tumor block (*see* **Note 4**) onto silane-coated slides.
2. Dry at 37°C overnight in an incubator (*see* **Note 5**).
3. Dewax using citroclear or xylene for 15 min before passing through graded alcohols into water before placing in PBS for 5 min (block endogenous peroxidase if using StreptABC; *see* **step 8**).
4. Pretreat sections with 12.5 mg protease type XXIV/100ml PBS for 20 min at 37°C (*see* **Note 6**).
5. Place in TBS (for APAAP) or PBS (for StreptABC) for 5 min, rinse, and apply JC70a (10mg/mL) primary antibody for 30 min.
6. Follow with standard APAAP or StreptABC immunohistochemistry (*see* **Note 7**).
7. APAAP (Dako, UK): This method uses a soluble enzyme antienzyme antibody complex (calf intestinal APAAP) to act on new fuschin substrate. The primary and final antibody complex is bridged by excess rabbit antimouse antibody, which binds to the primary mouse antibody with one Fab leaving an Fab site free to bind the tertiary complex. Repeated rounds of secondary and tertiary antibodies amplifies staining intensity. The enzyme hydrolyzes the naphthol esters in the substrate to phenols, which couple to colorless diazonium salts in the chromogen to produce a red color. Endogenous alkaline phosphatase is inhibited by the addition of 5 m*M* levamisole, which does not inhibit calf intestinal alkaline phosphatase. Incubation steps of 30 min with washing 3× in TBS with two rounds of intensification of 10 min should be used. Rinse in tap water for 2–3 min, tap dry, and mount in aqueous mountant.
8. StreptABC (Dako, UK): This method uses the high affinity of streptavidin for biotin. After application of the primary antibody, a biotinylated goat antimouse antibody at 1/40 is overlayed. The tertiary antibody complex of streptavidin-biotin-horseradish peroxidase (HRP) is then applied. The open sites on the streptavidin complex to the biotin on the secondary antibody. The brown endproduct is formed as 3′3-diamino benzidine HCl (DAB) is oxidized when it donates electrons to activate the HRP/H_2O_2 reaction. Blocking of endogenous peroxidase in paraffin-embedded tissues is performed by incubation of the section for 20 min in 0.5% H_2O_2 in methanol. Incubation of antibodies for 30 min with three washing steps of 5 min in PBS should be used. Rinse in tap water for 2–3 min, tap dry and mount in aqueous mountant.

9. Confirm satisfactory staining using normal entrapped vasculature as an internal positive control.
10. An optional parallel negative control section using an IgG_1 isotype antibody can also be run.

3.2. Assessment of Tumor Angiogenesis

Most studies use two observers over a conference microscope. The two investigators identify the hot spots and then independently quantify the vessels. Occasional discrepancies are resolved by consensus. Nevertheless, single observers can also accurately quantify tumor angiogenesis. Most solid tumors such as colorectal and breast cancers elicit a similar pattern of neovascularization and do not pose problems in counting. However, papillary tumors, including those of the bladder and thyroid, are difficult to interpret since each papillary formation, characteristic of its morphology, contains a vascular core. Care should thus be taken when assessing tumors with similar growth pattern.

3.2.1. Identification of Hot Spots

The three hot-spot areas containing the maximum number of *discrete* microvessels should be identified by scanning the entire tumor at low power (×40 and ×100) (**Fig. 1**). This is the most subjective step of the procedure, because the experience of the observer determines the success in identifying the relevant hot spots *(37)*. Poor selection will in turn lead to an inability to classify patients into different prognostic groups.

Therefore, it is recommended that inexperienced observers be trained, ideally by comparing their hot spots with those chosen by an experienced investigator, and this performed continually on different series until there is >90% agreement. Training can be completed by assessing sections from a series already known to contain prognostic information *(33)*.

Inexperienced observers tend to be drawn to areas with dilated vascular channels, often within the sclerotic body of the tumor. This is a function of the human eye, which is more sensitive to detecting vascular area than microvessel number. It is important to overcome this natural tendency and identify the maximum number of *discrete* microvessels. The central areas of the tumor often contain dilated vessels, and these regions, together with necrotic tumor, should be ignored. Vascular lumina or the presence of erythrocytes are not a requirement for a vessel to be considered countable. Indeed, many of the microvessels have a collapsed configuration. Although the hot spot areas can occur anywhere within the tumor, they are generally at the tumor periphery making it important to include the normal-tumor interface in the representative area to be

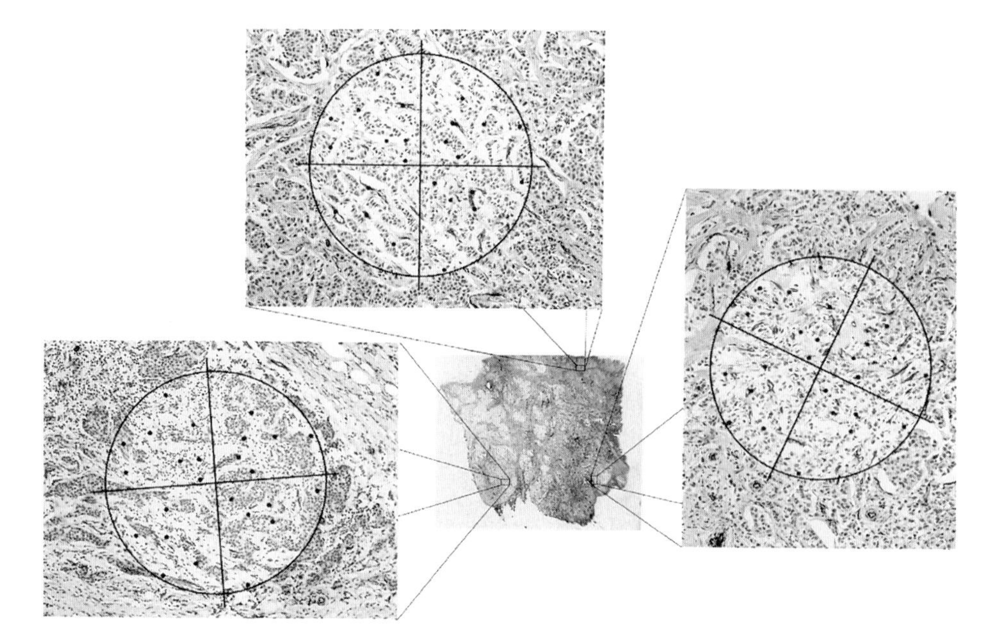

Fig. 1. The tumor is scanned at low power (×40–100) (*center*), and the three areas that contain the highest number of discrete microvessels are selected.

assessed. Vessels outside the tumor margin by one ×200–250 field diameter and immediately adjacent benign tissue should not be counted. The procedure takes 2–5 min.

3.2.2. Chalkley Counting

Once selected, a 25-point Chalkley point eyepiece graticule (*see* **Note 8**) is used to quantitate vascularity*(34)*. The graticule covers only a proportion of the microscope field area, so a magnification of ×200–250 is used (*see* **Note 9**). The graticule should then be oriented over each hot spot region so that the maximum number of graticule points are on or within areas of highlighted vessels (**Fig. 2**). It is worthwhile trying several graticule positions because indices can vary greatly depending on its alignment. Particular care should also be taken in the occasional case (<1% breast cancers) where antiCD31 has stained tumors with an intense plasma cell infiltrate. This can mimic a hot spot and obscure the underlying tumor vasculature. Plasma cells can otherwise be disregarded on morphological grounds (*see* **Note 10**). The mean of the three Chalkley counts is then generated for each tumor and used for statistical analysis. The procedure takes 2–3 min.

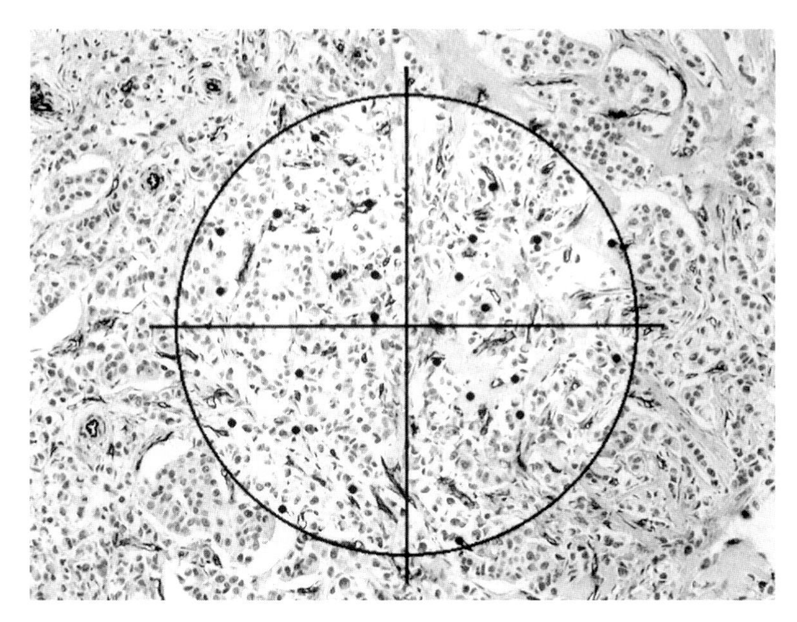

Fig. 2. Areas of tumor containing the highest number of discrete microvessels are examined at high power (×200–250) and the Chalkley point graticule is then rotated in the eyepiece so the maximum number of graticule dots coincide with the vessels or their lumens, which are then recorded.

3.3.3. Intratumoral Microvessel Density

For this index counting can be performed using a magnification of between ×200–400. Any endothelial cell or EC cluster separate from adjacent microvessels, tumor cells, or matrix elements is considered a countable vessel. Those which appear to be derived from the same vessel should also be counted if distinct. Again, vessel lumens and erythrocytes are not included in the criteria defining a microvessel. There is no cutoff for vessel caliber. The procedure takes 3–6 min.

4. Potential Improvements to Current Methods

4.1. Vascular Grading

Vascular grading facilitates assessment of angiogenesis in tissue sections and is akin to semiquantitative tumor grading. This vascular grading is based on the subjective appraisal by trained observers over a conference microscope ***(6,22)***. Significant correlations between vascular grade and both microvessel density ($p = 0.002$) and Chalkley count ($p = 0.0001$) have been demonstrated.

The method is reproducible *(38)*, but delineating criteria is difficult owing to the subjective nature of the system, and a considerable investment in time would be required to align the cutoffs required for multicenter studies. However, although there is some loss of power associated with translation of numerical to categorical data, the overall time savings engendered by this technique make it an attractive proposition. Further validation in a large series of randomized patients is warranted, nevertheless, to determine its prognostic utility before being applied in quantitative studies.

4.2. Novel Angiogenic Antigens

Instead of highlighting all the tumor-associated endothelium, an alternative approach would be to identify selectively only the vasculature that is undergoing active neovascularization. This helps to quantify tumor angiogenesis accurately and might also have important implications for antivascular targeting *(39)*. A number of antibodies have been identified that recognize antigens reported to be upregulated in tumor associated endothelium compared to normal tissues. These include EN7/44, endoglin, endosialin, and E-9 *(33,40)*. However, to date no studies assessing their utility have been performed.

4.3. Tumor Vascular Architecture

The vascular morphology of tumors is different within tumors of similar and different histological types *(41)*. Particular vascular patterns might help distinguish benign from malignant lesions *(42,43)* and be a prognostic marker; in ocular melanomas, a closed back-to-back loop vascular pattern was associated with death from metastasis *(44)*, and in lung carcinomas distinct patterns of neovascularization might potentially respond differently to anticancer treatments *(45)*.

4.4. Automation

Numerous studies have reported automation of the counting procedure by computer image analysis systems *(4,17,19,32,37,46–53)*. These systems have several drawbacks, not including the capital and running costs and those they share with manual methods. An endothelial marker, which gives sensitive and specific capillary staining, is essential for the system to accurately discriminate blood vessels from surrounding tissue elements. A high background signal interferes with this process. Although partially automated systems with area- and shape-filters using defined color tolerances are available, most systems are not fully automated, require a high degree operator interaction, and like manual counting suffer from observer bias. Also, software able to identify hot spots is not available but when developed will require motorized stages at additional expense. Thus, computer image analysis systems are currently more costly, time consuming (up to 40 min per case), and no more accurate than a trained observer. They are unsuitable for routine diagnostic practice. Nevertheless, data from these studies have demonstrated that most vascular indices, including microvessel density, vessel perimeter, and vascular area, are significantly correlated, suggesting that they are equivalent indices of angiogenesis *(32)*. Previously it had been hypothesized that microvessel density might not be the most important vascular parameter. A large vascular perimeter or area might be better measures of angiogenesis because these measures may better reflect the functional aspects of endothelial surface and the volume of blood available for interaction with the tumor *(32)*.

5. Potential Novel Techniques

The microvessels highlighted by immunohistochemistry in tissue section are the end result of a dynamic multistep process. The evolving neovasculature is the result of a complex interplay between extracellular matrix remodeling, endothelial cell migration and proliferation, and capillary differentiation and anastomosis *(54–56)*. Although it is not possible to measure these continuous processes, evidence suggests that a number of molecules involved in these events might be surrogate end points of angiogenesis. Thus, partly due to many of the inherent and methodological difficulties of vascular counts, these alternative strategies for quantifying tumor angiogenesis have also been pursued. These have particular drawbacks in that fresh specimens are required and the methodologies have yet to be standardized.

5.1. Angiogenic Factors and Receptors

Angiogenesis is the result of the net change in the balance of angiogenic stimulators and inhibitors (i.e., gain of promoters and/or loss of inhibitors). There are now numerous reports documenting up-regulation of several angiogenic factors and their receptors using a variety of techniques, including immunohistochemistry and *in situ* hybridization in a range of tumor types *(57–72)*. However, only a few have correlated these data to clinicopathological parameters or survival. Some, such as vascular endothelial growth factor (VEGF) have shown a significant relationship between their tumor levels and microvessel density (in breast *[73]*, brain *[74]*, cervix *[70]*, lung *[75]*, stomach *[76]*, and colon *[57]* cancers) and expression of KDR, a VEGF receptor, has also been correlated with high vessel counts in advanced stage colon carcinomas *(57)*. Furthermore, in a multivariate analysis VEGF expression levels gave independent prognostic information in breast carcinomas *(77)*. Similarly, thymidine phosphorylase in some studies has also been reported to be associated with microvessel density *(78,79)* and survival *(79)*.

A particular use of angiogenic factor measurement in patient sera, urine, or cerebrospinal fluid is the ability to perform serial measurements. Although some of the these studies have shown a relationship between angiogenic factor expression as a measure of tumor angiogenesis and patient survival ***(80–82)*** none of the current techniques are sensitive or specific enough to use for quantifying tumor angiogenesis ***(83,84)***. Different tumors use different angiogenic factors (breast carcinomas coexpress VEGF and thymidine phosphorylase (TP) ***[85]***, whereas they are reciprocally expressed in bladder cancers ***[86,87]***), and it is more likely that specific factor profiles for individual tumor types will be a more accurate measure of tumor angiogenesis.

5.2. Cell Adhesion Molecules

Increasing evidence suggests that many of the endothelial cell adhesion molecules of the immunoglobulin, selectin, and integrin superfamilies, which have physiological roles in immune trafficking and tumor metastasis, also play a major role in angiogenesis. Clinical studies showing melanoma patients with upregulated selectins on endothelium and a significantly poorer prognosis are validating the interest in cell adhesion molecules and their cognate ligands in tumor angiogenesis ***(88)***. Indeed soluble cell adhesion molecules are readily identified in sera of cancer-bearing patients although their relationship to tumor angiogenesis is yet unknown ***(89,90)***. Similarly, integrins, including $\alpha_v\beta_3$, have also been shown to have been upregulated in human breast carcinomas compared to normal or benign breast ***(91–93)*** and might also be a potential surrogate marker for angiogenesis.

5.3. Proteolytic Enzymes

Several studies have demonstrated that proteolytic enzymes including the plasminogen activators ***(94)*** and the matrix metalloproteinases ***(95)*** that are important in tumor cell invasion and migration are also important in angiogenesis. Indeed, a positive correlation between microvessel density and both urokinase plasminogen activator (uPA) and plasminogen activator inhibitor (PAI) has been reported ***(96)***. Thus the poor prognosis in tumors ***(97–104)*** associated with elevated levels of the uPA in the system are likely to be partly due to the angiogenic activity of these tumors. Measurement of proteases, particularly components of the urokinase system, might give some indication of the angiogenic activity of a tumor.

6. Summary

Continuing research into angiogenesis using quantitative data will not only broaden our understanding of the angiogenic process but will have several potential clinical applications beyond its use for prognosis. It might help in stratifying patients for cytotoxic therapy ***(105)***, aid monitoring and prediction of their response ***(106)***, and, with the advent of antiangiogenesis and vascular targeting, treatment could be stratified and altered based on these angiogenic measurements. The next few years will provide the data as to the reliability of quantitation of angiogenesis in tissue sections. During this time it is also probable that basic research will describe several candidate molecules that might become objective, sensitive, and specific enough to supersede the presently used assays.

7. Notes

1. One block is justified by the prevailing evidence, which shows a high concordance in vessel number between different blocks ***(8,29,30)***.
2. Studies within these microscope magnification/field area ranges derived prognostic information.
3. 8 µm cryostat sections can also be used but the area of tumor assessed is less representative.
4. The tumor block should be selected by examining H&E-stained slides.
5. If sections continually float off after antigen retrieval, drying at 56°C overnight will increase tissue adherence.
6. Alternative antigen retrieval techniques include pressure cooking and microwaving.
7. APAAP is preferred in tissues such as liver and kidney, which contain high endogenous biotin.
8. This can be obtained from Graticules Ltd, Morely Road, Botany Trading Estate, Tonbridge Wells, Kent, TN9 1ZN, UK. NB. Graticule size will depend on the microscope eyepiece diameter.
9. The field area would be too small were a higher magnification to be used. *See* **Subheading 1., paragraph 4**.
10. If anti-CD34 is used, positive stromal cells can also interfere with interpretation.

References

1. Folkman, J. (1971) Tumor angiogenesis: Therapeutic implications. *N. Engl. J. Med.* **285,** 82–86.
2. Brem, S., Cotran, R., and Folkman, J. (1972) Tumor angiogenesis: a quantitative method for histological grading. *J. Natl. Cancer Inst.* **48,** 347–356.
3. Mlynek, M., van Beunigen, D., Leder, L.-D., and Streffer, C. (1985) Measurement of the grade of vascularisation in histological tumor tissue sections. *Br. J. Cancer* **52,** 945–948.
4. Svrivastava, A., Laidler, P., Davies, R., Horgan, K., and Hughes, L. (1988) The prognostic significance of tumor vascularity in intermediate-thickness (0.76-4.0mm thick) skin melanoma. *Am. J. Pathol.* **133,** 419–423.
5. Porschen, R., Classen, S., Piontek, M., and Borchard, F. (1994) Vascularization of carcinomas of the esophagus and its correlation with tumor proliferation. *Cancer Res.* **54,** 587–591.
6. Weidner, N., Semple, J. P., Welch, W. R., and Folkman, J. (1991) Tumor angiogenesis and metastasis—correlation in invasive breast carcinoma. *N. Engl. J. Med.* **324,** 1–8.

7. Fox, S. B. (1997) Tumor angiogenesis and prognosis. *Histopathology* **30,** 294–301.
8. Van Hoef, M. E., Knox, W. F., Dhesi, S. S., Howell, A., and Schor, A. M. (1993) Assessment of tumor vascularity as a prognostic factor in lymph node negative invasive breast cancer. *Eur. J. Cancer* **29A,** 1141–1145.
9. Hall, N. R., Fish, D. E., Hunt, N., Goldin, R. D., Guillou, P. J., and Monson, J. R. (1992) Is the relationship between angiogenesis and metastasis in breast cancer real? *Surg. Oncol.* **1,** 223–229.
10. Sightler, H., Borowsky, A., Dupont, W., Page, D., and Jensen, R. (1994) Evaluation of tumor angiogenesis as a prognostic marker in breast cancer. *Lab. Invest.* **70,** 22A (abstract).
11. Vesalainen, S., Lipponen, P., Talja, M., Alhava, E., and Syrjanen, K. (1994) Tumor vascularity and basement membrane structure as prognostic factors in T1-2M0 prostatic adenocarcinoma. *Anticancer Res.* **14,** 709–714.
12. Rutgers, J. L., Mattox, T. F., and Vargas, M. P. (1995) Angiogenesis in uterine cervical squamous cell carcinoma. *Int. J. Gynecol. Pathol.* **14,** 114–118.
13. Ohsawa, M., Tomita, Y., Kuratsu, S., Kanno, H., and Aozasa, K. (1995) Angiogenesis in malignant fibrous histiocytoma. *Oncology* **52,** 51–54.
14. Barnhill, R., Busam, K., Berwick, M., Blesssing, K., Cochran, A., Elder, D., et al. (1994) Tumor vascularity is not a prognostic factor for cutaneous melanoma. *Lancet* **344,** 1237–1238.
15. Axelsson, K., Ljung, B., Moore II, D., Thor, A., Chew, K., Edgerton, S., et al. (1995) Tumor angiogenesis as a prognostic assay for invasive ductal breast carcinoma. *J. Natl. Cancer Inst.* 87.
16. McCarthy, S. A., Kuzu, I., Gatter, K. C., and Bicknell, R. (1991) Heterogeneity of the endothelial cell and its role in organ preference of tumor metastasis. *Trends Pharmacol. Sci.* **12,** 462–467.
17. Wakui, S., Furusato, M., Itoh, T., Sasaki, H., Akiyama, A., Kinoshita, I., et al. (1992) Tumor angiogenesis in prostatic carcinoma with and without bone marrow metastases: a morphometric study. *J. Pathol.* **168,** 257–262.
18. Carnochan, P., Briggs, J.C., Westbury, G., and Davies, A.J. (1991) The vascularity of cutaneous melanoma: a quantitative histological study of lesions 0.85–1.25 mm in thickness. *Br. J. Cancer* **64,** 102–107.
19. Visscher, D., Smilanetz, S., Drozdowicz, S., and Wykes, S. (1993) Prognostic significance of image morphometric microvessel enumeration in breast carcinoma. *Anal. Quant. Cytol.* **15,** 88–92.
20. Visscher, D., DeMattia, F., and Boman, S. (1994) Technical factors affecting image morphometric microvessel density counts in breast carcinomas. *Lab. Invest.* **70,** 168A (abstract).
21. Ottinetti, A. and Sapino, A. (1988) Morphometric evaluation of microvessels surrounding hyperplastic and neoplastic mammary lesions. *Breast Cancer Res. Treat.* **11,** 241–248.
22. Weidner, N., Folkman, J., Pozza, F., Bevilacqua, P., Allred, E. N., Moore, D. H., et al. (1992) Tumor angiogenesis: a new significant and independent prognostic indicator in early-stage breast carcinoma. *J. Natl. Cancer Inst.* **84,** 1875–1887.

23. Bosari, S., Lee, A. K., DeLellis, R. A., Wiley, B. D., Heatley, G. J., Silverman, M. L. (1992) Microvessel quantitation and prognosis in invasive breast carcinoma. *Hum. Pathol.* **23,** 755–761.
24. Bundred, N., Bowcott, M., Walls, J., Faragher, E., and Knox, F. (1994) Angiogenesis in breast cancer predicts node metastasis and survival. *Br. J. Surgery* **81,** 768 (abstract).
25. Li, V., Folkerth, R., Watanabe, H., Yu, C., Rupnick, M., Barnes, P., et al. (1994) Microvessel count and cerebrospinal fluid basic fibroblast growth factor in children with brain tumors. *Lancet* **344,** 82–86.
26. Parums, D., Cordell, J., Micklem, K., Heryet, A., Gatter, K., Mason, D. (1990) JC70: a new monoclonal antibody that detects vascular endothelium associated antigen on routinely processed tissue sections. *J. Clin. Pathol.* **43,** 752–757.
27. Barnhill, R. L., Fandrey, K., Levy, M. A., Mihm, M. J., and Hyman, B. (1992) Angiogenesis and tumor progression of melanoma. Quantification of vascularity in melanocytic nevi and cutaneous malignant melanoma. *Lab. Invest.* **67,** 331–337.
28. Sahin, A., Sneige, N., Singletary, E., and Ayala, A. (1992) Tumor angiogenesis detected by Factor-VIII immunostaining in node-negative breast carcinoma (NNBC): a possible predictor of distant metastasis. *Mod. Pathol.* 5:17A (abstract).
29. Martin, L., Holcombe, C., Green, B., Winstanley, J., and Leinster, S. (1996) Vascular heterogeneity in breast cancer assessed by microangiography and immunohistochemistry. *Br. J. Surg.* **83,** 702.
30. de Jong, J., van Diest, P., and Baak, J. (1995) Heterogeneity and reproducibility of microvessel counts in breast cancer. *Lab. Invest.* **73,** 992–926.
31. Horak, E. R., Leek, R., Klenk, N., LeJeune, S., Smith, K., Stuart, N., et al. (1992) Angiogenesis, assessed by platelet/endothelial cell adhesion molecule antibodies, as indicator of node metastases and survival in breast cancer. *Lancet* **340,** 1120–1124.
32. Fox, S. B., Leek, R. D., Weekes, M. P., Whitehouse, R. M., Gatter, K. C., and Harris, A. L. (1995) Quantitation and prognostic value of breast cancer angiogenesis: comparison of microvessel density, Chalkley count and computer image analysis. *J. Pathol.* **177,** 275–283.
33. Vermeulen, P. B., Gasparini, G., Fox, S. B., Toi, M., Martin, L., McCulloch, P., et al. (1996) Quantification of angiogenesis in solid human tumors—an international consensus on the methodology and criteria of evaluation. *Eur. J. Cancer* **32A,** 2474–2484.
34. Chalkley, H. (1943) Method for the quantative morphological analysis of tissues. *J. Natl. Cancer Inst.* **4,** 47–53.
35. Fox, S. B., Leek, R., Smith, K., Hollyer, J., Greenall, M., and Harris, A. (1994) Tumor angiogenesis in node negative breast carcinomas-relationship to epidermal growth factor receptor and survival. *Breast Cancer Res. Treat.* **29,** 109–116.
36. Dickinson, A. J., Fox, S. B., Persad, R. A., Hollyer, J., Sibley, G. N., and Harris, A. L. (1994) Quantification of angiogenesis as an independent predictor of prognosis in invasive bladder carcinomas. *Br. J. Urol.* **74,** 762–766.
37. Barbareschi, M., Weidner, N., Gasparini, G., Morelli, L., Forti, S., Eccher, C., et al. (1995) Microvessel quantitation in breast carcinomas. *Appl. Immunochem.* **3,** 75–84.

38. Fox, S. B., Leek, R., Bliss, J., Gusterson, B., Mansi, J., Gatter, K., and Harris, A. (1997) Tumor angiogenesis is associated with bone marrow micrometastasis in breast cancer patients. *J. Natl. Cancer Inst.* **89,** 1044–1049.
39. Burrows, F. J. and Thorpe, P. E. (1994) Vascular targeting—a new approach to the therapy of solid tumors. *Pharmacol. Ther.* **64,** 155–174.
40. Fox, S. B. and Harris, A. (1997) Markers of tumor angiogenesis: clinical applications in prognosis and anti-angiogenic therapy. *Invest New Drugs*, in press.
41. Warren, B. (1979) The vascular morphology of tumors. *In*: *Tumor Blood Circulation* (Peterson, H., ed). CRC Press, Boca Raton, Fla, pp. 1–47.
42. Smolle, J., Soyer, H. P., Hofmann-Wellenhof, Smolle-Juettner, F. M., and Kerl, H. (1989) Vascular archictecture of melanocytic skin tumors. *Path. Res. Pract.* **185,** 740–745.
43. Cockerell, C. J., Sonnier, G., Kelly, L., and Patel, S. (1994) Comparative analysis of neovascularisation in primary cutaneous melanoma and Spitz nevus. *Am. J. Dermatopathol.* **16,** 9–13.
44. Folberg, R., Rummelt, V., Parys-Van Ginderdeuren, V., Hwang, T., Woolson, R., Pe'er, J., and Gruman, L. (1993) The prognostic value of tumor blood vessel morphology in primary uveal melanoma. *Ophthalmology* **100,** 1389–1398.
45. Pezzella, F., Pastorino, U., Tagliabue, E., Andreola, S., Sozzi, G., Gasparini, G., et al. (1996) Non-small-cell lung carcinoma tumor growth without morphological evidence of neo-angiogenesis. *Am. J. Pathol.* **151(5),** 1417–1423.
46. Simpson, J., Ahn, C., Battifora, H., and Esteban, J. (1994) Vascular surface area as a prognostic indicator in invsasive breast carcinoma. *Lab. Invest.* **70,** 22A.
47. Brawer, M. K., Deering, R. E., Brown, M., Preston, S. D., Bigler, S. A. (1994) Predictors of pathologic stage in prostatic carcinoma. The role of neovascularity. *Cancer* **73,** 678–687.
48. Furusato, M., Wakui, S., Sasaki, H., Ito, K., and Ushigome, S. (1994) Tumor angiogenesis in latent prostatic carcinoma. *Br. J. Cancer* **70,** 1244–1246.
49. Bigler, S., Deering, R., and Brawer, M. (1993) Comparisons of microscopic vascularity in benign and malignant prostate tissue. *Human Pathol.* **24,** 220–226.
50. Williams, J. K., Carlson, G. W., Cohen, C., Derose, P. B., Hunter, S., and Jurkiewicz, M. J. (1994) Tumor angiogenesis as a prognostic factor in oral cavity tumors. *Am. J. Surg.* **168,** 373–380.
51. Wesseling, P., Vandersteenhoven, J. J., Downey, B. T., Ruiter, D. J., and Burger, P. C. (1993) Cellular components of microvascular proliferation in human glial and metastatic brain neoplasms. A light microscopic and immunohistochemical study of formalin-fixed, routinely processed material. *Acta Neuropathol. (Berl)* **85,** 508–514.
52. Charpin, C., Devictor, B., Bergeret, D., Andrac, L., Boulat, J., Horschowski, N., et al. (1995) CD31 Quantitative immunocytochemical assays in breast carcinomas. *Am. J. Clin. Pathol.* **103,** 443–448.
53. Van der Laak, J., Westphal, J., Schalkwijk, L., Pahplazt, M., Ruiter, D., de Waal, R., and de Wilde, P. (1998) An improved procedure to quantify tumor vascularity

using true color image analysis: comparison with the manual hot-spot procedure in a human melanoma xenograft model. *J. Pathol.* **184,** 136–143.
54. Paweletz, N. and Knierim, M. (1989) Tumor-related angiogenesis. *Crit. Rev. Oncol. Hematol.* **9,** 197–242.
55. Blood, C. H. and Zetter, B. R. (1990) Tumor interactions with the vasculature: angiogenesis and tumor metastasis. *Biochim. Biophys. Acta* **1032,** 89–118.
56. Bicknell, R. and Harris, A. L. (1991) Novel growth regulatory factors and tumor angiogenesis. *Eur. J. Cancer* **27,** 781–785.
57. Takahashi, Y., Kitadai, Y., Bucana, C. D., Cleary, K. R., and Ellis, L. M. (1995) Expression of vascular endothelial growth factor and its receptor, KDR, correlates with vascularity, metastasis and proliferation of human colon cancer. *Cancer Res.* **55,** 3964–3968.
58. Brown, L. F., Berse, B., Jackman, R. W., Tognazzi, K., Guidi, A. J., Dvorak, H. F., et al. (1995) Expression of vascular permeability factor (vascular endothelial growth factor) and its receptors in breast cancer. *Hum. Pathol.* **26,** 86--91.
59. Moghaddam, A., Zhang, H. T., Fan, T. P., Hu, D. E., Lees, V. C., Turley, H., et al. (1995) Thymidine phosphorylase is angiogenic and promotes tumor growth. *Proc. Natl. Acad. Sci. USA* **92,** 998–1002.
60. Anandappa, S. Y., Winstanley, J. H., Leinster, S., Green, B., Rudland, P. S., and Barraclough, R. (1994) Comparative expression of fibroblast growth factor mRNAs in benign and malignant breast disease. *Br. J. Cancer* **69,** 772–776.
61. Reynolds, K., Farzaneh, F., Collins, W. P., Campbell, S., Bourne, T. H., Lawton, F., et al. (1994) Association of ovarian malignancy with expression of platelet-derived endothelial cell growth factor. *J. Natl. Cancer Inst.* **86,** 1234–1238.
62. Garver, R. J., Radford, D. M., Donis, K. H., Wick, M. R., and Milner, P. G. (1994) Midkine and pleiotrophin expression in normal and malignant breast tissue. *Cancer* **74,** 1584–1590.
63. Janot, F., el-Naggar, A. K., Morrison, R. S., Liu, T. J., Taylor, D. L., and Clayman, G. L. (1995) Expression of basic fibroblast growth factor in squamous cell carcinoma of the head and neck is associated with degree of histologic differentiation. *Int. J. Cancer* **64,** 117–123.
64. Gomm, J. J., Smith, J., Ryall, G. K., Ballic, R., Turnbull, L., and Coombes, R. C. (1991) Localisation of basic fibroblast growth factor and transforming growth factor β1 in the human mammary gland. *Cancer Res.* **51,** 4685–4692.
65. Zarnegar, R. and DeFrances, M. C. (1993) Expression of HGF-SF in normal and malignant human tissues. *EXS* **65,** 181–199.
66. Daa, T., Kodama, M., Kashima, K., Yokoyama, S., Nakayama, I., and Noguchi, S. (1993) Identification of basic fibroblast growth factor in papillary carcinoma of the thyroid. *Acta Pathol. Japn.* **43,** 582–589.
67. Wong, S. Y., Purdie, A. T., and Han, P. (1992) Thrombospondin and other possible related matrix proteins in malignant and benign breast disease. *Am. J. Pathol.* **140,** 1473–1482.

68. Schultz-Hector, S. and Haghayegh, S. (1993) Basic fibroblast growth factor expression in human and murine squamous cell carcinomas and its relationship to regional endothelial cell proliferation. *Cancer Res.* **53,** 1444–1449.
69. Alvarez, J. A., Baird, A., Tatum, A., Daucher, J., Chorsky, R., Gonzalez, A. M., and Stopa, E. G. (1992) Localisation of basic fibroblast growth factor and vascular endothelial cell growth factor in human glial neoplasms. *Mod. Pathol.* **5,** 303–307.
70. Guidi, A. J., Abu, J. G., Berse, B., Jackman, R. W., Tognazzi, K., Dvorak, H. F., and Brown, L. F. (1995) Vascular permeability factor (vascular endothelial growth factor) expression and angiogenesis in cervical neoplasia. *J. Natl. Cancer Inst.* **87,** 1237–1245.
71. Zagzag, D., Brem, S., and Robert, F. (1988) Neovascularization and tumor growth in the rabbit brain. A model for experimental studies of angiogenesis and the blood-brain barrier. *Am. J. Pathol.* **131,** 361–372.
72. Visscher, D. W., DeMattia, F., Ottosen, S., Sarkar, F. H., and Crissman, J. D. (1995) Biologic and clinical significance of basic fibroblast growth factor immunostaining in breast carcinoma. *Mod. Pathol.* **8,** 665–670.
73. Toi, M., Kondo, S., Suzuki, H., Yamamoto, Y., Inada, K., Imazawa, T., et al. (1996) Quantitative analysis of vascular endothelial growth factor in primary breast cancer. *Cancer* **77,** 1101–1106.
74. Plate, K. H., Breier, G., Millauer, B., Ullrich, A., and Risau, W. (1993) Up-regulation of vascular endothelial growth factor and its cognate receptors in a rat glioma model of tumor angiogenesis. *Cancer Res.* **53,** 5822–5827.
75. Mattern, J., Koomagi, R., and Volm, M. (1996) Association of vascular endothelial growth factor expression with intratumoral microvessel density and tumor cell proliferation in human epidermoid lung carcinoma. *Br. J. Cancer* **73,** 931–934.
76. Maeda, K., Chung, Y. S., Ogawa, Y., Kang, S., Takatsuka, S., Ogawa, M., et al. (1996) Prognostic value of vascular endothelial growth factor expression in gastric carcinoma. *Cancer* **77,** 858–863.
77. Toi, M., Hoshina, S., Takayanagi, T., and Tominaga, T. (1994) Association of vascular endothelial growth factor expression with tumor angiogenesis and with early relapse in primary breast cancer. *Japn. J. Cancer Res.* **85,** 1045–1049.
78. Toi, M., Hoshina, S., Taniguchi, T., Yamamoto, Y., Ishitsuka, H., and Tominaga, T. (1995) Expression of platelet derived endothelial cell growth factor/thymidine phosphorylase in human breast cancer. *Int. J. Cancer* **64,** 79–82.
79. Fox, S. B., Westwood, M., Moghaddam, A., Comley, M., Turley, H., Whitehouse, R. M., et al. (1996) The angiogenic factor platelet-derived endothelial cell growth factor/thymidine phosphorylase is up-regulated in breast cancer epithelium and endothelium. *Br. J. Cancer* **73,** 275–280.
80. Nguyen, M., Watanabe, H., Budson, A. E., Richie, J. P., Hayes, D. F., and Folkman, J. (1994) Elevated levels of an angiogenic peptide, basic fibroblast growth factor, in the urine of patients with a wide spectrum of cancers. *J. Natl. Cancer Inst.* **86,** 356–361.

81. Toi, M., Taniguchi, T., Yamamoto, Y., Kurisaki, T., Suzuki, H., and Tominaga, T. (1996) Clinical-significance of the determination of angiogenic factors. *Eur. J. Cancer* **32A,** 2513–2519.
82. Salven, P., Teerenhovi, L., and Joensuu, H. (1997) A high pre-treatment serum vascular endothelial growth factor concentration is associated with poor outcome in non-Hodgkin's lymphoma. *Blood* **90,** 3167–3172.
83. O'Brien, T. S., Smith, K., Cranston, D., Fuggle, S., Bicknell, R., and Harris, A. L. (1995) Urinary basic fibroblast growth factor in patients with bladder cancer and benign prostatic hypertrophy. *Br. J. Urol.* **76,** 311–314.
84. Chow, N. H., Chang, C. J., Yeh, T. M., Chan, S. H., Tzai, T. S., and Lin, J. S. (1996) Implications of urinary basic fibroblast growth factor excretion in patients with urothelial carcinoma. *Clin. Science* **90,** 127–133.
85. Toi, M., Yamamoto, Y., Inada, K., Hoshina, S., Suzuki, H., Kondo, S., and Tominaga, T. (1995) Vascular endothelial growth factor and platelet-derived endothelial growth factor are frequently co-expressed in highly vascularized breast cancer. *Clin. Cancer Res.* **1,** 961–964.
86. O'Brien, T., Cranston, D., Fuggle, S., Bicknell, R., and Harris, A. L. (1995) Different angiogenic pathways characterize superficial and invasive bladder cancer. *Cancer Res.* **55,** 510–513.
87. O'Brien, T., Fox, S. B., Dickinson, A., Turley, H., Westwood, M., Moghaddam, A., et al. (1996) Expression of the angiogenic factor thymidine phosphorylase/platelet derived endothelial cell growth factor in primary bladder cancers. *Cancer Res.* **56,** 4799–4804.
88. Schadendorf, D., Heidel, J., Gawlik, C., Suter, L., Czarnetzki. (1995) Association with clinical outcome of expression of VLA-4 in primary cutaneous malignant melanoma as well as P-selectin and E-selectin on intratumoral vessels. *J. Natl. Cancer Inst.* **87,** 366–371.
89. Kageshita, T., Yoshii, A., Kimura, T., Kuriya, N., Ono, T., Tsujisaki, M., et al. (1993) Clinical relevance of ICAM-1 expression in prmary lesions and serum of patients with malignant melanoma. *Cancer Res.* **53,** 4927–4932.
90. Banks, R. E., Gearing, A. J., Hemingway, I. K., Norfolk, D. R., Perren, T. J., and Selby, P. J. (1993) Circulating intercellular adhesion molecule-1 (ICAM-1), E-selectin and vascular cell adhesion molecule-1 (VCAM-1) in human malignancies. *Br. J. Cancer* **68,** 122–124.
91. Brooks, P. C., Clark, R. A., and Cheresh, D. A. (1994)Requirement of vascular integrin alpha v beta 3 for angiogenesis. *Science* **264,** 569–571.
92. Brooks, P. C., Montgomery, A. M., Rosenfeld, M., Reisfeld, R. A., Hu, T., Klier, G., and Cheresh, D. A. (1994) Integrin alpha v beta 3 antagonists promote tumor regression by inducing apoptosis of angiogenic blood vessels. *Cell* **79,** 1157–1164.
93. Brooks, P. C., Stromblad, S., Klemke, R., Visscher, D., Sarkar, F. H., and Cheresh, D. A. (1995) Antiintegrin alpha v beta 3 blocks human breast cancer growth and angiogenesis in human skin. *J. Clin. Invest.* **96,** 1815–1822.
94. Pepper, M. and Montesano, R. (1990) Proteolytic balance and capillary morphogenesis. *Cell Diff. Dev.* **32,** 319–328.

95. Fisher, C., Gilbertson, B. S., Powers, E. A., Petzold, G., Poorman, R., and Mitchell, M. A. (1994) Interstitial collagenase is required for angiogenesis in vitro. *Dev. Biol.* **162,** 499–510.
96. Fox, S. B., Stuart, N., Smith, K., Brunner, N., and Harris, A. L. (1993) High levels of uPA and PAI-1 are associated with highly angiogenic breast carcinomas. *J. Pathol.* **170,** 388A(suppl).
97. Grøndahl- Hansen, J., Christensen, I. J., Rosenquist, C., Brünner, N., Mouridsen, H. T., Danø, K., and Blichert, T. M. (1993) High levels of urokinase-type plasminogen activator and its inhibitor PAI-1 in cytosolic extracts of breast carcinomas are associated with poor prognosis. *Cancer Res.* **53,** 2513–2521.
98. Janicke, F., Schmitt, M., Pache, L., Ulm, K., Harbeck, N., Hofler, H., and Graeff, H. (1993) Urokinase (uPA) and its inhibitor PAI-1 are strong and independent prognostic factors in node-negative breast cancer. *Breast Cancer Res. Treat.* **24,** 195–208.
99. Foekens, J. A., Schmitt, M., van, P. W., Peters, H. A., Bontenbal, M., Janicke, F., and Klijn, J. G. (1992) Prognostic value of urokinase-type plasminogen activator in 671 primary breast cancer patients. *Cancer Res.* **52,** 6101–6105.
100. Duffy, M. J., Reilly, D., O'Sullivan, C., O'Higgins, N., Fennelly, J. J., and Andreasen, P. (1990) Urokinase-plasminogen activator, a new and independent prognostic marker in breast cancer. *Cancer Res.* **50,** 6827–6829.
101. Schmitt, M., Janicke, F., Moniwa, N., Chucholowski, N., Pache, L., and Graeff, H. (1992) Tumor-associated urokinase-type plasminogen activator: biological and clinical significance. *Biol. Chem. Hoppe Seyler* **373,** 611–622.
102. Spyratos, F., Martin, P. M., Hacene, K., Romain, S., Andrieu, C., Ferrero, P. M., et al. (1992) Multiparametric prognostic evaluation of biological factors in primary breast cancer. *J. Natl. Cancer Inst.* **84,** 1266–1272.
103. Sumiyoshi, K., Serizawa, K., Urano, T., Takada, Y., Takada, A., and Baba, S. (1992) Plasminogen activator system in human breast cancer. *Int. J. Cancer* **50,** 345–348.
104. Ganesh, S., Sier, C. F. M., Heerding, M. M., Griffioen, G., Lamers, C. B. H. W., and Verspaget, H. W. (1994) Urokinase receptor and colorectal cancer survival. *Lancet* **344,** 401–402.
105. Protopapa, E., Delides, G. S., and Revesz, L. (1993) Vascular density and the response of breast carcinomas to mastectomy and adjuvant chemotherapy. *Eur. J. Cancer* **29A,** 1141–1145.
106. Fox, S. B., Engels, K., Comley, M., Whitehouse, R. M., Turley, H., Gatter, K. C., and Harris, A. L. (1997) Relationship of elevated tumor thymidine phosphorylase in node positive breast carcinomas to the effects of adjuvant CMF. *Ann. Oncol.* **8,** 271–275.

3

In Vivo Matrigel Migration and Angiogenesis Assays

Katherine M. Malinda

1. Introduction

Angiogenesis, the process of new blood vessels forming from preexisting vessels, is an important feature in developmental processes, wound healing, and pathologic conditions such as cancer and vascular diseases. Owing to the importance of angiogenesis, a relatively simple and rapid in vivo method to determine the angiogenic potential of compounds is desirable to augment in vitro findings.

One such quantitative method is the murine Matrigel plug assay, which can measure both angiogenesis and antiangiogenesis. Matrigel, an extract of the Englebreth-Holm-Swarm tumor composed of basement membrane components, is liquid at 4°C and forms a gel when warmed to 37°C *(1)*. When plated on Matrigel, human umbilical vein endothelial cells (HUVEC) undergo differentiation into capillary-like tube structures in vitro *(2,3)*. In vivo, Matrigel is either injected alone or mixed with potential angiogenic compounds and injected subcutaneously into the ventral region of mice, where it solidifies forming a "Matrigel plug." When known angiogenic factors, such as basic fibroblast growth factor (bFGF), are mixed with the Matrigel and injected, endothelial cells migrate into the plug and form vessels. The level of angiogenesis is typically viewed by embedding and sectioning the plugs in paraffin and staining using Masson's Trichrome, which stains the Matrigel blue and the endothelial cells/vessels red **(Fig. 1)**. These vessels contain erythrocytes, indicating that they form functional capillaries. Additionally, these capillaries stain positively for factor VIII-related antigen *(4,5)*. In unsupplemented Matrigel, few cells invade the plug. Strongly angiogenic compounds result in yellowish

From: *Methods in Molecular Medicine, Vol. 46: Angiogenesis Protocols*
Edited by: J. C. Murray © Humana Press Inc., Totowa, NJ

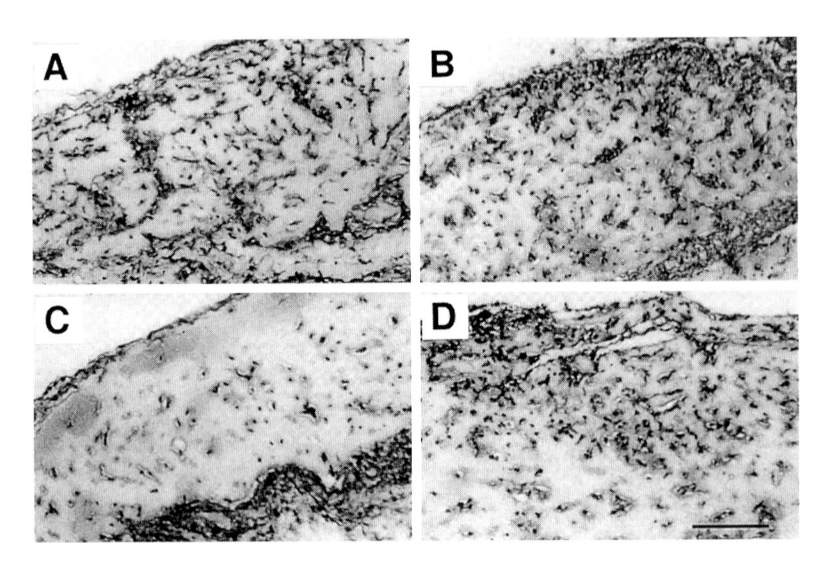

Fig. 1 Sections of Matrigel plugs stained with Masson's trichrome. All sections are oriented with the side underlying the skin at the top of the image. **(A, B)** Representative fields of plugs containing 5 μg/mL Tβ4. **(C)** Field showing Matrigel alone. **(D)** Representative field of a plug with 10 ng/mL ECGS (positive control). Sections with Tβ4 contain many more cells than Matrigel alone and have cells with a similar morphology to those in the ECGS control. Bar = 100 μm. With permission from Malinda, K. M., Goldstein, A. L. and Kleinman, H. K. (1997). Thymosin β4 stimulates directional migration of human umbilical vein endothelial cells. *FASEB J.* **11,** 474–481.

plugs, so initial indications of activity can be made at the time when plugs are removed from the mice.

The assay also can be utilized when putative antiangiogenic compounds are being tested. In this assay, Matrigel is premixed with bFGF (angiogenic compound), and the test substances are then added. Thus, test antiangiogenic substances inhibit the formation of vessels induced by bFGF in the Matrigel plug. In this case, the plugs removed from the mouse are relatively colorless and when viewed by Masson's trichrome staining, should contain few endothelial cells. Because of the strong vascular response to bFGF, it is also possible to measure hemoglobin levels with the Drabkin assay *(4)*.

One important consideration in using either of the above assays is the degree of variability that will be observed. Differences in the mice and in the basement membrane preparations will affect the background levels of blood vessel formation one observes. Age and gender of the mice selected also can result in different results between experiments. Vessel formation in young mice (6 mo old) is reduced as compared to mice 12–24 mo old *(4)*. Additionally, if the

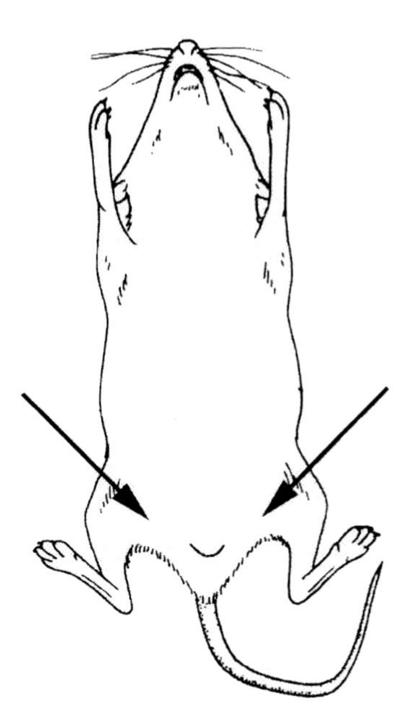

Fig. 2 Diagram of the ventral side of a mouse. Arrows show optimal sites for Matrigel injection.

Matrigel is injected into different sites in the mouse, variability can result. Lower angiogenic response is observed if the material is injected into the dorsal surface of the animal, while one of the best areas in terms of angiogenic response is the ventral side of the mouse in the groin area close to the dorsal midline (**Fig. 2**). Regardless of these potential problems, this assay is one of the best for the rapid screening of potential angiogenic and antiangiogenic compounds.

2. Materials

2.1. General Materials

1. C57Bl/6N female mice. Three or four mice are needed for each test group.
2. Additional cages and supplies, one for each test group.
3. Matrigel (Collaborative Biomedical, Bedford, MA) store at –20°C. Enough will be needed to inject 0.5 or 1 mL per mouse.
4. 3 mL syringes (Monoject).
5. 25-gauge needles (Monoject)
6. 14 mL Falcon round bottom tubes (#2059) (Becton Dickinson, Rutherford, NJ).
7. 10X Phosphate buffered saline (PBS) (Biofluids, Inc., Rockville, MD) diluted to 1X in sterile, distilled H_2O.
8. Endothelial cell growth supplement [bovine brain extract containing acid fibroblast growth factor (aFGF) and bFGF (ECGS)] (Collaborative Biomedical,

Becton-Dickinson, Rutherford, NJ): Prepare 1 mg/mL solution just before use in sterile, distilled H_20 and store unused portion in aliquots at –20°C. (Loses activity after 1 mo according to manufacturer.) Or bFGF (Collaborative Biomedical, Bedford, MA). Prepare solution 1 mg/mL just before use in sterile, distilled H_20 and store unused portion in aliquots at –20°C. Final concentration used in experiment is either 10 ng/mL or 100 ng/mL ECGS or 150 ng/mL bFGF.
9. Table top vortex (Vortex Genie 2, Scientific Industries).
10. Sterile pipet tips.
11. Sterile 5 mL disposable pipettes.
12. Scalpel or dissecting scissors

2.2. Embedding and Quantitation

1. Scintillation vials or specimen vials.
2. 10% formalin in PBS.
3. Histology service for paraffin embedding, sectioning, and staining with Masson's trichrome stain.
4. Microscope equipped with a video system linked to a computer with NIH Image software (free to the public).
5. Spreadsheet software.

3. Methods

3.1. Screening Putative Angiogenic Compounds

1. Thaw Matrigel on ice overnight at 4°C. At no time should Matrigel be warmed to room temperature during handling.
2. Mix Matrigel on vortex; place on ice (*see* **Notes 1** and **2**).
3. Pipet equal amounts of Matrigel for each test condition in separate 14 mL tubes on ice. Each mouse injection requires 0.5–1 mL with three or four mice for each condition (*see* **Notes 2** and **3**).
4. Tubes should be included for the positive and negative controls. Negative control is Matrigel alone; positive controls contain either 10 ng/mL and 100 ng/mL ECGS or 150 ng/mL bFGF added to Matrigel.
5. Equalize the volumes with cold PBS so that, after test substances have been added, all tubes contain Matrigel at equal dilutions (*see* **Note 4**). Mix on the vortex.
6. Load syringes with each solution and inject 0.5 or 1 mL subcutaneously into the ventral area of each mouse (*see* **Fig. 2**; **Notes 5** and **6**).
7. Place animals of each test group in separate labeled cages.
8. After 7–10 d, sacrifice the animals and remove the Matrigel plugs. Plugs appear as bumps on the ventral side of the animal and are removed using a sharp pair of scissors or a scalpel. Remove the plug with some surrounding tissue for orientation when histology is performed.

3.2. Screening Putative Antiangiogenic Compounds

1. Thaw Matrigel on ice overnight at 4°C.
2. Mix Matrigel on vortex; place on ice (*see* **Note 1**).

3. Pipet equal amounts of Matrigel for each test condition in separate 14 mL tubes on ice. Each mouse injection requires 0.5–1 mL, three or four mice for each condition (*see* **Notes 2** and **3**).
4. Add 150 ng/mL bFGF to all tubes except the one used for the negative control containing Matrigel alone.
5. Equalize the volumes with cold PBS so that, after test substances have been added, all tubes contain Matrigel at equal dilutions (*see* **Note 4**). Mix on the vortex.
6. Load syringes with each solution and inject 0.5 or 1 mL subcutaneously into the ventral area of each mouse (*see* **Fig. 2**; **Notes 5** and **6**).
7. Place animals of each test group in separate labeled cages.
8. After 7–10 d, sacrifice the animals and remove the plugs. Plugs appear as bumps on the ventral side of the animal and are removed using a sharp pair of scissors or a scalpel. Remove the plug with some surrounding tissue for orientation when histology is performed.

3.3. Fixation and Quantitation by Microscopy

1. Place plugs in labeled scintillation or specimen vials filled with 10% formalin. Allow plugs to fix overnight before embedding in paraffin.
2. Section and stain with Masson's trichrome, which stains the endothelial cells/vessels dark red and the Matrigel blue (*see* **Note 7**).
3. Visually inspect the sections to determine if the test substance is stimulatory and if further quantitation is necessary (**Fig. 1**).
4. Capture images at 10× magnification in the area underlying the skin (*see* **Note 8**).
5. Adjust the image to selectively highlight the gray level corresponding to the endothelial cells/vessels by using the *density slice function* of the software and the light level of the microscope.
6. Use the *measure option* to measure the area occupied by all cells selected by the *density slice function* (*see* **Notes 9** and **10**).
7. Count at least three random fields of the area underlying the skin on each slide.
8. Load the numbers into a spreadsheet and calculate the total area occupied by endothelial cells in the three fields and then average these areas among the three or four replicate mouse plugs.
9. Compare the resulting numbers for statistical significance.

4. Notes

1. Bubbles form when Matrigel is vortexed. Minimize vortexing to keep bubbles to a minimum or injection problems will occur. Also, allowing the tubes to stand on ice for 10 min will reduce the number of bubbles.
2. All reagents should be kept as sterile as possible to prevent infection of mice.
3. Always use gloves when working with reagents.
4. Keep dilution of Matrigel to a minimum. If too dilute, it will not gel when injected.
5. Use caution when handling animals. Only approved personnel should handle animals. Contact your animal safety officer for the appropriate safety regulations you must follow. Wear gloves, masks, gowns, hair cover, and boots when appropriate.

6. Inject solutions slowly; a bump should form under the skin. If the skin is punctured during injection, the material will flow out. Allow the Matrigel to gel for about 10 sec before removing needle at the end of the injection, otherwise injected material will leak out.
7. Fixation according to these methods can inactivate epitopes so that immunohistochemical analysis may not work. An alternative is to make frozen sections of the tissue and eliminate the fixation and paraffin steps.
8. An alternative method of quantitating vessel in-growth is to measure the hemoglobin content using, for example, a Drabkin reagent 525 kit (Sigma, St. Louis, MO).
9. Selecting only a portion of the image may be required, to avoid counting areas that contain bubbles, cracks or other artifacts that refract the light and may be picked up by the density slice feature. If this is necessary, use the drawing tool to draw a box, note the size in the "info" window, and draw the same sized box on all sections for the entire experiment.
10. Since animals will have different responses to the injected material, it is not unusual to experience variability in the responses of animals to different test substances.

References

1. Kleinman, H. K., McGarvey, M. L,. Liotta, L. A., Gehron-Robbey, P., Tryggvasson, K., and Martin, G. R. (1982) Isolation and characterization of type IV procollagen, laminin, and hepara sulfate proteoglycan from the EHS sarcoma. *Biochemistry* **24,** 6188–6193.
2. Kubota, Y., Kleinman, H. K., Martin, G. R., and Lawley, T. J. (1988) Role of laminin and basement membrane in the morphological differentiation of human endothelial cells into capillary-like structures. *J. Cell Biol.* **107,** 1589–1598.
3. Grant, D. S., Kinsella, J. L., Fridman, R., Auerbach, R., Piasecki, B. A., Yamada, Y., Zain, M., and Kleinman, H. K. (1992) Interaction of endothelial cells with a laminin A chain peptide (SIKVAV) in vitro and induction of angiogenic behavior in vivo. *J. Cell Physiol.* **153,** 614–625.
4. Passaniti, A., Taylor, R. M., Pili, R., Guo, Y., Long, P. V., Haney, J. A., Pauly, R. R., Grant, D. G., and Martin, G. R. (1992) A simple, quantitative method for assessing angiogenesis and antiangiogenic agents using reconstituted basement membrane, heparin, and fibroblast growth factor. *Lab. Invest.* **67,** 519–528.
5. Kibbey, M. C., Corcoran, M. L., Wahl, L. M., and Kleinman, H. K. (1994) Laminin SIKVAV peptide-induced angiogenesis in vivo is potentiated by neutrophils. *J. Cell Phys.* **160,** 185–193.

4

Alginate Microbead Release Assay of Angiogenesis

Clifford Y. Ko, Vivek Dixit, William W. Shaw, and Gary Gitnick

1. Introduction

Recently, the acceleration (and retardation) of blood vessel growth has been an increasingly frequent subject of study. With its potential application to a wide range of clinical disease processes, investigation certainly remains essential and promising. While in vitro investigation is traditional, well-controlled, and objective, studying angiogenesis in vivo can be quite difficult for a number of reasons. One major reason is the inherent tissue differences associated with blood vessel growth. Because all tissues are different, certain tissues tend to be inherently more vascular than others. As such, the growth (and concentration) of blood vessels occurs at different rates and proportions depending on that specific tissue. In the past several years, most in vivo angiogenesis work has been performed in the sclera as it allows for relatively easy access and the possibility of repeated observation. The sites to which investigation of angiogenesis might be applied, however, are invariably quite different and therefore additional tissues such as solid organs, fascia, muscle, and skin need to be studied as well. How can this be performed?

The application of angiogenic growth factors has been used to induce blood vessel growth. Growth factors, however, are also not easy to use. The half-life of most growth factors is very short (i.e., minutes). While previous investigations have sprinkled or dripped growth factors onto the study sites, those methods have led some to question their efficacy or accuracy. Furthermore, it is well known that once a tissue is manipulated by dissection, cutting, suturing, and so forth, angiogenesis invariably occurs as part of wound healing or tissue ischemia or both. As a result, if angiogenesis is demonstrated in vivo after dissection, the role of the angiogenic growth factor is uncertain, because the

From: *Methods in Molecular Medicine, Vol. 46: Angiogenesis Protocols*
Edited by: J. C. Murray © Humana Press Inc., Totowa, NJ

outcome may have resulted only from the tissue manipulation, wound healing, or ischemia.

This chapter describes a technique using a slow-release system for angiogenic growth factor, one that permits induction of angiogenesis virtually anywhere the system is applied. We utilize a microbead alginate delivery system ***(1–3)***. The advantage of this system is that it can be constructed with any type of growth factor [we have used it jointly with a nerve growth factor, brain-derived neurotrophic factor [BDNF], epidermal growth factor, and others], and the system is quite simple to apply in vivo. We have also found that subsequent molecular, histological, and/or immunological investigation of the tissue can be performed quite easily. This system, which does utilize tissue manipulation during its application, has been demonstrated to induce angiogenesis beyond that which would occur solely by tissue manipulation or ischemia, etc. Moreover, this system produces consistent results for a given tissue.

2. Materials

1. Calcium alginate: Keltone TM LV (Kelco, San Diego, CA). The alginate, derived primarily from *Macrocystis pyrifera*, contains approx 61% mannuronic acid and 39% guluronic acid.
2. NaCl: 0.9% solution.
3. $CaCl_2$: 1.5% solution.
4. Ultraviolet (UV) light for sterilization.
5. Compressed air supply.
6. Droplet-generating apparatus: A schematic representation of the droplet-generating apparatus used for this application is shown in **Fig. 1**. It consists of a syringe attached to an air-jet nozzle device. The nozzle consists of a solid brass cylinder that has been drilled along its axis to allow the placement of a 22- or 23-gauge needle through it. The hole is tapered from the top to the bottom in such a way that at the bottom of the nozzle there is a 1-mm circular clearance around the needle once it is in place. The top of the needle should be snug, so as to prevent movement of the needle once it is in place. The needle is not beveled but cut at right angles, with care to ensure that the cut edges are smooth. Placement screws are located near the middle of the nozzle to hold the needle in place. Also, these screws can be used to adjust the placement of the needle so that it is exactly in the center of the hole. The nozzle also consists of a second lateral hole that is drilled at a right angle, near the top part of the nozzle, that connects with the longer tapered hole that runs along the axis of the nozzle. The lateral hole is connected to a compressed air supply such that a constant stream through the air-jet nozzle flows coaxially to the needle. After the syringe is attached to the air jet it is placed in a syringe pump as shown (**Fig. 1**). The material to be encapsulated is pushed through the needle and as soon as it emerges it is plucked by the coaxial air-flow and is collected in the $CaCl_2$ solution. The mixture extruded through the droplet-generating apparatus forms microbeads of approx 300–700 μm diameter.

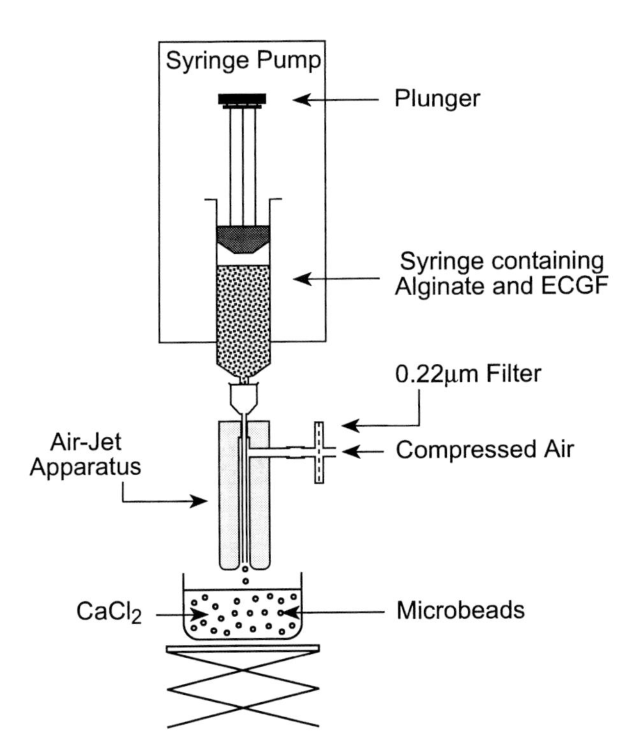

Fig. 1. Schematic demonstrating the microencapsulation apparatus used to create microbeads for the alginate microbead release system.

Larger and smaller diameter microbeads may also be prepared by using a smaller gauge needle and by adjusting air-flow rate through the nozzle. The optimal air-flow rate with our device is approximately 65–70 mL/sec. (*See* **Note 1**).

3. Methods

3.1. Alginate Preparation

1. Prepare a 3% solution of sodium alginate in sterile 0.9% NaCl.
2. Centrifuge the solution at 20,000*g* for 3 h to remove all particulate debris.
3. Aliquot the sodium alginate solution into sterile tubes and leave overnight under UV light to insure complete sterilization.

3.2. Calcium Alginate-Angiogenic Growth Factor Release System

The amount of growth factor used will depend on the activity of the preparation. The total amount of alginate used will be in a volume of 4 mL. This amount should be kept in mind when calculating the final concentration of growth factor (e.g., if the final concentration requires 1 mg/mL, then 4 mg is needed). Once this is calculated, the required amount of growth factor is dissolved in 0.25 mL of 0.9% NaCl solution. The freshly dissolved angiogenic growth factor is then added to 3.75 mL sodium alginate.

To construct microbeads, the mixture is extruded through the droplet-generating apparatus to form microdroplets of approximately 300–700 μm diameter. In this reaction, the free-flowing aqueous sodium alginate microdroplets immediately form solid calcium alginate hydrogel microbeads upon exposure to 1.5% calcium chloride solution. The microbeads are quickly rinsed in a minimal amount of calcium chloride solution.

3.3. In Vivo Placement of Alginate Release System

Our in vivo work thus far has been in the male Wistar rat weighing approx 300 g. We have placed the alginate release system in a variety of sites; however, probably the easiest and most reliable site is the fasciovascular groin flap *(4–5)*. The fasciovascular tissue is a translucent vascular film that adheres to the vessels and surrounding groin connective and adipose tissues. The blood supply to this tissue is supplied by the superficial inferior epigastric vessels, which branch directly from the femoral vessels.

1. Make a transverse incision in the groin crease.
2. Carefully dissect down to the femoral vessels. Superior and slightly superficial to these vessels is the groin fasciovascular tissue. Isolation of this fascial and fat tissue with the superficial epigastric vessels is possible with meticulous dissection. Take care not to transect the epigastric vessels.
3. The pedicled flap produced is isolated to an approximate size of 3 by 4 cm.
4. Alginate microbeads containing growth factor are placed within the flap, either by embedding them into the flap or by wrapping the flap around the beads and suturing it closed like a sack. Other shapes may be easier to use in certain sites (*see* **Note 2**).

4. Notes

1. When using the air jets, it is critical that the needle is placed at the center of the nozzle opening; failing to do so will result in irregularly shaped microbeads. In addition, alginate can sometimes adhere to the sides of the nozzle, thus clogging it. The air-flow rate should also be monitored carefully. A slow rate of flow may form a large droplet that can sometimes clog tip of the nozzle. Also, if the air-flow rate is too high it may cause the $CaCl_2$ to splash up to the nozzle tip, thus clogging it. The viscosity of the alginate is also a concern sometimes. A lower viscosity alginate will flow faster through the nozzle and the air-flow rate should be regulated to control microbead shape and size. We have found a 3% alginate solution to be optimal for our purpose. Finally, the tip of the needle should optimally be placed approx 1 cm from the surface of the $CaCl_2$. Placing it higher (i.e., further from the $CaCl_2$ solution) will result in irregularly shaped microbeads. Placing it lower risks splashing the $CaCl_2$ on to the nozzle tip, causing it to clog.

It should be noted that this is a very convenient procedure for immobilizing growth factors. The procedure is carried out at room temperature using gentle, buffered solutions. Thus care should also be taken to see that the $CaCl_2$ is not ice cold.

2. Variations in microbead shape: In our initial experience with the alginate release system, we utilized the microbead shape. However, when we started using the system in vivo, we found that the beads were sometimes a bit too cumbersome to use in certain situations, as they often would migrate with gravity or easily dislodge with even the slightest tissue manipulation. Therefore, we started constructing other shapes out of the identical alginate material. We have made more practical delivery-system shapes, including a filamentous (string-shaped) and a flat-sheet shaped design. We have found that both shapes can be more functional in vivo in some anatomic regions. The filamentous configuration is advantageous because it can be wrapped around certain structures, while the flat-sheet system can be laid securely adjacent to appropriate tissues to target release.

   To make the alginate filament system, the mixture of 3% alginate and growth factor is extruded through a 25-gauge needle into the 1.5% calcium chloride solution. To ensure uniform diameter of the string, a syringe pump apparatus is utilized to apply uniform pressure to the syringe. To make the flat sheet alginate system, our engineering department constructed a special device made of Lucite that fits securely onto a syringe and extrudes a uniformly flat alginate sheet through a slit at its endpoint. The slide dimensions are 25 mm × 0.25 mm. Again, a mixture of 3% alginate and growth factor is extruded through this device into the 1.5% $CaCl_2$ solution. To facilitate string and sheet formation, we found it advantageous to submerge the device tip in $CaCl_2$ solution during construction.

References

1. Ko, C. Y., Dixit, V., Shaw, W., and Gitnick, G. (1995) In vitro slow release profile of endothelial cell growth factor immobilised within calcium alginate microbeads. *Art Cells, Blood Subs. and Immob. Biotech.* **23,** 143–151.
2. Dixit, V., Darvasi, R., Arthur, M., Brezina, M., Lewin, K., and Gitnick, G. (1990) Restoration of liver function in Gunn rats without immunosuppression using transplanted micro-encapsulated hepatocytes. *Hepatology* **12,** 1342–1349.
3. Ko, C. Y., Dixit, V., Shaw, W., and Gitnick, G. (1997) Extensive in vivo angiogenesis from the controlled release of endothelial cell growth factor: implications for cell transplantation and wound healing. *J. Controlled Rel.* **44,** 209–214.
4. Ko, C. Y., Dixit, V., Shaw, W., and Gitnick, G. (1995) Succesful xenotransplantation of microencapsulated hepatocytes in the rat fasciovascular groin flap. Presented at the Tenth World Congress of the International Society for Artificial Organs, Taipei, Taiwan. Abstract book: **42.**
5. Borud, L. J., Shaw, W., Passaro, Jr. E., Brunicardi, F. C., and Mullen, Y. (1994) The fasciovascular flap: a new vehicle for islet transplantation. *Cell Transplant.* **3,** 509–514.

5

Disc Angiogenesis Assay

Anthony C. Allison and Luis-F. Fajardo

1. Introduction

The aim of our research was to develop a quantitative assay for angiogenesis in mammals, especially the mouse. This is a convenient experimental animal because of its small size, which allows compact housing and experimentation with angiogenic factors or inhibitors in limited supply. Mouse genetics is an advanced discipline, resulting in the availability of many inbred strains and histocompatible tumors. Recombinant growth factors and other proteins are usually of human or mouse origin, and the desirability of using proteins of the experimental animal under study has been demonstrated. Mice genetically engineered to overproduce or not produce particular growth factors or receptors are valuable experimental tools.

Growth factors can accelerate the proliferation and/or migration of endothelial cells and surrounding connective tissue cells, and it is desirable to quantify these processes independently. The disc angiogenesis assay arose from our observations of the growth of blood vessels into a polyvinyl alcohol sponge implanted sub-cutaneously in the mouse *(1)*. The design was improved to allow quantification of vascular in-growth from the circumference of a disc, as well as incorporation of labeled thymidine into endothelial and other connective tissue cells *(2)*. In this way effects of angiogenic factors or inhibitors on the proliferation and migration of endothelial cells and surrounding connective tissue cells could be monitored separately.

2. Materials

1. Polyvinyl alcohol sponge (Kanebo, PVA, Rippey Co., Santa Clara, CA); 2 mm thickness.
2. Cell-impermeable filters (0.45 micron, HAWP 013, Millipore).

From: *Methods in Molecular Medicine, Vol. 46: Angiogenesis Protocols*
Edited by: J. C. Murray © Humana Press Inc., Totowa, NJ

3. #1 Millipore glue (XX70 000 00, Millipore).
4. Acetate copolymer (Elvax, Dupont).
5. [Methyl-^{3}H]-Thymidine (6.7 Ci/mmol, Dupont).
6. Photographic emulsion (NTB-2, Eastman Kodak Corp).
7. Leuconyl blue (BASF; distributed by Wyandotte Co., Hotland, MI).

3. Methods

3.1. Preparation of Sponge Disc

1. A disc of 13 mm diameter and 2 mm thickness is prepared from sterile materials, under sterile conditions within a laminar-flow hood. Alternatively, nonsterile discs can be sterilized by a variety of methods.
2. The flat circular sides of the disc are covered with the cell-impermeable filters and sealed with glue (**Fig. 1**). This leaves only an external 2 mm rim for penetration or exit of cells.
3. When indicated, and prior to placement of the filters, a hole 3 mm in diameter is bored in the center of the disc.
4. A pellet of sponge material 2 mm in diameter, containing an angiogenic agonist or antagonist to be tested, is coated with acetate copolymer and placed in the central hole. Each pellet of this size can hold up to 20 µL of the material to be tested. Test substances we have used, providing a strong angiogenic stimulus, include recombinant human epidermal growth factor (EGF; usually 20 µg per disc) and recombinant human basic fibroblast growth factor (bFGF; usually 20 µg per disc).
5. Care should be exercised to avoid physical or chemical damage to the sponges, or to the angiogenic or inhibitory agents once loaded into the disc. Sponges tolerate dry heat up to 120°C, but can be damaged by boiling water.

3.2. Implantation of Sponge Disc

1. The mouse is anesthetized by an intraperitoneal injection of ketamine hydrochloride (50 mg/kg), xylazine (5 mg/kg), and acepromazine maleate (1 mg/kg).
2. The shaved skin surface is sterilized with 70% ethanol, and a 1.5 cm long incision into the subcutis is made at least 1 cm away from the desired location of the disc. Blunt dissection is used to produce a tunnel toward the site of implantation.
3. Phosphate-buffered saline (PBS) is dripped into the area with a Pasteur pipette. Holding it gently with forceps, the PBS-moistened disc is then inserted through the wound and into the tunnel up to the desired site.
4. The skin wound is closed with 3-4 metal clips.
5. The animals are housed as usual, with standard chow and water *ad libitum*, during the period of the experiment, 7–30 d, most often 14 d (*see* **Notes 1–4**).

3.3. Assay for Cellular Proliferation in Sponge Disc

Uptake of tritiated thymidine (^{3}HTdR) is used to evaluate DNA synthesis, and therefore proliferation of endothelial and other cells *(3)*. Total incorpora-

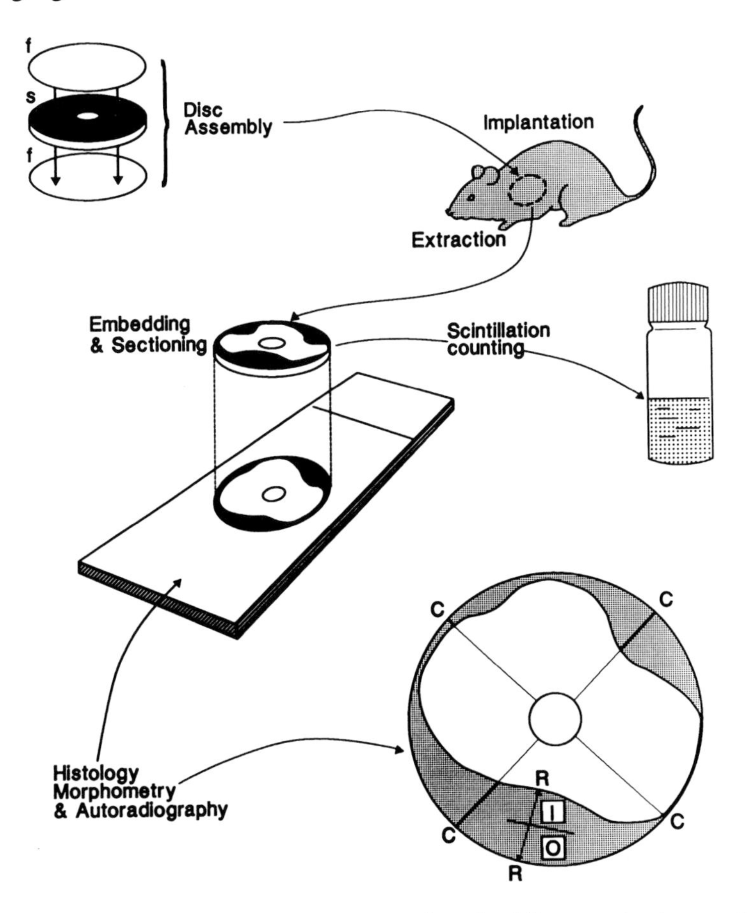

Fig. 1. Disc assembly, implantation, removal, embedding, sectioning, and analysis (quantitative and qualitative). *See* **Subheading 3.**

tion of label is measured by scintillation counting while autoradiography identifies the different cell types synthesizing DNA.

1. Mice are injected intraperitoneally with [methyl-^{3}H]-thymidine (66 μCi/25 g body weight) at 24, 18, and 12 h before sacrifice (total = 200 μCi). Alternatively radiolabel can be delivered continuously by osmotic minipumps (Alza Co., Palo Alto, CA) implanted subcutaneously, for 7–14 d.
2. After sacrifice, tissues are removed, fixed, and paraffin-embedded.
3. Serial 6 μm-thick sections are cut.
4. For determination of total radioactivity, sections are subjected to alkali solubilization and beta radiation activity determined in a scintillation counter. These values are representative of DNA synthesis of all cells in the sample.

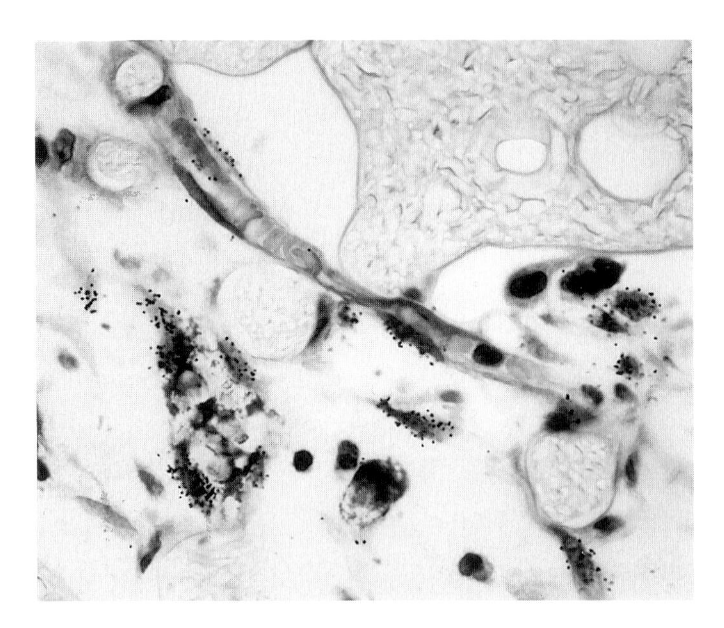

Fig. 2. Autoradiograph of paraffin section from disc stimulated with 20 mg of bFGF. The diagonally-oriented capillary in the center has at least two endothelial cells labeled by ^{3}HTdR. The labeled cells outside of the capillary are fibroblasts. Inner zone of growth. Counter-stained with H&E ×460.

5. For autoradiography, sections are prepared by dipping the slide-mounted sections in photographic emulsion. These slides are held at 4°C for 35 days and developed. The developed slides are counterstained with hematoxylin and eosin. In these autoradiographic preparations, we perform differential counting of labeled vs. unlabeled cells using standard-size microscopic fields, two in the outer and two in the inner zone of each disc. Cells containing 5 or more grains are considered labeled. (In our experience, after 5 wk exposure the great majority of labeled cells contain more than 10 grains per cell, while background is less than 1 grain per 100 μm^2; **Fig. 2.**) Raw data were analyzed using RSI statistical analysis software. Having obtained means, standard errors, and standard deviations for each group of discs, comparisons of experimental and control groups are carried out using Student's *t* test. Other methods for identifying cells that have proliferated are available.

3.4. Identification of Blood Vessels in Sponge Discs by Light Microscopy

1. An injection of Leuconyl blue (0.4 mL of a 40% solution in PBS) is given intracardially to the mouse 15 min prior to euthanasia ***(4)***.
2. Mice are killed with a large intra-peritoneal dose of anesthetic followed by quick cervical dislocation, or by placing the animal(s) in a concentrated CO_2 chamber.

3. A careful incision is made in the skin overlying the implanted disc, which is then dissected gently from the surrounding tissues, and removed.
4. The filter is carefully dissected away from the disc with a sharp blade to permit effective fixative penetration and dehydration.
5. The dissected disc is placed in 10% neutral formalin for 48–72 h. The disc is the embedded flat in paraffin, in such an orientation that initial sections will be taken from the uncovered side of the disc.
6. Multiple, 6-µm thick planar sections are cut (**Fig. 1**). For light microscopy and measurement of radial growth, sections are stained with hematoxylin and eosin; for measurement of total area growth, with toluidine blue. Sections are also cut for autoradiography and scintillation counting.

3.4. Quantification of Angiogenesis

3.4.1. Centripetal Vessel Growth

Direct measurement of blood vessels can be performed by various methods, including point counting on histologic sections *(5)* and determination of intravascular volume, e.g., with radioactive isotopes *(6)*. Such methods are tedious and impractical when examining a large number of discs. We have devised a simpler, indirect procedure based upon the centripetal growth of new vessels:

1. Project an image of the section at 100× magnification onto a horizontal surface (we use a Bausch and Lomb microprojector).
2. Identify the innermost blood vessel.
3. Measure in centimeters the radial distance between the leading tip of the innermost vessel and the rim of the disc (**Fig. 1**).

We have named this end-point "centripetal vessel growth" (CVG). The CVG is directly proportional to the total vessel growth and is therefore a reliable, indirect measure of such growth *(1,2)*.

3.4.2. Total Growth Area

We have shown in **Subheading 3.4.1.** that the total growth area within the disc is consistently proportional to the area occupied by blood vessels (**Fig. 3**). Therefore, any measurement of the total growth area is an accurate, indirect measurement of the area occupied by blood vessels. A method of determining the total growth area is to measure the radial fibrovascular growth along two perpendicular diameters (four radii) using the projection apparatus (**Subheading 3.4.1., step 1**; **Fig. 1**). The average value of the four measurements is then obtained for each disc.

A precise system of measuring total growth area uses 6-µm paraffin sections stained with toluidine blue to obtain high contrast (**Fig. 1**). Toluidine blue uniformly stains the area of growth and not the sponge trabeculae. The sections

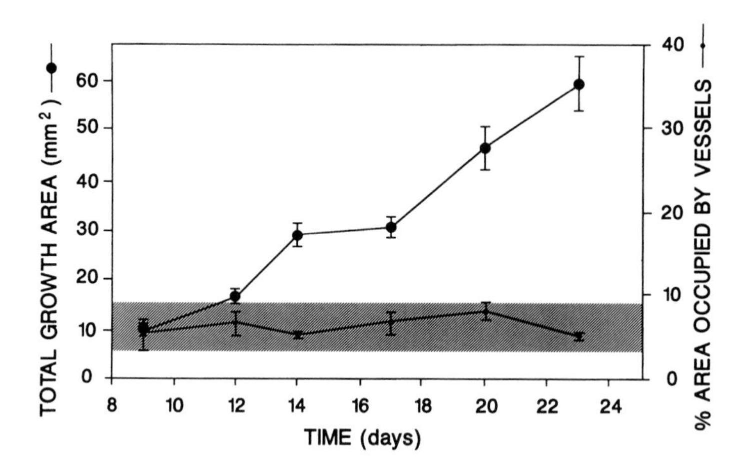

Fig. 3. Vessel growth is directly proportional to total fibrovascular growth (*see* **Subheading 3.4.**). As the latter increases continuously (*upper line*) during the period of observation, the proportion of the growth area occupied by blood vessels (*lower line*) remains constant. Total growth area was measured by computer-assisted digital image analysis, and vascular area by point counting (*see* text). Growth stimulated by 20 mg of EGF. Bars, standard error of mean.

(mounted on slides without cover slips) are placed on the stage of a microscope fitted with a computer-assisted digital image analysis system using the NIH Image program, which measures the entire area of growth automatically, in pixels. Threshold values are standardized for a particular experiment, and pixel values are converted to mm^2.

3.4.3. Differential Cell Counting

Differential counting of the various cells that comprise the fibrovascular growth is performed with microscope reticules. The relative proportions of various cell types, as well as the absolute numbers of the cells per given area are established by differential counting within standard-size areas of the various zones of growth in the disc assay (**Fig. 1**). The space occupied by vessels and other elements is determined using the intersection points of microscope grids. Such procedures, which are time consuming, were used initially to analyze the biology of fibrovascular growth and to validate the more often used quantitative methods described in the previous paragraphs.

3.6. Characteristics of Vascular Growth in the Disc Assay

Centripetal growth into the disc is composed of blood vessels and surrounding stroma (**Figs. 1, 4**) and is usually asymmetric. This asymmetry may be related to the density of mother vessels (venules) in the surrounding host tis-

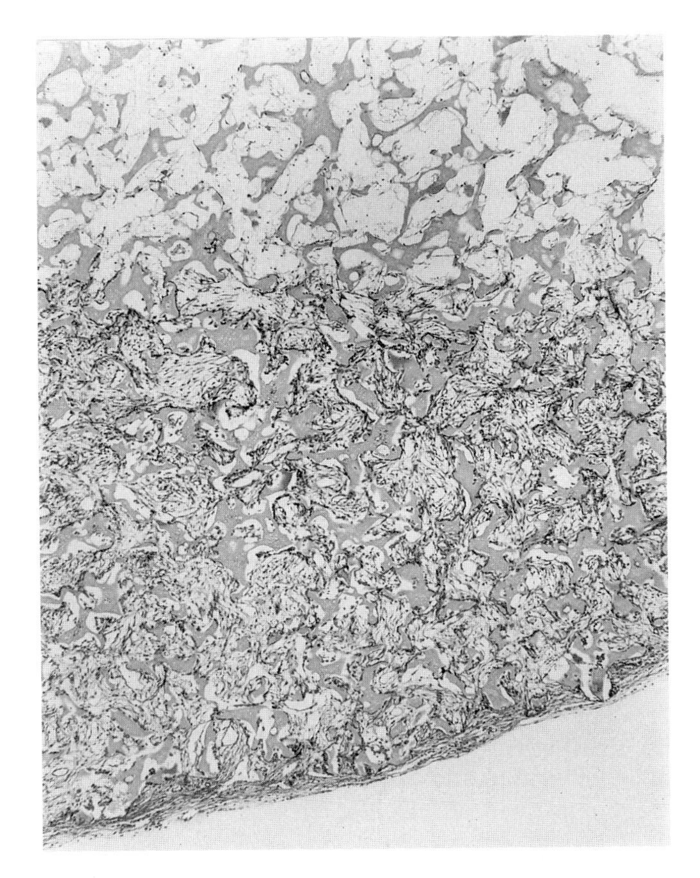

Fig. 4. Section of disc showing empty polyvinyl sponge in the upper one third of the photograph (*inner portion of disc*). The lower two thirds (*outer, or peripheral portion of disc*) contain vessels and fibroblasts among the sponge trabeculae. The rim of the disc is at the bottom. Disc stimulated with 20 mg of EGF, 14 d after implantation. H&E ×50.

sue, from which the vessel sprouts are derived. The stroma contains fibroblasts, macrophages, neutrophils, and occasional lymphocytes, in order of frequency of occurrence. The proportions of these elements vary considerably between the inner zone of the disc and the outer zone (as divided by an arbitrary circular line into two bands). Some time-dependent variation is observed, but this is not substantial. There is no significant difference in the proportion of cell types between discs with spontaneous growth (i.e., occurring without stimulant) and discs in which growth has been stimulated by endothelial growth factors. By using metachromatic stains a moderate number of mast cells can be demonstrated in the tissues of the host immediately around the disc, at d 20 and beyond. Neither the 14- nor the 27-d discs contain mast cells within the regions of angiogenesis.

Throughout the entire area of growth the majority of cells are fibroblasts; the second most common cells are endothelial; and other cells (neutrophils, macrophages etc.) comprise the rest. By 14 d of growth stimulated by 20 µg of epidermal growth factor (EGF) the proportions of these three categories of cells are 53%, 25%, and 22%, respectively. However in the outer zone of the disc, the cellularity is high, while in the inner zone it is low (**Fig. 4**). Although nonendothelial cells predominate in both zones (67–81% at d 14 of spontaneous growth), the proportion of endothelial cells (EC) is higher in the inner than in the outer zone (33% versus 19% at 14 d of spontaneous growth). The space occupied by the various elements has been determined by point counting with a grid in standard histology sections *(5)*. The sponge trabeculae occupy a constant proportion of the disc (around 30%). At 14 d of EGF-stimulated growth, fibroblasts comprise 18% and blood vessels only 5%; a large portion of space (>46%) is not occupied by vessels, cells, or trabeculae.

The area of the disc occupied by blood vessels is a consistent percentage of the total growth area regardless of the size of the latter. In other words, as the area grows from d 9 to d 23, the portion occupied by blood vessels grows in parallel to the growth of the total fibrovascular area. Hence, measurement of the total area is an accurate index of the size of the blood vessel area (**Fig. 3**). Electron microscopic observations on the cell types in the disc, and on the stroma, have been published (**2**; **Fig. 5**).

3.7. Normal Growth Pattern

We have studied sequentially the fibrovascular growth in the disc assay. The typical growth curve induced by EGF is shown in **Fig. 6**. The total fibrovascular growth area increases continuously, with some plateau between 12 and 15 days. The total incorporation of ^{3}HTdR (also shown in **Fig. 6**) increases exponentially, reaching maximum at d 12, and then declines continuously to d 23. This decline in the number of cells undergoing DNA synthesis, in the face of continuing tissue growth, indicates that beyond d 12 cell migration makes a proportionately greater contribution than cell proliferation. From multiple ^{3}HTdR autoradiography experiments it is clear that most of the cells synthesizing DNA are located in the inner zone, as opposed to the outer zone. For example, on the 14th d of spontaneous growth 49% of the EC in the inner zone are labeled (vs 7% in the outer zone). The same is true for fibroblasts (35% vs 6%).

3.8. Applications of the Disc Assay

This system has been used to test the effects of putative agonists and antagonists of angiogenesis, pharmacological and otherwise. We have performed sev-

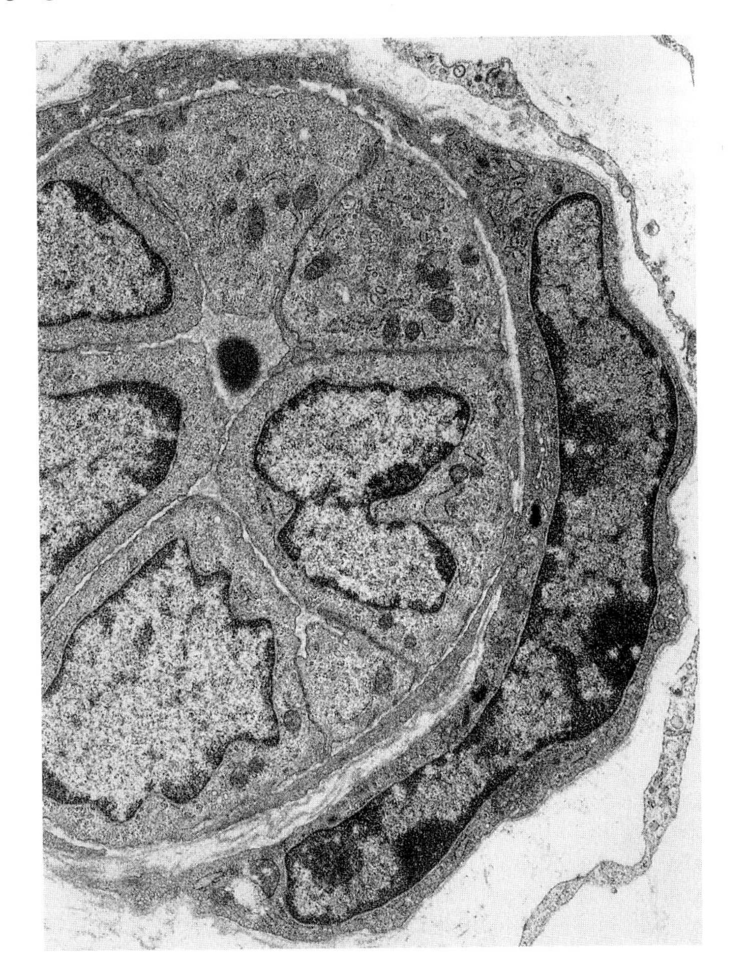

Fig. 5. Vessels from the inner zone of growth ECs are pyramidal, with an outer base invested by a well-defined basement membrane. The lumen is small and contains a small part of a tangentially sectioned red cell. Most of the vessel is encircled by a pericyte (*right*) with a flattened nucleus and prominent rough endoplasmic reticulum. This cell is also covered by a basement membrane. The disc contained 20 µg basic FGF. Stained with uranyl acetate and lead citrate. ×12,600.

eral experiments using the assay for this purpose, some of which are described in the following subheadings.

3.8.1. Effects of Angiogenic Agonists

Initially we tested the effects of known agonists. EGF, or prostaglandin (PG)E_1 were placed in the center of the discs and shown to increase fibrovascular growth. These agonists also increased DNA synthesis by ECs. Heparin,

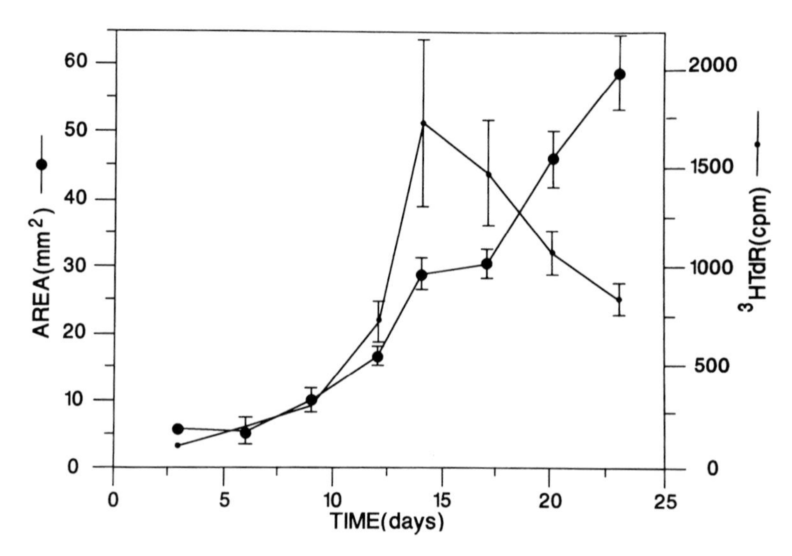

Fig. 6. Total growth area (measured by digital image analysis) is compared with total cell proliferation (as determined by ^{3}HTdR incorporation) in discs containing 20 mg of EGF. While total growth area is continuously increasing, ^{3}HTdR incorporation decreases from d 14 on. Each point corresponds to 5–10 discs. Bars, standard error of mean.

used alone, increased the area of fibrovascular growth without increasing DNA synthesis, whereas heparin plus bFGF increased DNA synthesis as well as the growth area (**Fig. 7**) (*see* **Note 5**).

3.8.2. Inhibitors of Angiogenesis

The disc assay was then applied to study the effects of postulated inhibitors of angiogenesis, either local (hyperthermia, X-radiation) or systemic (glucocorticoids, cyclooxygenase inhibitors). The hypothesis that hyperthermia—as used in cancer therapy—interferes with angiogenesis was based on our in vitro observation that capillary EC are sensitive to hyperthermia, which is lethal for EC in a dose-dependent fashion *(7)*. The results left no doubt that hyperthermia, within the dose range used in clinical cancer therapy, inhibits angiogenesis *(8)*. Hence, at least some of the antitumor effects of hyperthermia are the result of interference with tumor microcirculation.

In another study *(9)* effects of X-radiation on angiogenesis were analyzed. A dose-response relationship was observed when X-radiation was administered on d 11 after implantation and the disc was removed on d 20 *(9)*. These and previous observations point to ECs as important targets of ionizing radia-

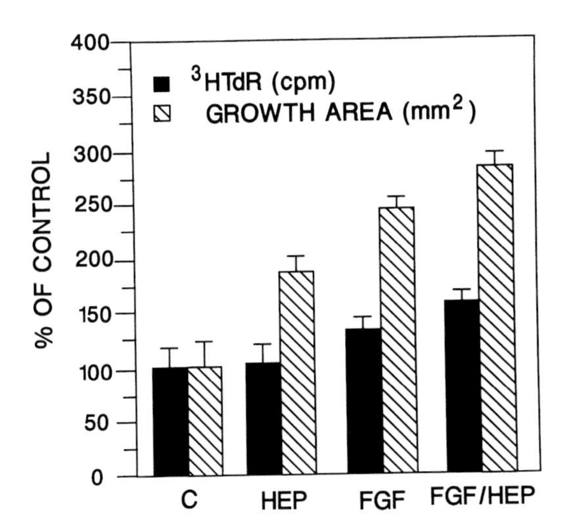

Fig. 7. Heparin promotes angiogenesis mainly by stimulating cell migration. Comparison of total growth area (*hatched bars*) with cell proliferation (expressed as ^{3}HTdR incorporation; *solid bars*) indicates that little or no cell proliferation contributes to total growth stimulated by heparin alone. When bFGF is added, both growth and proliferation are increased. C, spontaneous growth; HEP. porcine heparin (Sigma, 50 mg per disc); FGF, basic fibroblast growth factor (20 mg per disc). A total of 200 mCi of ^{3}HTdR was infused subcutaneously over 14 d by an Alza osmotic minipump. Both growth and cell proliferation are expressed as percentage of control values. All samples analyzed at 14 d of growth. Standard errors of mean are indicated.

tion in the stroma, especially during the period of active proliferation of these cells induced by growth factors.

Effects of systemically administered small molecular inhibitors of angiogenesis were also studied. Glucocorticoids are potent inhibitors of EC proliferation in culture and of angiogenesis in vivo ***(10)***. In the disc assay gluco-corticoids were found to be highly effective in counteracting the angiogenic effects of EGF and bFGF (b-fibroblast growth factor). For reasons outlined in the discussion we postulated that E- and I-type prostaglandins might be transducers of the angiogenic effects of bFGF. It followed that a cyclooxygenase (PGH synthase) inhibitor (ketorolac) might block the in vivo angiogenic effect of bFGF. Using the disc assay model, this was found to be the case; ketorolac inhibited bFGF-stimulated angiogenesis, presumably because it interfered with the synthesis *de novo* of PGE. However, it did not affect angiogenesis stimulated by preexisting PGE (**Fig. 8**), showing that the drug does not block signal transduction mechanisms following the binding of PGs to receptors. There is considerable current interest in the possibility that selective inhibitors of the

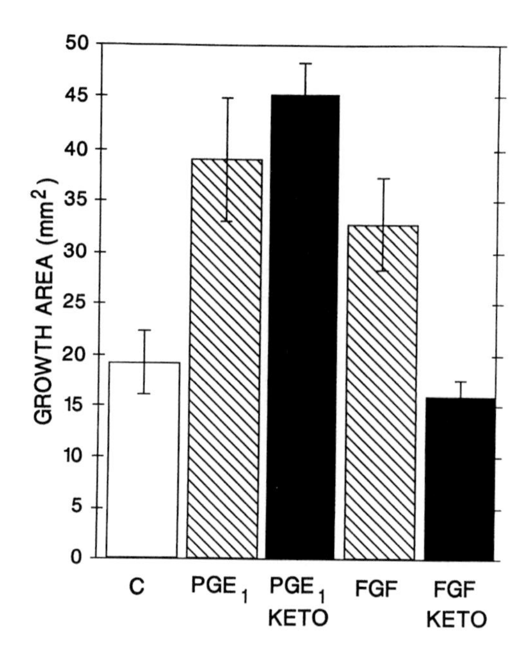

Fig. 8. Effect of a cyclooxygenase inhibitor (ketorolac) on angiogenesis. In vitro studies have indicated that capillary EC proliferation stimulated by fibroblast growth factor requires the presence of PGs. As shown here in the disc assay, the same occurs in vivo: ketorolac (4 mg daily by gavage) inhibits the 14 d angiogenesis stimulated by bFGF (20 mg; last 2 bars), presumably because it interferes with the synthesis of new PGE. However, ketorolac does not affect angiogenesis stimulated by already synthesized PGE (PGE_1; 75 mg in the center of the disc) as shown in the 2nd and 3rd bars. The control disc in the first bar contained no agonist or antagonist and thus indicates spontaneous growth. C, control. PGE_1, prostaglandin E_1. FGF: basic fibroblast growth factor; Keto. ketorolac. Standard errors of mean are indicated.

type 2 isoform of PGH synthase may inhibit angiogenesis without damaging the stomach or small intestine.

The synthetic flavonoid, flavone-8 acetic acid (FAA), has antitumor effects in the mouse, which have been attributed to effects on microvessels ***(11)***. Tested in the disc assay in a wide range of concentrations (10–60 μg) FAA had no demonstratable effect on angiogenesis (unpublished).

3.8.3. Tumor necrosis factor-α (TNF-α)

TNF-α has been reported to suppress EC proliferation and the formation of microvascular sprouts in vitro, and to have toxic effects on stimulated EC. It has also been reported to induce angiogenesis in the rat and rabbit corneas.

Because of these discrepant effects, and the induction by TNF-α of necrosis in some experimental animal tumors, we investigated systematically the effects of the cytokine on angiogenesis in mice ***(12)***. We used recombinant 17 kDa mouse TNF-α, which binds to and activates both the p55 and p75 receptor (human TNF-α binds selectively to the p55 receptor). Two sets of experiments employing the disc assay in 136 BALB/c mice revealed a bimodal angiogenic response to mrTNF-α. At low concentrations (0.01–1 ng/disc) mrTNF-α induced angiogenesis: both vessel growth and EC proliferation ($p<0.01$ as compared with spontaneous growth). At high concentrations (1–5 μg/disc), or delivered at the high rate of 16–60 ng/h by osmotic minipumps, mrTNF-α also inhibited angiogenesis in regionally-placed discs.

These findings validate those of previous investigators that appeared contradictory. They also help to explain the effects of TNF-α when used as a cancer chemotherapeutic agent. Since the majority of tumor cells studied are resistant to the direct cytotoxic effects of TNF-α, it has been proposed that the observed tumor necrosis produced by the cytokine may be due to effects on tumor vasculature. TNF-α induces the expression on EC of tissue factor and has other procoagulant effects. In high concentrations TNF-α is cytotoxic for stimulated EC. Such effects could rapidly lead to tumor necrosis. Our findings suggest that sustained high levels of TNF-α could also suppress angiogenesis, whereas lower concentrations could actually augment the process.

All these findings suggest that high concentrations of TNF-α would be required to exert antitumor effects in vivo. The cytokine has shocklike, procoagulant, and other toxic effects, which limits the systemic concentrations that can be attained. Under these conditions TNF-α has not been found to have clinically useful antitumor activity. However, intraarterial administration to humans by isolated limb perfusion of high doses of TNF-α, together with melphalan, interferon-γ, and hyperthermia, was reported to show efficacy in patients with sarcomas and melanomas ***(13)***.

3.8.4. Transforming Growth Factor β1 (TGF-β1)

TGF-β1 is a multifunctional acid-stable dimer of 25 kDa produced by many cell types. Again, it has been reported by some authors to induce angiogenesis, or potentiate angiogenesis induced by vascular endothelial growth factor, whereas others have reported that TGF-β inhibits angiogenesis. We have investigated the effects of TGF-β1 in the disc assay ***(14)***. TGF-β in concentrations from 1–500 ng had no demonstrable effect on angiogenesis whereas large doses (1–2 μg) induced significant angiogenesis. Antibody against TGF-β decreased spontaneous vascular growth to a level below that observed in controls containing irrelevant IgG, suggesting that this cytokine is involved in normal angiogenic processes such as wound healing. Discs containing TGF-β showed more collagen and a greater accumulation of neutrophils than control discs or discs containing other cytokines.

It was concluded that TGF-β is angiogenic in vivo when it reaches a relatively high threshold level. This angiogenic effect of TGF-β may be indirect, requiring the recruitment of leukocytes that produce angiogenic factors. TGF-β is one of the cytokines contributing to the "spontaneous" angiogenesis which occurs in the disc assay when growth factors are not added.

3.9. Advantages and Disadvantages of the Disc Assay

Each of the angiogenesis assays has its own advantages and disadvantages, and each one has an application for which it is best suited. For instance, observation of living individual capillaries as they form and become patent is best done in the rabbit ear chamber ***(15)*** or its successors, such as the rat air pouch ***(16)*** and the hamster cheek pouch ***(17)***. Sequential recording of the growth of multiple capillaries into a biological material is best done in the cornea (of rabbits, rats, and others) ***(18)***. The measurement of vascular growth is best performed in the disc assay.

3.9.1. Advantages

The disc consists of easily obtainable and inexpensive materials. Many discs can be used in one experiment. Being small, they are implanted easily in various animal species, particularly rodents. The discomfort associated with the corneal assays does not occur with the disc assay. It is well tolerated and causes little or no skin inflammatory reaction that could augment or interfere with the process of angiogenesis. The animals do not scratch the incision wound or the site of implant.

The disc assay is a mammalian system, unlike the chorioallantoic membrane, and therefore more relevant to human physiology and pathology. In the disc assay there is always a moderate, well-characterized and measurable spontaneous vascular growth, while in other systems there is little or no spontaneous vascular growth. Therefore in the disc assay one can test not only angiogenic agonists, but also antagonists, the latter by measuring the decrease in spontaneous growth.

Unlike in some angiogenesis models, where microvessels grow with little or no stroma, in the disc assay vessels and stromal elements grow concomitantly. The latter replicates the mesenchymal proliferation that occurs in physiological states (e.g., proliferative endometrium, wound healing) and in pathological conditions (e.g., chronic inflammation, neoplasms), characterized by angiogenesis (*see* **Notes 6** and **7**).

The most important advantage of the disc assay is the fact that vascular growth can be quantified easily and reproducibly. This is not the case in most other systems. For instance, measurement of vessel growth in the cornea involves complex and lengthy procedures ***(18)***. Discriminating between new and preexisting vessels in the hamster pouch can be difficult, while in the disc assay all vessels are new by definition. The entire circumference of the disc is available in a section, and all cellular elements can be identified with appropriate stains. In the disc assay multiple vessels can be studied and measured, and multiple discs can be used for each point. In systems designed for visual monitoring of living vessels (e.g., rabbit ear chamber), only a few vessels can be measured and, due to cost, there may not be enough vessels, or animals, to obtain statistically significant data. Finally, the growth curves and the relative proportions of vessels and fibroblasts are now well established for the disc assay.

3.9.2. Disadvantages

The system is not designed for, nor does it allow, progressive observations of living blood vessels. External and gross inspection cannot be performed. Each disc provides information for only one time point. Histological embedding, sectioning, and staining are required.

4. Notes

1. Mice tolerate the implants well and do not scratch incisions made on the dorsum or the flank.
2. Consistency in the handling of the animals and placement of the discs has been found to be critical to the success of a given experiment.
3. We have used mice of either sex, most often of the Swiss Webster or BALB/c strains, approximately 8 wk old and weighing 20–30 g. Other investigators in various laboratories have used the disc assay in rats or rabbits (personal communications).
4. Good fibrovascular growth occurs whether the disc is placed subcutaneously in the abdomen, or in the thorax.
5. Recombinant EGF and bFGF, have been tested in the disc assay and found to increase EC proliferation and angiogenesis. X-radiation and hyperthermia both inhibit angiogenesis in the disc angiogenesis assay, and this may be one of the ways in which they exert antitumor effects.
6. The disc assay can be used to examine the effects of fibrogenic agents or inhibitors. The system allows separate quantification of the migration and proliferation of fibroblasts and the amount of collagen synthesized, e.g., by extraction of hydroxyproline from the disc. Subtypes of collagen may also be determined immunologically.

7. The assay can also be used to examine the role of different stromal components in EC proliferation and migration, as suggested by in vitro studies *(23)*. The sponge of the disc assay can be coated before implantation with fibronectin, vitronectin, proteoglycan, various types of collagen, and other connective tissue components.

References

1. Fajardo, L. F., Kowalski, J., Kwan, H. H., Prionas, S. D., and Allison, A. C. (1988a) The disc angiogenesis system. *Lab. Invest.* **58,** 718–724.
2. Kowalski, J., Kwan, H., Prionas, S. D., Allison, A. C., and Fajardo, L. F. (1992) Characterization and applications of the disc angiogenesis system. *Exp. Mol. Path.* **56,** 1–19.
3. Hobson, B. and Denekamp, J. (1984) Endothelial proliferation in tumors and normal tissues: continuous labeling studies. *Br. J. Cancer* **49,** 405–413.
4. Reinhold, H. S., Hopewell, J. W., and Rijsoort, A. V. (1983) A revision of the Spalteholz method for visualizing blood vessels. *Int. J. Microcirc. Clin. Exp.* **2,** 47–52.
5. Chalkley, H. W. (1943) Method for the quantitative morphologic analysis of tissues. *J. Natl. Cancer Inst.* **4,** 47–53.
6. Mahadevan, V., Hart, I. R., and Lewis, G. P. (1989) Factors influencing blood supply in wound granuloma quantitated by a new in vivo technique. *Cancer Res.* **49,** 415–419.
7. Fajardo, L. F., Schreiber, A. B., Kelly, N. I., and Hahn, G. M. (1985) Thermal sensitivity of endothelial cells. *Radiat. Res.* **103,** 276–285.
8. Fajardo, L. F., Prionas, S. D., Kowalski, J., Kwan, H. H., and Allison, A. C. (1988b) Hyperthermia inhibits angiogenesis. *Radiat. Res.* **114,** 297–306.
9. Prionas, S. D., Kowalski, J., Fajardo, L. F., Kaplan, I., Kwan, H. H., and Allison, A. C. (1990) Effects of x-irradiation on angiogenesis. *Radiat. Res.* **124,** 43–49.
10. Ingber, D. E., Madri, J. A., and Folkman, J. (1986) A possible mechanism for inhibition of angiogenesis by angiostatic steroids: induction of capillary basement membrane dissolution. *Endocrinology* **119,** 1768–1775.
11. Zwi, L. J., Baguley, B. C., Gavin, J. B., and Wilson, W. R. (1990) The use of vascularized spheroids to investigate the action of flavone acetic acid on tumor blood vessels. *Br. J. Cancer* **62,** 231–237.
12. Fajardo, L. F., Kwan, H. H., Kowalski, J., Prionas, S. D., and Allison, A. C. (1992) Dual role of tumor necrosis factor-α in angiogenesis. *Am. J. Pathol.* **140,** 539–599.
13. Eggermont, A. M. M., Schraffordt, Koops, H., Kroan, B. B. R., Lienhard, D., Klausner, J. M., and Lejeune, F. J. (1994) The European experience in TNFα isolated limb perfusion for nonresectable extremity soft tissue carcinomas. *Eur. Cytokine Network* **5,** 224–226.
14. Fajardo, L. F., Prionas, S. D., Kwan, H. H., Kowalski, J., and Allison, A. C. (1996) Transforming growth factor β1 induces angiogenesis in vivo with a threshold pattern. *Lab. Invest.* **74,** 600–608.

15. Sandison, J. C. (1924) A new method for the microscopic study of living growing tissues by the introduction of a transparent chamber in the rabbit's ear. *Anat. Rec.* **28,** 281–287.
16. Reinhold, H. S., Blachiewicz, B., and Berg-Blok, A. (1979) Reoxygenation of tumors in "sandwich" chambers. *Eur. J. Cancer* **15,** 481–489.
17. Warren, B. A. and Shubik, P. (1966) The growth of the blood supply to melanoma transplants in the hamster cheek pouch. *Lab. Invest.* **15,** 464–478.
18. Proia, A. D., Chandler, D. B., Haynes, W. L., Smith, C. F., Suvarnamani, C., Erkel, F. H., and Klintworth, G. K. (1988) Quantitation of corneal neovascularization using computerized image analysis. *Lab. Invest.* **58,** 473–479.
19. Young, C. L., Adamson, T. C., III, Vaughan, J. H., and Fox, I. R. (1986) Immunohistologic characterization of synovial membrane lymphocytes in rheumatoid arthritis. *Arthritis Rheum.* **27,** 32–39.
20. Hirata, S., Matsuhara, T., Saura, R., Tateishi, H., and Hirohata, K. (1989) Inhibition of in-vitro vascular endothelial cells proliferation and in-vivo vascularization by low-dose methotrexate. *Arthritis Rheum.* **32,** 1065–1069.
21. Allison, A. C. and Kowalski, W. J. (1989) Prostaglandins as transducers of proliferation signals in microvascular endothelial cells and the pharmacological control of angiogenesis. In *Vascular Endothelium-Receptors and Transduction Mechanisms* (Catravas, J. D., Gillis, C. N., and Ryan, U. S., eds.), Plenum Press, New York, pp. 99–110.
22. Page, R. C. (1986) Gingivitis. *J. Clin. Periodontol.* **13,** 345–353.

6

Sponge Implant Model of Angiogenesis

Silvia P. Andrade

1. Introduction

The process of capillary growth, angiogenesis, is an integral part of wound healing and repair mechanisms. When it occurs during these conditions, it is tightly controlled and strictly delimited. However, in tumor growth and in a variety of vascular diseases, unrestrained angiogenesis can contribute significantly to the pathology and persistence of these manifestations *(1)*. Research on angiogenesis was initiated with the development of several bioassays that have permitted direct observations of the microvasculature in the living animal. The bioassays have been used for a variety of purposes; for example, to detect angiogenesis activity in malignant and normal cells and tissue, to screen purified test substances for angiogenic activity, and to elucidate the cellular events that accompany vessel growth. The response observed after the introduction of an appropriate stimulus such as mechanical injury or injection of neoplastic tissue implants has allowed the cataloging of the main events of the angiogenic cascade as well as the characterization of pro- and antiangiogenic factors.

The in vivo assays now available to study blood vessel formation include implantation of a variety of synthetic matrices in which determination of several quantitative and qualitative components of the fibrovascular tissue that infiltrate the sponge compartment can be assessed. The common principle underlying this technique is that injury caused by introduction of the device elicits within the area circumscribed by the implant a response that mimics the stages of wound repair; therefore, the angiogenic response in inflammatory tissue can be evaluated *(2,3)*. Additionally, spongy implants have also been used as a framework to host different tumor cell lines in rodents *(4–6)* for studying tumor angiogenesis. The advantage of implantation technique for the pur-

From: *Methods in Molecular Medicine, Vol. 46: Angiogenesis Protocols*
Edited by: J. C. Murray © Humana Press Inc., Totowa, NJ

pose of investigating tumor-induced angiogenesis is that assessment of the relative contributions of the tumor cells to early changes in the implant blood-flow can be detected even before visible growth of the tumor mass is evident.

The development of vasoactive regulatory systems and pharmacological reactivity of the neovasculature have also been investigated by means of sponge implantation technique *(6–9)*. Yet, the cannulated sponge implant model has emerged as an alternative biological route for site-specific and systemic drug delivery in cases where repeated injections at the conventional sites are not feasible. For example, the tails and skin of experimental animals are too fragile to stand daily injections. Several parameters derived from physiological, biochemical, and morphological approaches have been used for defining the components of the repair processes as well as for quantitating the implant fibrovascular infiltration of the host tissue. By employing radioactive isotopes and fluorogenic-dyes washout techniques, measurement of blood-flow in the wound compartment can be performed, which in turn indicates the functional state of the neovasculature and the interaction between the angiogenic site and the systemic circulation *(3,8)*. Biochemical determination of several components of the fibrovascular tissue, such as wet and dry weight, DNA, protein, extracellular matrix components, hemoglobin, enzyme activity and others, can provide assessment of cellular proliferation kinetics and extracellular matrix components involved in the process *(7,10–14)*. Utilizing morphological or morphometric approaches, the sequence of histological changes and vascular density can be determined *(3,8,10,11)*. **Figure 1** shows a schematic representation of several approaches and parameters that might be analyzed in the cannulated sponge model of angiogenesis. Analysis and modulation of various aspects of the inflammatory response that accompanies implantation are facilitated by the readily accessible location of the device and have contributed to new insights and strategies towards better understanding the angiogenic process. This chapter is an account of the usefulness of the sponge implant model of angiogenesis and a detailed description of the methodology.

2. Materials

1. Sponge matrix: A number of different sponge matrices have been used for inducing fibrovascular growth and as host to implanted tumor cells. The synthetic materials are mainly polyvinyl alcohol, cellulose acetate, polyester, polyether, and polyurethane alone or in combination. In our laboratory we use sponge discs made of polyether polyurethane. This type of material possesses the following characteristics: uniform pore size and intercommunicating pore structure, ability to resist chemical treatment, and biocompatibility.
2. Polythene tubing for cannula, 1.4 mm internal diameter, 1.2 cm long.

Cannulated sponge disc

Applications
- Characterization of fibrovascular tissue components
- Study of tumour and inflammatory angiogenesis
- Alternative biological route for drug delivery

Parameters
- Biochemical -enzyme activity
  -extracellular matrix components
  -hemoglobin content
  -etc
- Morphological -histology
  -morphometry
  -immunohistochemistry
- Physiological -blood flow
  -neovascular reactivity and permeability

Fig. 1. Applications and parameters that may be analyzed using the cannulated sponge implant model of angiogenesis.

3. Polythene tubing for plug, 1.2 mm internal diameter, 0.6 cm long.
4. 5-0 silk sutures for attachment of the cannula to the center of the sponge discs, for holding the sponge disc in place following implantation, and for closing the surgical incision.
5. Clippers.
6. Ethanol solution (70% in distilled water).
7. Sterilized surgical gauze.
8. Blades.
9. Curved scissors.
10. Forceps.

3. Methods

3.1. Preparation of Cannulated Sponge Implants

1. Circular sponge discs are cut from a sheet of sponge using a cork-borer. Usually the diameter and thickness of the disc depend on the animal used. For mice and rats, the recommended minimum dimensions are 8 mm × 4 mm and 12 mm × 6 mm, respectively.

1. Needle puncture to exteriorize the cannula
2. Midline dorsal incision to introduce the cannulated sponge disc
3. Cannulated sponge disc

Fig. 2. Schematic representation of the cannulated sponge disc and its arrangement following implantation.

2. A segment of polythene tubing (1.4 mm internal diameter) is secured to the interior of each sponge disc (i.e., midway through its thickness) by three 5-0 silk sutures, in such a way that the tube is perpendicular to the disc face (*see* **Fig. 2**).
3. The open end of the tube is sealed with a removable plastic plug made of a smaller polythene tubing (1.2 mm internal diameter). This cannula allows accurate injection of tracers, tests substances, and withdrawal of fluid into and from the interior of the implant.
4. The cannulated sponge discs are sterilized by boiling in distilled water for 10 min, then placed in sterile glass petri dishes, and irradiated overnight under ultraviolet light in a laminar flow hood.

3.2. Anesthesia

1. Rats are anesthetized with intramuscular injection of 0.5 mL/kg of 0.315 mg/kg fentanyl citrate and 10 mg/mL fluanisone.
2. Mice are anesthetized using the combination of fentanyl citrate and fluanisone acetate plus 5 mg/mL of midazolam hydrochloride each at a dose of 0.5 mL/kg. Ether inhalation can also be used.

3.3. Surgical Procedure for Sponge Disc Implantation

Implantation of sponge discs is performed with aseptic techniques following induction of anesthesia:

1. The hair on the dorsal side is shaved and the skin wiped with 70% (v/v) ethanol in distilled water.
2. A 1 cm midline incision is made, and through it a subcutaneous pocket is prepared by blunt dissection using curved scissors.
3. The sterilized sponge implant is inserted into the pocket, its cannula pushed through a small incision previously made on the cervical side of the pocket.
4. The base of the cannula is sutured to the animal skin to immobilize the sponge implant.
5. The cannula is plugged with a smaller piece of sealed polythene tubing to prevent infection.
6. The midline incision is closed by three interrupted 5-0 silk stitches.
7. The animals are kept singly with free access to food and water after recovery from anesthesia.

The cannulated sponge discs can be left *in situ* for periods ranging from days to weeks. A schematic representation of the cannulated sponge disc and its arrangement after subcutaneous implantation is shown in **Fig. 2.** (*See* **Notes 1–5**.)

3.4. Estimation of Blood Flow in the Sponge Implant by the ^{133}Xe Washout Technique

The sequential development of blood-flow in the implanted sponges, originally acellular and avascular, can be determined by measuring the washout rate of ^{133}Xe injected into the implants, a technique developed to measure blood flow and thus monitor the vascular changes indirectly. This is based on the principle that the amount of a locally-deposited radioactive tracer decreases at a rate proportional to the blood flow at the site of the injection. The decrease in radioactive counts should be exponential, and $t_{1/2}$ (time taken for the radioactivity to fall to 50% of its original value) for the washout inversely related to the local blood flow *(15)*:

1. Anesthetize the animals as in **Subheading 3.2.**.
2. Inject ^{133}Xe (10 µL containing approx 1×10^6 cycles per second) into the implant via the cannula. Immediately after injection, plug the cannula to prevent evaporation of the tracer.
3. Monitor the washout of radioactivity from the implant using a collimated gamma-scintillation detector for 6 min. The detector, a sodium iodide thallium activated crystal (1 in × 1 in) should be positioned directly above the site of injection.
4. Record radioactivity for 40-sec periods and print on a scalar ratemeter.

5. The radioactivity-vs-time data should be fitted to an exponential decline curve to derive $t_{1/2}$ (half-time in minutes), after deduction of background radioactivity.

A particular advantage of the ^{133}Xe-$t_{1/2}$ assay is that it allows nondestructive, and thus repeatable, measurements of blood flow in the same animal over the period of neovascularization of the sponge. This combination of techniques requires fewer animals and also allows an estimate of variability in individual animals.

3.5. Estimation of Blood Flow by Efflux of Sodium Fluorescein from an Implanted Sponge

Low-molecular fluorochrome-complexed tracers or fluorogenic dyes provide additional methods for detecting new blood vessels. Compared with radioactive isotope compounds, the advantages of fluorescent dyes are obvious; fluorescence is relatively nontoxic, nonradioactive and inexpensive ***(16)***. The measurement of fluorochrome-generated emission in the bloodstream following its application to the sponge implant compartment at various intervals postimplantation reflects the degree of local blood-flow development and the interaction of the angiogenic site with the systemic circulation ***(8)***. This approach can be used to study sponge-induced angiogenesis quantitatively and to investigate the pharmacological reactivity of the neovasculature. Measurements of the extent of vascularization of sponge implants can be made by estimating $t_{1/2}$ (min) of the fluorescence peak in the systemic circulation following intraimplant injection of sodium fluorescein at fixed time intervals (for example; d 1, 4, 7, 10 and 14) postimplantation.

1. Anesthetize the animal.
2. Determine blood background fluorescence by piercing the extremity of the tail and collecting 5 µL of blood with a heparinized yellow tip. Transfer the blood sample to a centrifuge tube contained 1 mL of isotonic saline (0.9%).
3. At time 0, administer sodium fluorescein (50 µL of a sterile solution of 10% sodium fluorescein per kg weight) to anesthetized animals.
4. After 1 min collect the first blood sample following dye injection as in **step 2**.
5. At 3 min collect a second blood sample. Repeat this procedure every 2–3 min for 25–30 min.
6. Centrifuge blood samples for 10 min at 1400*g* (2000 rpm). Keep the supernatant for fluorescence determination (excitation 485nm/emission 519nm).
7. From the fluorescence values estimate the time taken for the fluorescence to peak in the bloodstream (absorption) and the time required for the elimination of the dye from the systemic circulation (elimination). These parameters are expressed in terms of half-time ($t_{1/2}$; time taken for the fluorescence to reach, or to decay to, 50% of the peak value in the systemic circulation).

3.6. Biochemical Analysis of Implanted Sponges

Quantitation of various biochemical parameters further supports the functional characterization of the fibrovascular tissue that infiltrates the implants and has been used to corroborate assessment of angiogenesis *(7,10–14)*:

1. Remove the implants at any time postimplantation as required.
2. Immediately upon removal, weigh, homogenize, and centrifuge the tissue in isotonic physiological solution (saline 0.9% or phosphate-buffered saline).
3. Store the supernatant of the homogenate at –20°C for later analysis.

3.7. Histological Analysis of Implanted Sponges

To further establish the sequential development of granulation tissue and blood vessels in the implants, several histologic techniques have been employed:

1. Kill the animals bearing implants and dissect the implants free of adherent tissue.
2. Fix implants in formalin (10% w/v in isotonic saline) and embed in paraffin.
3. Cut sections (5–8 μm) from halfway through the sponge thickness.
4. Stain and process for light microscopy studies.

Figure 3 shows examples of sponge implants at 7 and 14 d postimplantation. The implants can be photographed by transillumination using an inverted microscope, which allows qualitative and morphometric analysis of the vascular pattern within the sponge matrix (**Fig. 4**).

4. Notes

1. The attachment of the cannula to the center of the sponge disc is facilitated by making in one end of the polythene tubing 'tooth-like' structures (usually four), in such a way that the thread is secured in one of them and then sutured to the sponge.
2. To avoid leakage following injection of test substances via the cannula, it is important to hold the base of the cannula with a forceps. Sometimes it is necessary to force in the injected substances by applying pressure to the cannula.
3. To avoid infection, plugs should be changed every time they are removed from the cannula.
4. Administration of drugs should be performed 2–3 d postimplantation to allow time for encapsulation of the sponges.
5. To eliminate acute effects, vasoactive substances (vasodilators or vasoconstrictors) should be given 6–8 h prior to blood-flow measurement.

Acknowledgements

This work was supported by grants from CNPq, FAPEMIG and PRONEX-Brazil.

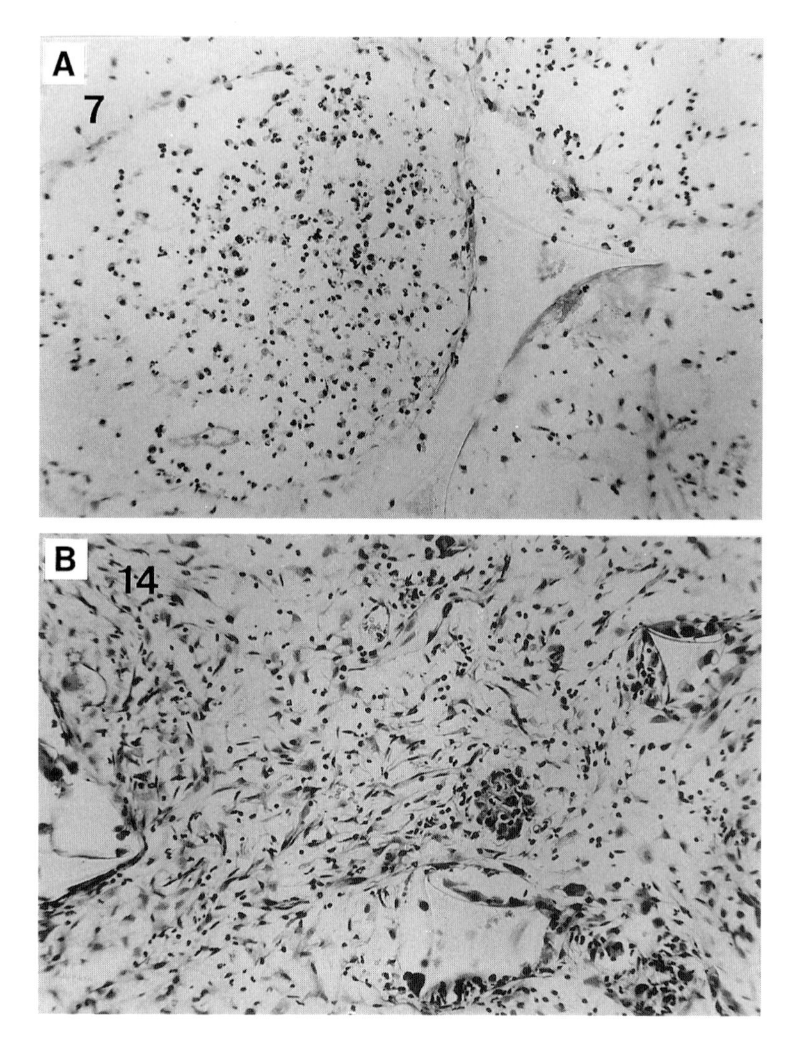

Fig. 3. Sequence of histological changes during angiogenesis in the rat sponge granuloma. Sections of implants removed on d 7 or 14 postimplantation, fixed in formalin, embedded in paraffin and 5 μm sections stained (H&E ×500). The sponge matrix is seen as triangular shapes. (A) At d 7, the pores of the implants are filled with a fibrous network, numerous polymorphonuclear leukocytes and spindle-shaped cells (fibroblasts). (B) By d 14 postimplantation, a uniformly dense, more highly organized fibrous matrix, with numerous spindle-shaped cells and capillaries interspersed within the sponge, can be seen.

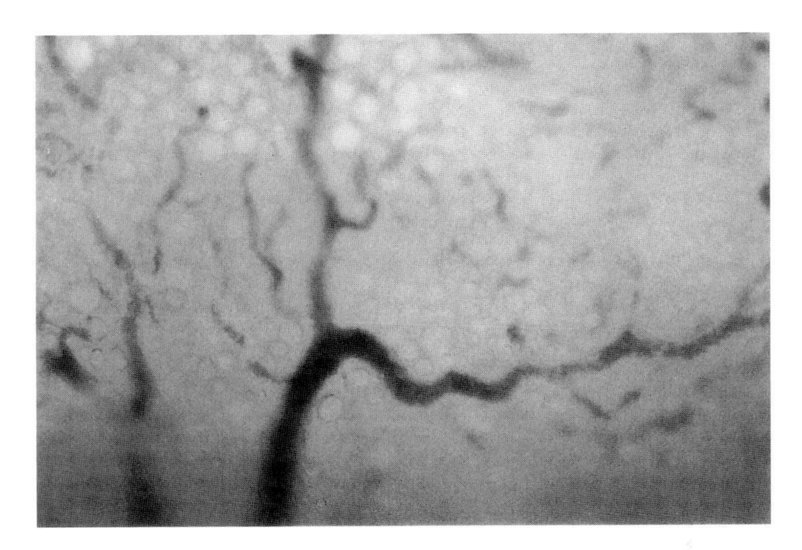

Fig. 4. Photomicrograph of the vascular pattern of a trans-illuminated sponge implant at d 7 postimplantation. Tortuous, dilated, and saccular capillaries are seen infiltrating the polyether-polyurethane sponge matrix, which appears as a beehive-like structure (magnification ×7). The implant was photographed with an Olympus-IM inverted microscope equipped with a 35 mm camera.

References

1. Folkman, J. and Shing, Y. (1992) Angiogenesis. *J. Biol. Chem.* **267,** 10,931–10,934.
2. Edwards, R. H., Sarmenta, S. S., and Hass, G. M. (1960) Stimulation of granulation tissue growth by tissue extracts; study by intramuscular wounds in rabbits. *Arch. Path.* **69,** 286–302.
3. Andrade, S. P., Fan, T. P. D., and Lewis, G. P. (1987) Quantitative in vivo studies on angiogenesis in a rat sponge model. *Br. J. Exp. Path.* **68,** 755–766.
4. Thiede, K., Momburg, F., Zangemeister, U., Schlag, P., and Schirrmacher, V. (1988) Growth and metastasis of human tumors in nude mice following tumor-cell inoculation into a vascularized polyurethane sponge matrix. *Int. J. Cancer* **42,** 939–945.
5. Mahadevan, V., Malik, S. T. A., Meager, A., Fiers, W., Lewis, G. P., Hart, I. R. (1990) Role of tumor necrosis factor in flavone acetic acid-induced tumor vasculature shutdown. *Cancer Res.* **50,** 5537–5542.
6. Andrade, S. P., Bakhle, Y. S., Hart, I., and Piper, P. J. (1992) Effects of tumor cells and vasoconstrictor responses in sponge implants in mice. *Br. J. Cancer* **66,** 821–826.
7. Andrade, S. P., Vieira, L. B. G. B., Bakhle, Y., Piper, P. J. (1992) Effects of platelet activating factor (PAF) and other vasoconstrictors on a model of angiogenesis in the mouse. *Int. J. Exp. Path.* **73,** 503–513.

8. Andrade, S. P., Machado, R. D. P., Teixeira, A. S., Belo, A. V., Tarso, A. M., and Beraldo, W. T. (1997) Sponge-induced angiogenesis in mice and the pharmacological reactivity of the neovasculature quantitated by a fluorimetric method. *Microvasc. Res.* **54,** 253–261.
9. Walsh, D. A., Hu, D. E., Mapp, P. I., Polak, J. M., Blake, D. R., and Fan, T. P. D. (1996) Innervation and neurokinin receptors during angiogenesis in the rat sponge granuloma. *Histochem. J.* **28,** 759–769.
10. Andrade, S. P., Cardoso, C. C., Machado, R. D. P., and Beraldo, W. T. (1996) Angiotensin-II-induced angiogenesis in sponge implants in mice. *Int. J. Microcirc. Clin. Exp.* **16,** 302–307.
11. Hu, D. E., Hiley, C. R., Smither, R. L., Gresham, G. A., and Fan, T. P. D. (1995) Correlation of 133Xe clearance, blood flow and histology in rat sponge model for angiogenesis. *Lab. Invest.* **72,** 601–610.
12. Buckley, A., Davidson, J. M., Kamerath, C. D., Wolt, T. B., and Woodward, S. C. (1985) Sustained release of epidermal growth factor accelerates wound repair. *Proc. Natl. Acad. Sci. USA* **82,** 7340–7344.
13. Plunkett, M. L. and Halley, J. A. (1990) An in vivo quantitative angiogenesis model using tumor cells entrapped in alginate. *Lab. Invest.* **62,** 510–517.
14. Ford-Hutchinson, A. W., Walker, J. R., and Smith, J. H. (1977) Assessment of anti-inflammatory activity by sponge implantation techniques. *J. Pharmacol. Meth.* **1,** 3–7.
15. Kety, S. S. (1949) Measurement of regional circulation by local clearance of radioactive sodium. *Am. Heart. J.* **38,** 321–331.
16. McGrath, J. C., Arribas, S., and Daly, C. J. (1996) Fluorescent ligands for the study of receptors. *Trends Pharmacol. Sci.* **17,** 393–399.

7

Hollow Fiber Assay for Tumor Angiogenesis

Roger M. Phillips and Michael C. Bibby

1. Introduction

It is becoming increasingly clear that the process of angiogenesis is important for tumor growth and metastasis and as such provides an exploitable target for therapeutic intervention. Consequently a number of useful model systems has been developed to investigate the angiogenic process associated with malignancy and to discover new therapeutic strategies.

The hollow fiber assay is currently being used by the National Cancer Institute (NCI) in the USA as part of their random drug-screening program for new anticancer drugs *(1)*. Its function is to act as a "filter" through which compounds emerging from the initial in vitro human cell line screen are passed, and to help ensure that only those compounds with promise for further development are selected for evaluation against human tumor xenografts. The hollow fiber assay as used by the NCI involves the short term culture (1 d) of cells inside biocompatible polyvinylidene fluoride fibers followed by implantation into mice at two anatomical sites, *viz*, intra-peritoneally (ip) and subcutaneously (sc). Drugs are administered ip for 4 d and the fibers removed 2 d after the last treatment. Cell survival is measured in vitro using a modified 3-[4,5-dimethylthiazol-2-yl]-2,5 diphenyl-2H-tetrazolium bromide (MTT) assay *(2)*.

Recent studies in this laboratory *(3)* have demonstrated that extension of implantation times beyond the 6-d period normally employed by NCI results in extensive vascularization of the fibers in the sc site and that this vascularization has significant implications for drug delivery to the tumor cells. In addition to being ideal for investigating the influence of blood vessel development on drug exposure parameters and subsequent chemosensitivity, the hollow fiber assay provides a useful model system for identifying and interpreting molecular events in the process of tumor angiogenesis.

From: *Methods in Molecular Medicine, Vol. 46: Angiogenesis Protocols*
Edited by: J. C. Murray © Humana Press Inc., Totowa, NJ

2. Materials

1. Polyvinylidene fluoride (PVDF) fibers: These should have an internal diameter of 1 mm and have a molecular weight cut off of 500,000 Da. They can be purchased from Spectrum Medical Ind. (Houston, USA).
2. Ethanol solution (70% in distilled water).
3. Scissors.
4. Autoclavable bags, boxes (stainless steel) and lid (approx dimensions: 250 mm × 50 mm × 40 mm)
5. Syringes (5 and 10 mL) plus 21-gauge needles
6. Tissue culture medium (e.g., RPMI 1640 supplemented with 20% fetal calf serum, sodium pyruvate [2 m*M*], penicillin/streptomycin [50 IU/mL, 50 μg/mL], and L-glutamine [2m*M*]).
7. Sterile forceps.
8. Polythene bag and tape.
9. Hemocytometer.
10. Ice box and ice bag.
11. Vari-mix rocker plate.
12. Demarcated steel tray and two small steel trays. Also an instrument tray for forceps, scissors, and smooth-jawed needle holders.
13. Bacti-Cinerator III Sterilizer (Fisher Scientific Co., USA).
14. Orbital shaker.
15. Thermolyne heating mats (Fisher Scientific Co., USA).
16. Cell culture consumables: Culture flasks, universal tubes, 6-well plates, and so forth.

3. Methods

3.1. Preparation of Hollow Fibers Prior to Tumor Cell Implantation

The fibers are supplied in 3-ft.-long segments, which are dehydrated. They must be cut to length, rehydrated, and sterilized before cells are inoculated. There are three steps involved in this procedure: ethanol saturation, water sterilization, and medium incubation.

3.1.1. Step1: Ethanol Saturation

1. Gently apply pressure to any closed, cut ends of the fiber to open.
2. Fill the syringe with 70% ethanol and flush each fiber individually. Place the fiber in a tray containing 70% ethanol.
3. Transfer the fibers to a glass chromatography column (600 mm tall) and fill with 70% ethanol.
4. Store vertically for a minimun of 72 h. Fibers can be stored in this way indefinitely. It is important that the fibers do not dry out; otherwise, the procedure outlined above must be repeated.

3.1.2. Step 2: Water Sterilization

1. Transfer the ethanol soaked fibers to a tray.
2. Place 100 mL distilled water into an autoclavable box.
3. Flush each fiber with fresh ethanol using a syringe and then flush each fiber with distilled water.
4. Do not remove the syringe yet but place the fiber in the box with the water and flush to eliminate air from the fiber (*see* **Note 1**). Remove the syringe only after air bubbles have been eliminated.
5. Place the lid on the box, secure with tape, and place into an autoclave bag and autoclave (131°C for 5 min). Note that the fibers melt at 143°C and should not be autoclaved at temperatures exceeding 132°C.

3.1.2. Step 3: Medium Incubation

The fibers are now sterile and from this point onward all work should be performed in sterile conditions using aseptic procedures.

1. Place two autoclave boxes (one containing the sterilized fibers and the other empty, but sterile) into the cabinet and open the lids.
2. Place 100 mL of growth medium in the empty box; open forceps and rest them on the inside of one of the lids.
3. Fill a syringe with medium and place a 21-guage needle onto the syringe. Using forceps to hold the fiber (*see* **Note 2**), insert the needle into the open end of the fiber and flush through with medium.
4. Move the fiber to the box containing medium. Place the fiber in the medium and flush to dispel air. Remove the needle and syringe.
5. Repeat for each fiber; close the box and secure with tape. Incubate fibers overnight at 37°C in an atmosphere containing 5% CO_2.

3.2. Inoculation of Hollow Fibers with Cell Lines

The next stage of the process requires good organization within the cabinet and a suggested layout of material is presented in **Fig. 1**.

1. Trypsinize cells (monolayer cultures approaching confluence in T-75 flasks) and resuspend cells in growth medium containing 20% fetal calf serum (FCS). (For some cell lines, growth can be enhanced by the inclusion of conditioned medium [5%] obtained from cell cultures at confluence.)
2. Count cells using a hemocytometer and adjust to required cell density using growth medium (±conditioned medium). The cell density required varies depending upon the cell line used but densities between 1×10^6 and 1×10^7 cells per mL are typical.
3. Store the cell suspension on ice whilst the cabinet is prepared for hollow fibers as follows: Place 2 mL of growth medium into each well of a 6-well culture plate. Also fill a universal tube with growth medium. Spray the cabinet with 70% etha-

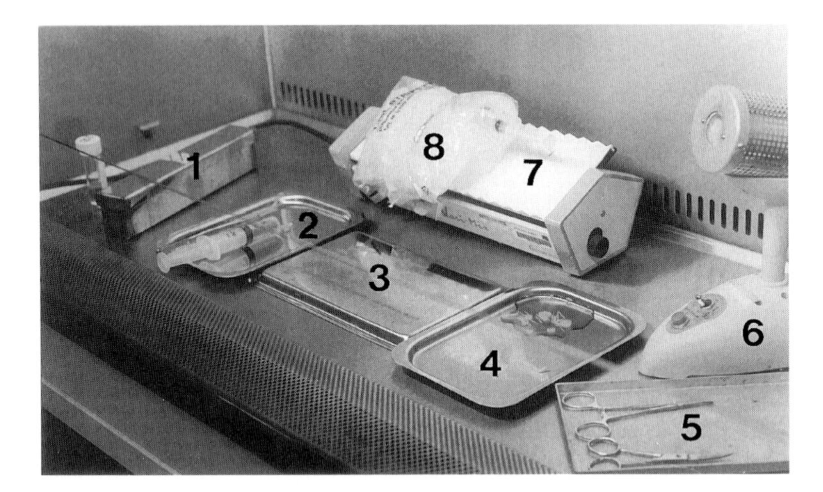

Fig. 1. Recommended layout of equipment for cell inoculation into hollow fibers: **1**, box of prepared fibers; **2**, small tray for instruments in use; **3**, demarcated tray; **4**, small tray; **5**, instrument tray; **6**, Bacti-Cinerator; **7**, Vari-mix rocker plate; **8**, ice-pack with syringe.

nol. Position the plate rocker and bacteriological incinerator at the rear of the cabinet.

4. Place a 10 mL syringe (with the plunger removed) into an ice pack and mould the ice pack around it. Secure the ice pack to the plate rocker.
5. Place the demarcated tray on the cabinet work-surface in front of you; position the small tray to the right of the demarcated tray and the second tray to the left of it. Place the instrument tray farthest right.
6. The next stages of the operation require the use of sterile gloves. It is good practice to keep one hand sterile (i.e., touching only sterile objects) and the other hand "dirty" to touch nonsterile objects.
7. With the right hand (sterile), pick up the 10 mL syringe and with the left hand (dirty) remove the paper and discard. Remove the cap from the medium (in the universal tube) and take up some medium using the syringe (held in the right hand). Pick up the needle in the left hand and attach to the syringe. Place the sterile syringe onto the instrument tray.
8. Repeat the procedure with the 5 mL syringe for the cell suspension and place into the syringe barrel in the ice pack on the rocker plate. Switch the rocker plate on.
9. Manipulating the fiber with the forceps, flush it through with fresh medium.
10. Flush the fiber with cell suspension, heat the smooth jawed needle holders in the incinerator for 3–5 sec, and clamp the loose end of the fiber to heat seal. Ensure the fiber is obviously full and heat-seal the needle end.
11. Lay the fiber containing the cell suspension along the demarcated tray, cover with medium and heat-seal at 1.5-cm intervals. Cut across the seals with scissors

to separate the segments. Each segment should have a tail of 2 mm for ease of handling (*see* **Notes 3** and **4**).

12. Place the segments in the wells of the 6-well plate and incubate at 37°C overnight in an atmosphere containing 5% CO_2.

3.3. In Vivo Implantation of Hollow Fibers and Qualitative Assessment of Angiogenesis

1. Prior to implantation in vivo, hollow fibers are cultured in vitro in T-75 flasks containing 30 mL medium for a period of 14 d. At this stage, the fibers are "mature" containing central necrotic cores and a rim of viable cells adjacent to the fiber wall. During this in vitro incubation, it is important that the fibers are constantly agitated (on an orbital shaker at 500 rpm).
2. Recipient mice are subjected to brief inhalation anesthesia and individual fibers are implanted sc in the flank by the use of a trocar. Routinely we have implanted three individual fibers through the same trocar incision, although single fibers can be implanted in different sites, e.g., in either flank or dorsally. To date we have implanted fibers containing human cancer cell lines into nude mice or murine tumor cells into syngeneic mice.
3. At various times after implantation (studies have been conducted up to 32 d post-implantation), mice are killed by cervical dislocation, a single incision is made through the skin, and the shin is peeled back to reveal the fibers *in situ*.
4. Photographic records are made of the implantation site, and the extent of neovascularization can be determined either by manual counting or image analysis (**Fig. 2**).

4. Notes

1. Flushing fibers: When the fibers are being washed through with water it is possible to trap air bubbles. Make sure that the fiber is fully immersed before removing the needle.
2. Handling fibers: It should be noted that handling the fibers with forceps, while trying to insert a needle into the small open end, takes a steady hand, good eyesight, and practice.
3. Sealing fibers: If seals are not properly formed they may open, resulting in leaking cells. It is important not only to practice the sealing procedure but also the cutting between the seals. This requires a steady hand and good eyesight. A magnifying lens may be helpful.
4. Washing completed fibers: It is important to ensure that fibers are well washed after seeding and sealing; otherwise cells can grow on the outside.

Acknowledgments

This work was supported by War on Cancer, Bradford, UK. The authors are grateful to staff at NCI for the provision of detailed protocols on their drug screening hollow fiber assay and to Tricia Cooper, Beryl Cronin, and Jennifer Pearce for technical assistance.

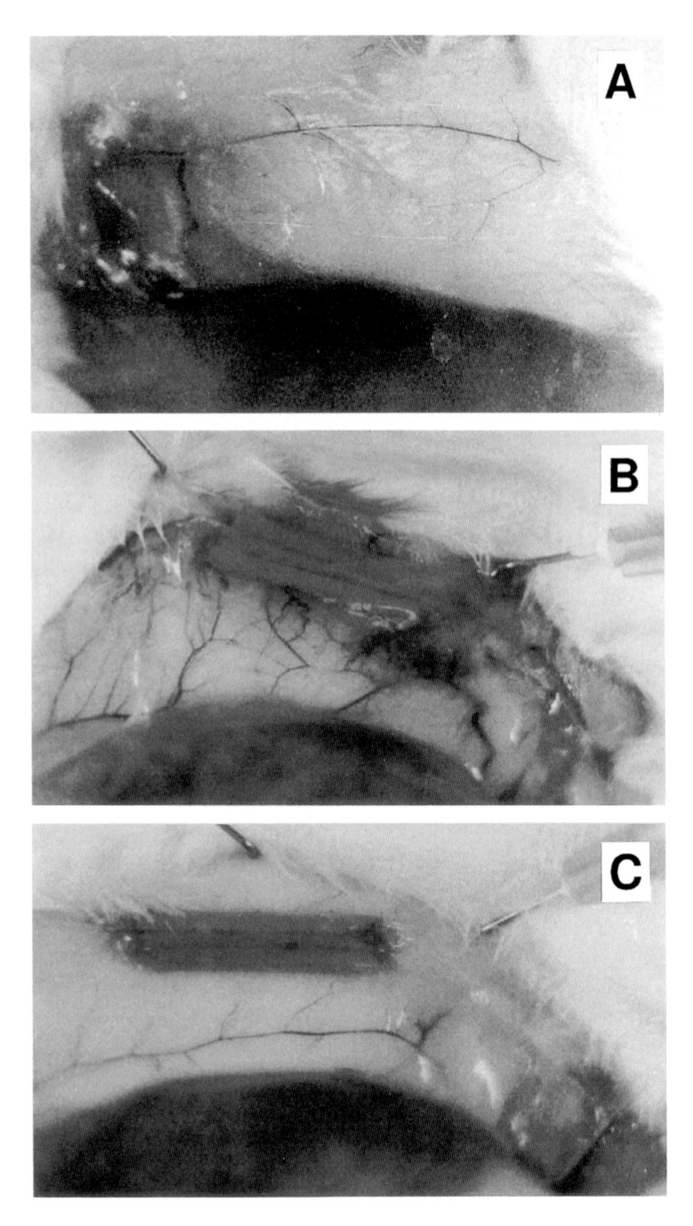

Fig. 2. Morphological appearance of blood vessel development to hollow fibers (containing MAC 15A cells) implanted sc in NMRI mice. (**A**) The normal appearance of subdermal vasculature in the absence of fibers. (**B**,**C**) Blood vessel development to hollow fibers containing tumor cells and culture medium respectively, 28 d after implantation.

References

1. Anderson, L. F. (1995) New drug screen assay uses fewer mice, while cutting costs. *J. Natl. Cancer Inst.* **87,** 1213–1214.
2. Hollingshead, M. G., Alley, M. C., Camallier, R. F., Abbott, B. J., Mayo, J. G., Malspeis, L., and Grever, M. R. (1995) In vitro cultivation of tumor cells in hollow fibers. *Life Sci.* **57,** 131–141.
3. Phillips, R. M., Pearce, J., Loadman, P. M., Bibby, M. C., Cooper, P. A., Swaine, D. J., and Double, J. A. (1998) Angiogenesis in the hollow fiber tumor model influences drug delivery to tumor cells: Implications for anticancer drug screening programs. *Cancer Res.* **58,** 5263–5266.

8

Dorsal Skinfold Chamber Preparation in Mice

Studying Angiogenesis by Intravital Microscopy

Axel Sckell and Michael Leunig

1. Introduction

Since 1924, when the first transparent chamber model in animals was introduced by Sandison *(1)*, many other chamber models have been described in the literature for studying angiogenesis and microcirculation in a wide variety of neoplastic and nonneoplastic tissues by means of intravital microscopy (for reviews *see* ***2–4***). Because angiogenesis is an active and dynamic process, one of the major strengths of chamber models is the possibility of monitoring angiogenesis in vivo continuously up to several weeks with high spatial and temporal resolution. In addition, after the termination of experiments, tissue samples can be excised easily and further examined by various in vitro methods (e.g., histology, immuno-histochemistry, and molecular biology).

The advantages of using mice as experimental animals are, for instance, the availability of a large number of different well-defined mouse strains, including transgenic or knockout mice and the wide variety of commercially generated agents suitable for mice (e.g., monoclonal antibodies, nanoparticles, and single gene products).

This chapter describes the protocol for the surgical preparation of the dorsal skinfold chamber in mice as well as the method to implant tumors in this chamber for further investigations of angiogenesis and other microcirculatory parameters. The model *(5,6)* presented here is the development of a similar model in hamsters *(7)*. In brief, take a fold of the depilated dorsal skin of an anesthetized mouse and cut out surgically a circular area of one skin layer (consisting of epidermis, dermis, subcutis, cutaneous muscle, and subcutaneous

From: *Methods in Molecular Medicine, Vol. 46: Angiogenesis Protocols*
Edited by: J. C. Murray © Humana Press Inc., Totowa, NJ

fatty tissue) completely. Then, fix the skinfold like a sandwich between the two titanium frames of the chamber and close the operation field with a sterile cover slip to avoid drying, infection, or mechanical damage of the inner layer, i.e. the cutaneous muscle, of the unprotected side of the opposite skin. For tissue implantation and/or other local treatments the chamber can easily be opened again by removing the cover slip and reclosed with a new sterile cover slip, respectively. The cutaneous muscle serves as site for implantation.

2. Materials

Except for commonly used devices, all materials necessary for the preparation of the dorsal skinfold chamber in mice are listed below with a detailed manufacturer's record. Evidently, all materials given are only suggestions and may be modified for personal preferences.

2.1. Dorsal Skinfold Chamber Preparation

2.1.1. Facilities and Apparatus

1. Laminar flow hood (Faster S.r.I., Ferrara, Italy).
2. Dry Sterilizer (Model IS-350, Inotech, Intergra Biosciences, CH-8304 Wallisellen, Switzerland).
3. Dissecting microscope (Wild, 6-50×, type 334790, CH-9435 Heerbrugg, Switzerland).
4. 2 Flat custom-made thermal pads (WISAG, CH-8057 Zurich, Switzerland).
5. Halogen lamp with two flexible swan-neck light transmission tubes (Intralux, 150H, Volpi AG, CH-8902 Urdorf-Zürich, Switzerland).
6. Custom-made skin spreading device. This consists of a heavy metal base and two flexible swan-neck tubes for applying tension to spread out the mouse skinfold, prior to the fixation of the back titanium frame of the dorsal skinfold chamber to the skinfold of the mouse (Workshop, Department of Clinical Research, University of Berne, CH-3010 Berne, Switzerland).

2.1.2. Drugs

1. Isotonic sodium chloride (0.9% NaCl solution injectable).
2. Anesthesia: Mixture consisting of isotonic sodium chloride, ketamine hydrochloride (Ketalar®, Parke-Davis, Morris Plains, NJ) and xylazine (XylaJect®, Phoenix Pharmaceutical, St. Joseph, MO).
3. Depilatory cream (Primex®, Mibella AG, CH-5033 Buchs, Switzerland).

2.1.3. Dorsal Skinfold Chambers

1. Custom-made dorsal skinfold chambers (Workshop, Institute for Surgical Research, Klinikum Grosshadern, University of Munich, D-81377 Munich, Germany; **Fig. 1**) consisting of 2 titanium frames, 3 screws (M2 × 6), 6 nuts (size 4),

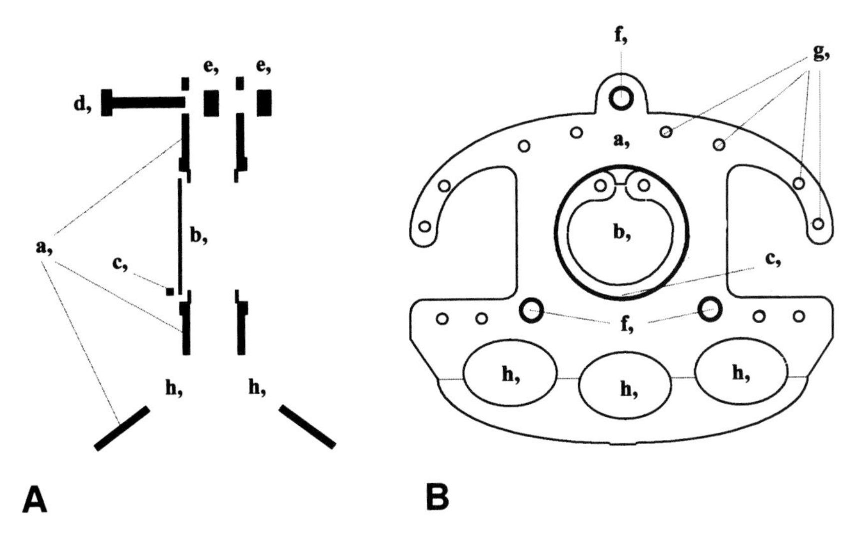

Fig. 1. Construction plan of the dorsal skinfold chamber. **(A)** Cross-section. **(B)** Lateral view: *a*, titanium frame; *b*, cover slip; c, tension ring; d, screw; e, nut; f, screw hole; g, bore holes for holding sutures; h, holes for weight reduction of the chamber.

and one tension ring to keep the sterile cover slip in position after closing of the chamber preparation.

2. Sterile cover slips (diameter 11.75 mm, circular, Assistent, D-97647 Sontheim, Germany).
3. Special pair of pliers (Garant®, Germany) to bring the tension ring in position to keep the chamber closed and to remove it again, respectively.
4. Wrench (CHR-VAN, size 4, SKG, Germany).

2.1.4. Surgical Instruments

1. Electric hair clipper (Electra®II, GH 204 or 201; Aesculap®, Aesculap AG and CO. KG, D-78532 Tuttlingen, Germany) equipped with a 1/20-mm cutting head (GH 700; Aesculap).
2. 2 Delicate hemostatic forceps (Baby-Mosquito, BH 115, Aesculap).
3. 1 Needle holder (Castroviejo, BM 2, Aesculap).
4. 1 Delicate dissecting forceps (Micro-Adson, BD 220, Aesculap).
5. 2 Microforceps (BD 331, Aesculap).
6. 1 Pair of dissecting scissors, fine patterns (Cottle-Masing, sharp, OK 365, Aesculap).
7. 1 Pair of microscissors (spring type) with round handles (FD 103, Aesculap®).
8. 1 Pair of microscissors (spring type) with flat handles and cross-serration (Vannas, FD 15, Aesculap).
9. Sterile scalpel blades (#15, BB 515, Aesculap).

10. Sutures: Polypropylene (4-0 U.S.P., 1.5 metric, DSM 13P, Sharpoint, Surgical Specialities Corporation, Reading, PA).

2.1.5. Other Materials

1. Mouse (body weight 25–30 g, age 6–12 wk); depending on the research goal, inbred, outbred, immune-competent, immune-deficient, and so on.
2. One cage per animal (*see* **Note 1**).
3. Syringes (1 mL).
4. 26G-Needles (26G3/8, 0.45 × 10).
5. Sterile nonwoven swabs (5 × 5 cm).
6. Sterile Q-tips (cotton pads on wooden sticks).
7. Fine black waterproof permanent pen.
8. Surgical masks.
9. Rubber gloves.
10. 70% Alcohol to disinfect skin of the mouse, surgical instruments, and rubber gloves.

2.1.6. Additional Equipment for Tissue Implantation

1. Custom-made device consisting of a slitted polycarbon tube (internal diameter 24 mm, length 120 mm) and a special mounting stage to fix the tube with the animal in it (Workshop, Institute for Surgical Research, Klinikum Grosshadern, University of Munich, D-81377 Munich, Germany).
2. Adhesive tape (Transpore™ 3M, Hypoallergenic, ≈1.2 mm × 9.1 m, 3M Medical-Surgical Division, St. Paul, MN).
3. Hanks' balanced salt solution (H-9269, 100 mL, Sigma-Aldrich Co., LTD, Irvine KA, UK) stored at 6°C.
4. Sterile Petri dishes (diameter: ≈100 × 20 mm).

3. Methods

3.1. Surgical Preparation of Dorsal Skinfold Chamber

A scheme of the setup for the surgical preparation of the dorsal skinfold chamber is shown in **Fig. 2**. All surgical procedures should be performed under aseptic conditions (*see* **Note 2**).

1. Anesthetize the mouse by an injection of a mixture of ketamine (100 mg/kg body weight) and Xylazine (10 mg/kg body weight) im into the limb (*see* **Note 3**).
2. For depilation of the entire dorsum of the mouse, carefully shave the anesthetized mouse with the electric hair clipper. Put the mouse on a thermal pad outside the hood and apply a thick layer (1–2 mm) of depilatory cream on the shaved skin area. After the cream was allowed to take effect for 5–10 min (*see* **Note 4**), it can be easily removed using a non-woven swab soaked in hand-warm germ-free water, wiping in the caudal-to-cranial direction (*see* **Notes 5** and **6**).

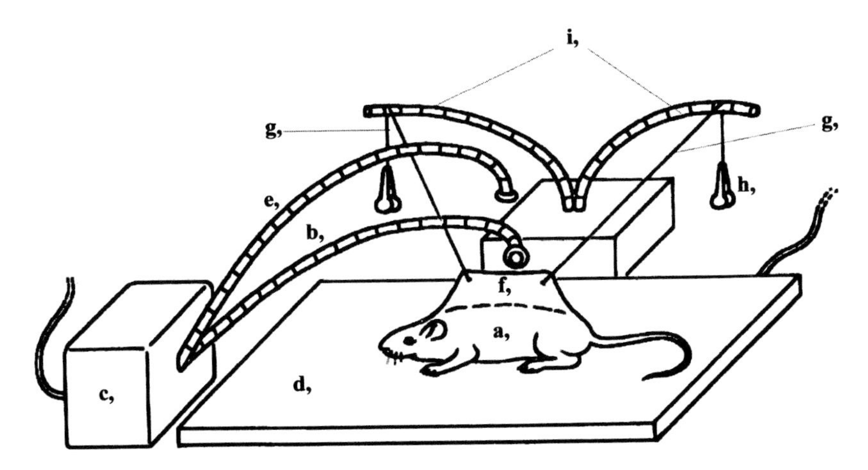

Fig. 2. Setup for the surgical preparation inside the hood: *a*, mouse; *b*, swan-neck light transmission tube #2 (for transillumination from behind); *c*, halogene lamp; *d*, thermal pad; *e*, swan-neck light transmission tube #1 (for epi-illumination from above); *f*, skinfold; *g*, holding thread; *h*, Baby-Mosquito; *i*, flexible swan-necks of the skin spreading device.

3. Dry the wet depilated skin with a dry sterile swab and disinfect the skin with a 70% alcohol-soaked swab.
4. Place the mouse between the light transmission tube (tube #2) of the halogen lamp (**Fig. 2**) and the surgeon, in a prone position on an opened sterile non-woven swab lying on the thermal pad inside the hood. The longitudinal axis of the mouse (with its head lined up to the left-hand side of the surgeon) should be parallel to the frontal plane of the surgeon. Illuminate the mouse from the top with the light transmission tube (tube #1) of the halogen lamp. Adjust tube #2 with its central light ~2 cm parallel to the surface of the thermal pad and perpendicular to the longitudinal axis of the mouse (**Fig. 2**).
5. Lift up a fold of the depilated dorsal skin. This fold, running from the sacrum to the neck of the mouse, should be located directly over, and parallel to, the spine. Adjust the fold under transillumination (tube #2) in such a way that both sides of the skinfolds become congruent (*see* **Note 7**). Then spread the skin in the upright position by fastening two holding threads (polypropylene 4-0; no knots but one Baby-Mosquito each fixed at their ends) at the edge of the skinfold (*see* **Note 8**) and hanging them over the two flexible swan-necks of the custom-made skin spreading device (**Fig. 2**; *see* **Note 9**).
6. Fix the first titanium frame of the chamber to the side of the skinfold facing away from the surgeon with two temporary holding sutures (polypropylene 4-0) through both of the borehole pairs, left and right of the apical screw (*see* **Notes 10–12**).

7. Remove the two holding threads completely and make two small incisions (using a pair of Cottle-Masing scissors or a #15 scalpel blade) perforating the entire skinfold to let the two lower screws of the titanium frame come through (*see* **Notes 13** and **14**). The fixation of the latter two screws using the Baby-Mosquitos helps to adapt the skinfold temporarily to the titanium frame facilitating further surgical preparations. The bent tip of the Baby-Mosquito should then point towards the mouse and not the surgeon.
8. Use the cranial Baby-Mosquito to turn the back titanium frame in a perpendicular position to the transilluminating light. The edge of the circular area being projected through the central window of the back titanium frame to the skin facing the surgeon can now be easily marked with a dotted line using a fine black permanent pen.
9. Place the mouse in right lateral position under the dissecting microscope (~6-fold magnification; *see* **Note 15**) with the skinfold pointing to the surgeon.
10. Using delicate dissecting forceps (Micro-Adson) and micro scissors (FD 103), remove all layers of the skin completely (epidermis, dermis, subcutis, cutaneous muscle, parts of the subcutaneous fatty tissue) along the marked dotted line (*see* **Notes 16** and **17**).
11. Stop possible bleeding along the edge of the wound gently by using sterile Q-tips slightly moistened with saline. Now allow 3–5 min to elapse to ensure that no further after-bleedings take place. Meanwhile, to avoid drying out of the operation field, perfuse the area with isotonic saline (*see* **Note 18**).
12. Absorb excess saline with dry Q-tips (*see* **Note 19**) and put the optical magnification on ~10-fold. *The next step is probably the most critical step of the chamber preparation*: Use microforceps (BD 331) and a pair of microscissors (Vannas) to carefully remove the last layer of subcutaneous fatty tissue which is connected to the underlying cutaneous muscle of the opposite skin (*see* **Notes 20–22**).
13. Close the chamber preparation like a sandwich with the second titanium frame: First connect the lower two screws to the corresponding holes of the second frame and then the apical one (*see* **Note 23**). If no air bubbles are visible between the cover slip and the underlying cutaneous muscle, screw on the nuts to finally fix the two frames of the chamber together (*see* **Notes 24** and **25**).
14. Perform four holding sutures (polypropylene 4-0) to spread and fix the edges of the sandwiched skinfold to all four pairs of boreholes left and right from the apical screw in both chamber frames. After stitching but before closing the knots of the two central holding sutures, cut the old temporary holding sutures from **step 6** and remove them completely (*see* **Note 26**).
15. Place the operated mouse in its cage and leave it on a thermal pad outside the hood at least until the mouse has regained consciousness (*see* **Note 27**).

3.2. Tissue Implantation into the Chamber Preparation

In the following, the explantation of a solid tumor from a donor mouse and the consecutive implantation of a chunk of this tumor into the dorsal skinfold preparation of a recipient mouse is described (*see* **Notes 28** and **29**):

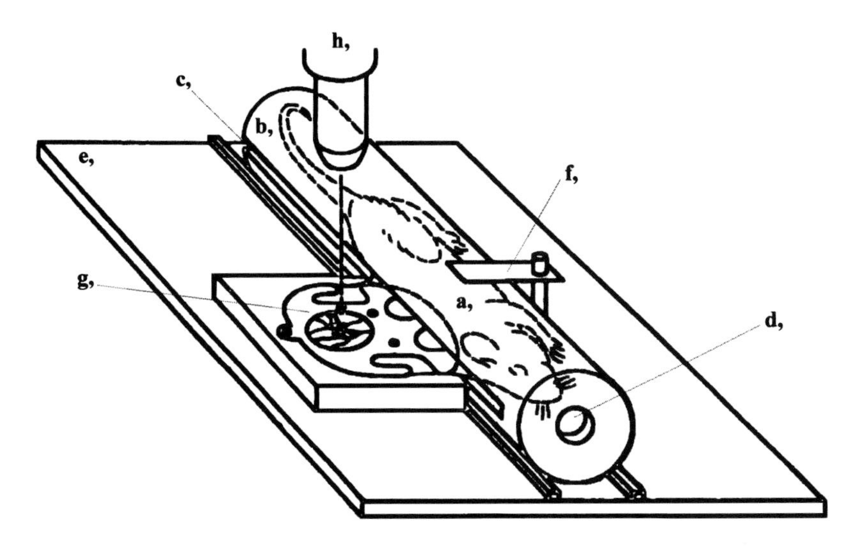

Fig. 3. A mouse fitted with a dorsal skinfold chamber inside a polycarbon tube fixed on the special mounting stage: *a*, mouse; *b*, slitted polycarbon tube; *c*, slit; *d*, breathing hole; *e*, special mounting stage; *f*, device to fix the polycarbon tube on the mounting stage; *g*, dorsal skinfold chamber preparation; *h*, objective of the intravital microscope.

1. Allow at least 48 h to elapse for the animals to recover completely from surgery before implantation of any tissue into the dorsal skinfold chamber preparation. Exclude all animals from further treatment and sacrifice them when there are signs of bleeding, inflammation, or any other irritation at the implantation site (*see* **Note 30**). Only chambers meeting criteria of intact microcirculation *(8)* should be used as sites for implantation.
2. Sacrifice the donor mouse bearing a solid subcutaneous tumor according to official Guidelines for Care and Use of Experimental Animals. For 2–3 min, completely insert the dead animal into a 70% alcohol solution for disinfection. Excise the desired tumor surgically under aseptic conditions in the hood and put it into a sterile Petri dish previously filled with cold (~6°C) Hanks' balanced salt solution.
3. The dissecting microscope is only needed for this step of the protocol (magnification ~10-fold). Remove the capsule and all hemorrhagic or necrotic parts of the tumor with the help of microforceps and a pair of microscissors. Cut the remaining tumor into small chunks of a diameter no greater than ~0.5–1 mm (*see* **Note 31**).
4. Put the nonanesthetized recipient mouse in the slitted polycarbon tube (**Fig. 3**). An adhesive tape fixed across the slit right behind the chamber jutting out will prevent the animal escaping from the tube (*see* **Note 32**). Then fix the chamber in a horizontal position in the special mounting stage, which will also serve as a stage to perform intravital microscopy of the implant at a later time (**Fig. 3**).

5. Remove the tension ring with the wrench. Use a 26G needle as lever to lift the cover slip a few millimeters, finally grasping and removing it with microforceps.
6. Transfer one of the tumor chunks with another set of sterile microforceps onto the cutaneous muscle in the center of the open chamber (*see* **Note 33**).
7. With a new, sterile cover slip, reclose the chamber preparation. Before inserting the tension ring to fix the cover slip in position, make sure that there are no persisting air bubbles. These air bubbles can be removed with a Q-tip (*see* **Notes 24** and **34**).
8. Cover the central window of the back titanium frame, which has no cover slip inserted, with a piece of adhesive hypoallergenic tape (*see* **Notes 35** and **36**) and release the mouse back into its cage after removing the other adhesive tape from the tube.
9. Intravital microscopy of the implanted tumor chunk can be performed now repeatedly in the conscious or anesthetized animal, by means of normal light and transillumination, or by means of epi-illumination from a mercury lamp and a fluorescent filter set in combination with appropriate fluorescent dyes injected iv into the animal.

4. Notes

1. For optimal quality of chamber preparation, one of the basic requirements is that the area of the dorsal skin associated with the chamber preparation lacks any injuries, scars, or other irritations. Therefore, prior to surgery, only mice from one brood should be held together in one cage, since mice from different broods tend to cause injury to one another. After chamber implantation the animals must be housed separately in single cages; otherwise, they may destroy each other's chamber preparations by scratching and biting.
2. Work under a laminar flow hood and wear a surgical mask as well as rubber gloves to minimize the possibility of bacterial contamination of the chamber preparation. Between preparation of two different animals, all surgical instruments should be first cleaned mechanically with sterile nonwoven swabs soaked in alcohol and then sterilized with a dry sterilizer. The gloves should be washed with alcohol and changed from time to time.
3. To avoid cooling down of the body temperature of the mouse, the anesthetized animal should be placed on a thermal pad (~37°C) whenever possible.
4. While the depilatory cream is taking effect, clean and prepare the surgical instruments for the following chamber preparation.
5. To avoid irritation of the skin, wipe gently but unsparingly with fresh water soaked swabs.
6. It is important to use depilatory cream as needed to remove all hairs. Otherwise during later transillumination microscopy, the remaining hair roots will show up as dark shadows, decreasing the optical quality of the region of interest.
7. Points of reference are the larger vessels of the skin, which run symmetrically to each other on the left and right side of the sagittal plane of the mouse.

8. The chamber should fit between the two holding sutures.
9. The weight of a Baby-Mosquito fixed at the end of each holding thread is heavy enough to keep the skinfold in an upright position.
10. If possible, place the central window of this frame in a manner that it is lying centrally between the two main vascular trunks coming from caudal and cranial.
11. The skin between the two sutures should be unstressed and the apical screw of the chamber frame should just jut over the upper edge of the skinfold.
12. Because the two holding sutures have to be removed at a later time (*see* **step 14**) and to avoid local skin necrosis do not make these sutures too tight.
13. The location of larger skin vessels may be controlled with transilluminating light. These vessels should not be cut or damaged by the incisions.
14. There should not be any tension on the skin area between the two screws and holding sutures.
15. The two Baby-Mosquitos may be used to adjust the skinfold and keep it level, parallel to the surface of the thermal pad.
16. Be sure to remove all macroscopic particles left inside or around the marked skin area before cutting. Loose hairs or small fibers of the nonwoven swab may be detected easily under the dissecting microscope and can be removed using microforceps.
17. Avoid hurting the underlying inside of the opposite skinfold. It may be advantageous to perform the initial incision in the center of the marked area. Then continue cutting towards and along the dotted line, respectively.
18. Allow enough time for bleeding to stop. The last layer of subcutaneous fatty tissue still protects the underlying cutaneous muscle of the opposite skin, which will later serve as the site of tissue implantation. After-bleeding at a later time onto the unprotected cutaneous muscle can easily destroy the chamber preparation.
19. Place the Q-tip close to the edge of the operation field. In doing so, touching of the vulnerable inside of the underlying skin is avoided.
20. Be sure to dissect the subcutaneous fatty tissue from the underlying cutaneous muscle by cutting, and not by pulling it away. Too much pulling may lead to disruption of small vessels of the muscle and thus to uncontrolled micro-bleeding.
21. **Steps 12** and **13** have to be performed as free of interruption as possible, to avoid drying and damage to the tissue layer (cutaneous muscle), which will later be used as a bed for implanting other tissues.
22. To save time, it may be advantageous to remove all remaining subcutaneous fatty tissue *in toto*, starting from caudal and toward cranial if you are cutting right-handed.
23. At the time of closure of the chamber preparation, the cover slip should already be inserted and fixed in the second titanium frame with the tension ring.
24. Usually the cutaneous muscle should stick to the cover slip solely by adhesion forces, automatically expelling remaining air. After closure of the two chamber frames small, persisting air bubbles may be carefully "pushed out" of the chamber from behind through the central window of the first titanium frame with a dry Q-tip. If you fail to remove all air bubbles, open the slit between both titanium frames for some millimeters, insert a few drops of saline between the cover slip

and the cutaneous muscle using a 26G needle to drive out remaining air bubbles, and close the chamber again.

25. The first nut should be screwed onto the apical screw. Make sure not to tighten the nuts too much. This could result in local skin necrosis or a deficient blood flow to and from the skin being part of the chamber preparation.
26. Make at least 6–8 knots in each of the four holding sutures, since the mice sometimes try to chew through sutures.
27. It will take a maximum of 1–2 d for the mouse to get completely accustomed to its new "knapsack." To allow the mouse to eat easily during these first days after surgery, some food may be put directly on the floor of the cage. After this time period the mouse should show normal behavior again, e.g. cleaning itself, eating, drinking, sleeping, playing, and climbing around in the cage.
28. To avoid immune reactions between the recipient animal and the tumor use either iso-grafted mouse carcinomas or immune-deficient mice as recipients (e.g., severe combined immunodeficient [SCID] mice).
29. Only fast-growing tumors are suitable for implantation since the mice must be sacrificed (on average) less than 30 d after the initial chamber implantation. Stimulated by the weight of the chamber, new skin will grow and lead to a lateral tipping over of the chamber preparation causing a reduced blood flow to and from the skinfold sandwiched between the two titanium frames. As a rule of thumb, solid tumors reaching visible size within 1–3 wk after sc implantation may be suitable for implantation into the chamber preparation.
30. Daily weight monitoring may help to appraise the general state of health of the animal. After an initial loss of weight (less than 10%) mice should stabilize again within the first 48 h after surgery. When bearing a tumor, a further loss of weight might be observed in these animals with increasing tumor volume over the time.
31. To avoid warming of the tumor chunks before implantation in different animals put the Petri dish on ice from time to time.
32. Since it may be stressful for mice to be inserted into a tube during experiments, it is recommended to leave a tube in the cage one or two weeks prior to the chamber implantation so that they become accustomed to it.
33. To avoid drying of the cutaneous muscle and to facilitate air-bubble-free reclosure of the chamber, moisten it with few drops of saline.
34. Be sure that the tumor chunk implanted does not move away from its position in the center of the chamber.
35. The growing tumor may sometimes provoke an itching stimulus at the site of implantation. Using a tape may prevent injuries to the skin and implant of the mice caused by scratching.
36. Avoid direct contact between the tape and the skin of the mouse.

References

1. Sandison, J. C. (1924) A new method for the microscopic study of living growing tissues by the introduction of a transparent chamber in the rabbit's ear. *Anat. Rec.* **28,** 281–287.

2. Menger, M. D. and Lehr, H. A. (1993) Scope and perspectives of intravital microscopy—bridge over from in vitro to in vivo. *Immunol. Today* **14,** 519–522.
3. Leunig, M. and Messmer, K. (1995) Intravital microscopy in tumor biology: Current status and future perspectives (review). *Int. J. Oncol.* **6,** 413–417.
4. Jain, R. K., Schlenger, K., Höckel, M., and Yuan, F. (1997) Quantitative angiogenesis assays: Progress and problems. *Nature Med.* **3,** 1203–1208.
5. Leunig, M., Yuan, F., Menger, M. D., Boucher, Y., Goetz, A. E., Messmer, K., and Jain, R. K. (1992) Angiogenesis, microvascular architecture, microhemodynamics, and interstitial fluid pressure during early growth of human adenocarcinoma LS174T in SCID mice. *Cancer Res.* **52,** 6553–6560.
6. Leunig, M., Yuan, F., Berk, D. A., Gerweck, L. E., and Jain, R. K. (1994) Angiogenesis and growth of isografted bone: Quantitative in vivo assay in nude mice. *Lab. Invest.* **71,** 300–307.
7. Endrich, B., Asaishi, K., Goetz, A. E., and Messmer, K. (1980) Technical report: A new chamber technique for microvascular studies in unanesthetized hamsters. *Res. Exp. Med.* **177,** 125–134.
8. Sewell, I. A. (1966) Studies of the microcirculation using transparent tissue observation chambers inserted in the hamster cheek pouch. *J. Anat.* **100,** 839–856.

9

Angiogenesis Assays Using Chick Chorioallantoic Membrane

David C. West, W. Douglas Thompson, Paula G. Sells, and Mike F. Burbridge

1. Introduction

The study of the angiogenic process and the search for novel therapeutic agents to inhibit, or stimulate, angiogenesis has employed a wide range of in vivo 'angiogenesis' assays (reviewed in **refs. 1–3**). These differ greatly in their difficulty, quantitative nature, rapidity, and cost. The classical in vivo models include the rabbit ear chamber, hamster cheek pouch, dorsal skin chamber, dorsal skin and air-sac model, anterior chamber/iris and avascular corneal pocket assay, and the chick embryo chorioallantoic membrane (CAM) assay. More recent methods involve the implantation of preloaded Matrigel or alginate plugs, or collagen or polyvinyl sponges *(1)*. Largely owing to its simplicity and low cost, the CAM is the most widely used in vivo model for the study of both angiogenesis and antiangiogenesis *(1,4)*

Originally the tool of the embryologist, the chick embryo chorioallantoic membrane was adapted as an angiogenesis assay by Folkman and coworkers *(5–9)*, initially to examine the angiogenic activity of tumor tissues. Two complementary methods were developed. Possibly the simplest and most widely used form of the assay is performed with the CAM *in situ*, the samples (up to two per egg) being applied through a window made in the shell. The alternative form of the assay, the shell-less CAM, involves transfer of the early, 3- or 4-days-old, embryo and its extraembryonic membranes to a glass Petri dish for further development. This gives a wider surface area, allowing multiple sample application on the same egg, and easier viewing and photography, but suffers from poor embryo viability. Consequently, several modifications have been introduced in an effort to increase the time-span of embryo viability

From: *Methods in Molecular Medicine, Vol. 46: Angiogenesis Protocols*
Edited by: J. C. Murray © Humana Press Inc., Totowa, NJ

(10–13). Variants of the shell-less model apply the sample to the yolk sac, or vitelline membrane, which is superseded by the chorioallantoic membrane from about d 7 of development. This form of assay has largely been employed to study inhibitors, toxicity, granulation tissue formation, and early vascular effects *(14–19)*.

The CAM develops over the first 11 d of the 21-d gestation period. At about 24 h the development of blood islands around the embryo can be seen. Capillaries appear in the yolk sac of fertilized hens' eggs at 48 hr and grow rapidly over the next 6–8 d. The embryonic yolk sac itself consists of four layers: ectoderm (bilayered vitelline membrane), somatic mesoderm, visceral mesoderm (including blood vessels), and endoderm *(16)*. The circulatory system originates from the mesoderm and develops as three distinct divisions, in the following sequence: the extraembryonic (for nutrition), the accessory or allantoic (for respiration and excretion), and the embryonic (for distribution within the embryo). By d 4, the yolk sac membrane is well vascularized and continues to develop until the chorioallantoic membrane supersedes it, about d 7. The endothelial cells of this developing vasculature have an abundance of secretory organelles, most likely related to the synthesis of the extracellular matrix and basement membrane *(20)*. After d 11, proliferation of endothelial cells in existing capillaries is complete, and they appear to have acquired the morphological characteristics of differentiated cells *(7,20)*.

During the first 10 d of incubation, the organs differentiate rapidly *(21)*. The heart undergoes a transition to the adult myosin phenotype on d 6 and is fully chambered by d 8 *(22)*. The spleen is recognizable by d 4 and of considerable mass by d 6 *(23)*. Between the d 4–5 of development, the mesonephric kidney begins to secrete urine into the allantois, which arises as a diverticulum of the hind gut *(24)*. The liver develops between d 3 and 9. Detectable levels of proteins concerned with blood clotting (fibrinogen and prothrombin) and globulin *(25)* appear on d 12 or 13, but complement is not detected until at least d 17.

The chick embryo does not have a fully developed immune system until late in its gestation. The thymus, which is first visible at d 5, is invaded by stem cells on d 6 or 7 *(26,27)* and differentiates into cortical and medullary zones on d 13. True lymphocytes are thought to be present by d 17 *(28–30)*, but responsiveness to T-cell mitogens does not develop until d 20 *(31)*. Cells bearing B-determinants first appear in the cloacal bursa, about d 12. Yolk sac phagocytic cells, probably of endothelial origin, have been described *(16)* and mast cells have been identified on d 3 of incubation *(32)*, but inflammatory responses are largely absent in the yolk sac membrane.

Given the immaturity of its immune system, it is surprising that one of the main limitations of the CAM assay is that samples may induce nonspecific inflammatory reactions, epithelial hyperplasia, or granulation tissue formation, all of which can stimulate a secondary angiogenesis reaction ***(16,33–35)***. However, similar indirect angiogenesis also occurs in most other in vivo assays. Nonspecific reactions have been reported for a variety of carrier vehicles (e.g., Millipore filters, fiberglass discs, gelatin and polymer sponges, filter paper, agarose or polyacrylamide discs, and Thermanox cover slips) and egg shell itself. In addition, a high concentration of salt crystals can induce a hyperosmotic damage to chorion epithelium, again inducing a nonspecific response ***(35,36)***. The focal application of angiogenic stimulators or inhibitors first described by Folkman ***(5)*** remains the most widely used method. However, in apparent angiogenic stimulation, the focal "spoke-wheel" vascular pattern of the CAM may be the effect of local buckling and contraction of the CAM, or a local distortion of supply vessels due to fibrosis, and may not result from a local increase in capillary numbers ***(37,38)***. In response to the problems of non-specific carrier and salt effects, and this potential localized physical distortion of the CAM in focal application, several workers have developed assay systems in which enough liquid sample is applied to cover the whole of the CAM surface ***(39–43)***. However, in these methods the degree of angiogenesis can only be assessed by chemical means (protein/collagen synthesis, through incorporation of radioactive amino acids, or DNA synthesis, using thymidine incorporation) or by image analysis, as the response affects the whole area of the CAM.

Numerous methods of quantifying the CAM angiogenesis response have been published in the last 25 years. Semiquantitative scoring methods depend on a subjective grading scale for response, but with experience and blind sample application this mode of assessment offers a convenient method to study large numbers of samples and make serial observations ***(1,3,4,13,44–47)***. Use of a numerical grading scale allows calculation of a coefficient of angiogenesis ***(13,44,45)***, and statistical analysis using the Wilcoxon Matched-Pairs Signed-Ranks test or Fisher's test. Determination of the number of converging vessels within a 1 mm radius of the application point ***(48,49)***, or the number of vessels intersecting 3 concentric rings, at 4, 5 and 6 mm diameter from the sample ***(50,51)***, have also been used as a measure of the angiogenic response. Several laboratories have developed semiautomatic image analysis methods based on these same principles ***(51–53)***.

Recently, a novel quantitative variant of the CAM assay has been proposed ***(54)***. In this method the soluble test substance is prepared in a collagen solution, which is then allowed to gel at 37°C. Prior to application to the 8 d CAM, the sample gel is placed between two parallel pieces of nylon mesh. New vessels grow vertically into the collagen gel and the nylon meshes align the vessels for rapid counting. It is claimed that new vessels can be quantified in less than 1 min for each sample.

Early methods were almost totally geared to assay of angiogenesis, but the interest in negative regulators of angiogenesis and antiangiogenic drugs has seen the development of assays for the inhibition of angiogenesis. The original focal and liquid CAM assays have been, or can be, used in this context by coapplication of both an angiogenic agent and the potential inhibitor. In addition, several methods have been developed based on inhibition of the developing vasculature of the yolk sac membrane and early CAM.

As the reader has no doubt gathered, almost every worker has his or her own variation on the CAM assay. In the following subheadings, we outline four major variants of the assay: (1) Focal sample application to the CAM *in situ*; (2) liquid sample application to the CAM *in situ*; (3) an inhibition assay on the yolk sac membrane/early CAM *in situ*, and (4) the shell-less yolk sac membrane method for early vascular effects and inhibition. These have been used routinely for at least the last five years, and in some cases for as long as 20 years.

2. Materials

2.1. Preparation of Eggs for In Situ Sample Application

1. Fertile hen eggs (*see* **Notes 1, 2**, and **3**).
2. Betadine iodine scrub or 70% ethanol.
3. 37°C humidified incubator (fan assisted preferred) (*see* **Note 4**).
4. 2 mL syringe with G21 needle.
5. Modeler's electric drill with selection of cutting tools (Zircon Drill, BDH Cat No. 0190) or a diamond tipped glass-slide marker.
6. Household vacuum cleaner (necessary to remove shell dust) (*see* **Note 5**).
7. Fine pointed forceps and a microspatula.
8. Scotch "Magic" adhesive tape.

2.2. Sample Preparation for Focal Application

1. 1% aqueous methylcellulose (viscosity of 2% aqueous solution at 25°C, 4000 centipoises, Sigma).
2. Sample (sterile if possible) (*see* **Note 6**).
3. 10 μL automatic pipette.
4. 2-mm-diameter Teflon rods (flat ended) inserted vertically in a Teflon, or Perspex, base. A 10 × 8 array of rods is suitable for most purposes (*see* **Note 7**).
5. Stainless steel dental needle, G30 or smaller.
6. Fine pointed forceps.

2.3. Collagen Synthesis Assay

1. [U- ^{3}H]- or [U- ^{14}C]- proline (Amersham/ ICN).
2. A sharp, sterile 1-cm diameter stainless steel cork-borer.
3. Ice-cold 5% and 20% aqueous trichloroacetic acid (TCA).
4. Refrigerated centrifuge capable of centrifuging 3 mL at 10,000*g*.

5. 1 *M* NaCl in phosphate-buffered saline.
6. Filter sterilized (0.2 µm) sterile collagenase solution (Sigma Type II, 1 mg/mL in Dulbecco's Modified Essential Medium (DMEM), containing 200 U/mL penicillin and 200 µg/mL streptomycin).
7. 37°C shaking waterbath or incubator.
8. Scintillation fluid and scintillation counter.

2.4. Visual Assessment of Angiogenesis

1. Stereomicroscope (Wild M32, continuously variable magnification). ×4–×8 objective and ×10 eyepiece most useful.
2. Fiberoptic light source (Schott K1500).
3. 35-mm camera mounted on microscope.
4. 4% paraformaldehyde in phosphate-buffered saline is prepared by dissolving 40 g of paraformaldehyde in 1 L of phosphate-buffered saline (PBS), by stirring at 60°C in a fume hood until dissolved. This is stored frozen until required. Alternatively, commercial formalin solution (40% w/v) may be diluted with phosphate-buffered saline or normal saline.
5. Syringe (5 mL) fitted with a 26-guage needle.

2.5. In Situ Inhibition Assay

1. 6-d-old hen eggs, prepared as in **Subheading 3.1.**
2. 1% aqueous methylcellulose (viscosity of 2% aqueous solution at 25°C, 4000 centipoises, Sigma).
3. Teflon slabs (50 × 100 × 3 mm; *see* **Note 7**).
4. 10 µL automatic pipette.
5. Sterile dissection forceps.
6. Liquid paraffin.
7. Tween 80 (Sigma).
8. Electric homogenizer.
9. Syringe (1 mL) fitted with a 26-guage needle.
10. Suitable dissecting microscope and camera system (optional).

2.6. Sample preparation for non- focal liquid application

1. Sterile (0.2 µm filtered) solution of sample in phosphate-buffered saline.
2. 1 mL syringe.

2.7. [^{3}H]- Thymidine Incorporation Assay

1. Methyl-[^{3}H]-thymidine (2 µCi/ml in phosphate-buffered saline).
2. Ice-cold normal saline.
3. Pointed scissors.
4. Homogenizer with a suitable head for a 5 mL volume.
5. Ice-cold 10% aqueous trichloroacetic acid.
6. Centrifuge to take "Universal" containers at 1000*g*.

7. Instagel Plus (Packard Biosciences), a thixotropic scintillation fluid.
8. Scintillation counter.

2.8. Preparation of Shell-less Eggs

1. Fertile hen's eggs (*see* **Notes 1** and **2**).
2. 70% ethanol.
3. 37°C humidified incubator (fan assisted preferred).
4. Laminar flow cabinet.
5. "Clingfilm hammocks," formed by draping non-PVC clingfilm (Perfa-Cling, UK) over a section of plastic drainpipe, about 7 cm diameter and 4 cm high, and securing with a rubber band.
6. Sterile petri dish.
7. Dissecting microscope.
8. Fiberoptic light source (Schott K1500).

2.9. Sample Application on Shell-less Eggs

1. 2 mm diameter methylcellulose discs (*see* **Subheading 2.2.**), filter paper discs (Whatman no. 2) or nitrocellulose discs.
2. Fine forceps.

3. Methods

3.1. Preparation of Eggs for In Situ Sample Application

1. On d 0, bring the fertile hens' eggs to room temperature, and wash with 2% v/v aqueous Betadine, drain, and place in a fan-assisted humidified incubator/egg incubator, at 37–37.5°C, preferably fitted with tilting shelves (*see* **Note 8**).
2. On d 3, remove the eggs from the incubator and swab with Betadine or 70% alcohol. Make a small hole the shell at the pointed end of the egg and, using a 2-mL syringe with a 21G needle, remove 2 mL of albumen. Return the eggs to the incubator and incubate horizontally with hole uppermost (*see* **Note 9**).
3. On d 4, swab the blunt end of the egg with Betadine or 70% alcohol and, using a 21G syringe-needle, or a fine-cutting drill bit, make a small hole/cut in the shell, at the blunt end of the egg, to puncture the air sac. Then, using the same fine-cutting drill bit, or a diamond-tipped glass slide marker, score a 1-cm square window (size depends on further procedures) in the shell between the two previous cuts. After scoring, remove the square of shell with sharp pointed forceps and a small spatula, exposing the CAM (**Fig. 1**; *see* **Note 10**). This procedure causes the membrane to drop as the egg contents fill the space left by collapse of the air space at the end of the egg and the withdrawal of albumen (*see* **Note 11**). Dispose of any nonviable eggs at this stage and, for each of the remainder, cover the square opening in the shell with adhesive tape. Return the eggs to the incubator, keeping them horizontally with the window uppermost, until sample application on d 10 (d 8 may also be used; *see* **Note 3**).

Fig. 1. Windowed egg showing exposed chorioallantoic membrane (CAM).

3.2. Focal Sample Application

1. Prepare a 1% w/v aqueous solution of methylcellulose and autoclave to sterilize. Allow to cool whilst stirring and store sterile solution at 4°C (*see* **Note 6**).
2. Sterilize Teflon rods by autoclaving or 70% ethanol.
3. On day of sample application (*see* **Note 15**), in a laminar flow cabinet or clean environment, mix equal volumes of sterile 1% methylcellulose and sample (10 μL for each sample application). Vortex the mix thoroughly, because methylcellulose is very viscous. Control discs of methylcellulose only should be prepared in parallel (*see* **Notes 12** and **13**).
4. Pipette 10 μL of sample mix on to the top of a Teflon rod, or slab. After all the samples have been applied to the Teflon rods, the samples are kept in the flow cabinet until they have completely dried to a thin 2-mm diameter disc (30–60 min; *see* **Note 14**).
5. On day of sample application, organize the surviving eggs into groups of at least five eggs per sample. (*See* **Note 15**.) Apply samples to each egg in turn. Peel back the adhesive tape to reveal the underlying CAM, and make a small puncture in the CAM with a fine dental needle (*see* **Notes 16, 17**). Peel the sample disc from its Teflon rod, using fine forceps, and place it on the CAM over the puncture. Reseal the egg window and replace the eggs in the incubator for a further 3 d (*see* **Note 15**).

3.3. Collagen Synthesis Assay

1. Prepare eggs and samples as in **Subheadings 3.1.** and **3.2.**
2. On d 10, immediately after application of the methylcellulose disc, apply 4 μL of aqueous [U- ^{3}H]- proline (1 μCi) on to the sample pellet.

3. After visual assessment of the angiogenic response on d 13 (**Subheading 3.4.**), carefully excise the membrane using fine scissors and forceps.
4. With a sharp, sterile 1-cm-diameter cork-borer, cut a circle of each membrane (1 cm diameter and approximately 5 mm radius from the point of sample application) and transfer each separately to 3 mL of ice-cold 5% TCA.
5. After 4 hr at 4°C, recover the membranes by centrifugation at 10,000*g* for 10 min, at 4°C.
6. Wash the membranes twice with 3 mL of ice-cold 5% TCA, followed by three washes with 3 mL 1 *M* NaCl in phosphate-buffered saline.
7. Digest washed membrane in 1 mL of sterile collagenase at 37°C, for 16 hr, with shaking or mixing.
8. Terminate the digestion and precipitate undigested protein by the addition of 1 mL of ice-cold 20% TCA.
9. After 4 hr at 4°C, centrifuge the digest at 10,000*g* for 20 min and count an aliquot of the supernatant, containing collagen peptides, in liquid scintillation counter to determine [^{3}H]- proline incorporation into collagen peptides. (*See* **Note 18**.)

3.4. Visual Assessment and Photography

1. On d 13 and/or d 14 (*see* **Note 19**), examine CAM for angiogenesis at site of sample application. Part, or all, of the methylcellulose disc remains to indicate the area.
2. Note the general condition of the CAM and embryo. Do not score dead or dying eggs (membrane appears opaque and it is evident that there is a poor blood supply to the CAM vessels).
3. Score the degree of angiogenesis: 0, no change; 1, changes in vessels (disorganized or bending), but not directed to the point of application; 2, a few microvessels converge; moderate microvascular and large vessel convergence; and 3, classical spoke wheel (**Fig. 2**; *see* **Note 20**).
4. The CAM can be photographed to provide a permanent record. A total magnification (objective × ocular lens) of ×40 to ×80 is generally most suitable (*see* **Note 21**).
5. In addition, or alternately, fix the CAM and excise it for photography. Inject ice-cold 4% paraformaldehyde–PBS under the CAM (approx 2 mL) and also flood the upper surface of the CAM. Leave at 4°C for at least 4 hr (overnight fixation is generally more practical). Excise the CAM, using small dissection scissors and forceps (*see* **Note 22**), into PBS and place the CAM on a clean dry microscope slide. Photograph under the dissecting microscope, as above.

Fig. 2. *(see opposite page)* Positive focal reactions: **(A)** Grade 4 reaction showing "spoke wheel" pattern, with bending of large vessels (in situ, magnification ×16); **(B)** weaker Grade 2 reaction, converging vessels on upper side of sample, largely microvascular, typical of vascular endothelial growth factor (VEGF; *in situ*, magnification ×32); and **(C)** paraformaldehyde-fixed CAM, showing a Grade 4 reaction (magnification ×32).

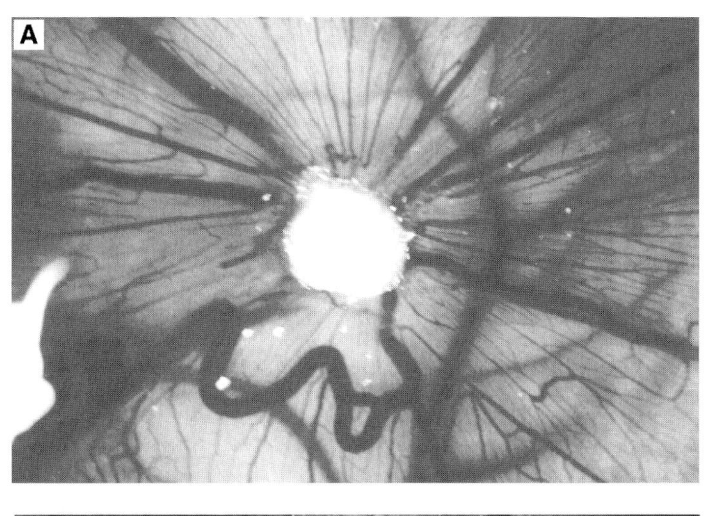
A

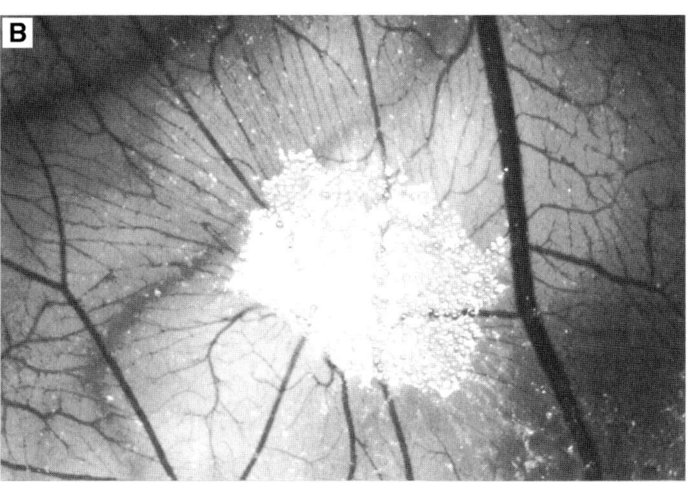
B

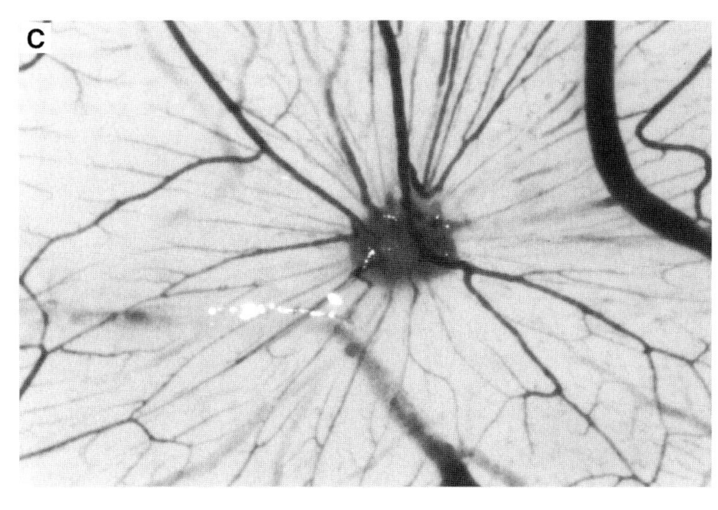
C

3.5. In situ Inhibition assay

1. Prepare a 1% w/v aqueous solution of methylcellulose and autoclave to sterilise. Allow to cool whilst stirring (*see* **Note 6**).
2. Sterilize Teflon rods by autoclaving or 70% ethanol.
3. Dissolve compounds to be tested in suitable solvents at 20 mM (*see* **Notes 6** and **23**).
4. Mix equal volumes of compound solution and 1% methylcellulose, and place 10 μl drops onto a Teflon slab (*see* **Note 7**). Empty discs (negative control) should be prepared in parallel. Allow to dry.
5. Using sterile forceps, gently peel the methylcellulose discs from the Teflon slabs, and place in the central area of the 15 mm diameter CAM of 6-d-old eggs (*see* **Notes 24**, **25**, and **26**).
6. Return eggs to the incubator for a further 48 hr.
7. In a homogenizer, mix equal volumes of liquid paraffin and 4% (v/v) aqueous Tween 80 to obtain a fine white emulsion (*see* **Note 27**).
8. Using a 1 mL syringe fitted with a 26-gauge needle, inject 1 mL of this emulsion into the chorioallantois, so that the vascular network of the CAM stands out against the white background.
9. Observe with the naked eye. An area in which relatively large vessels (those already present on d 6) cross an otherwise avascular zone of 3 mm or more in diameter is scored as positive (**Fig. 3**; *see* **Note 28**).
10. The antiangiogenic activity of each compound is given as the percentage of tested eggs presenting a positive response (*see* **Notes 24** and **29**).

3.6. Nonfocal (Liquid) Sample Application Technique

1. Prepare hen eggs as described in **Subheading 3.1.**
2. On d 10, organize the surviving eggs into groups of preferably no less than 10. Prepare samples and inject 0.3 mL of test, or control, sample through the adhesive tape onto the upper surface of the CAM (*see* **Note 30**).
3. On d 11, 18 hr after sample application (*see* **Note 31**), fix the CAM as outlined in **Subheading 3.4**. Dehydrate and mount *en face* on large slides, using DPX resin (*see* **Note 32**).

Visual analysis, ×30 magnification, shows general increase in vessel number and tortuosity with positive samples (**Fig. 4**). Subsequent image analysis of vessel number, branching, and length can be used to quantify these changes. Alternatively, measurement of the incorporation of methyl-[^{3}H]-thymidine into DNA may be used.

3.7. Assay of DNA Synthesis as Methyl-[^{3}H]-Thymidine Incorporation

1. On d 11 apply 0.5ml of methyl-[^{3}H]-thymidine (2μCi/ mL of normal) onto the CAM surface of each egg. Return to incubator for 20 min (*see* **Notes 33** and **34**).

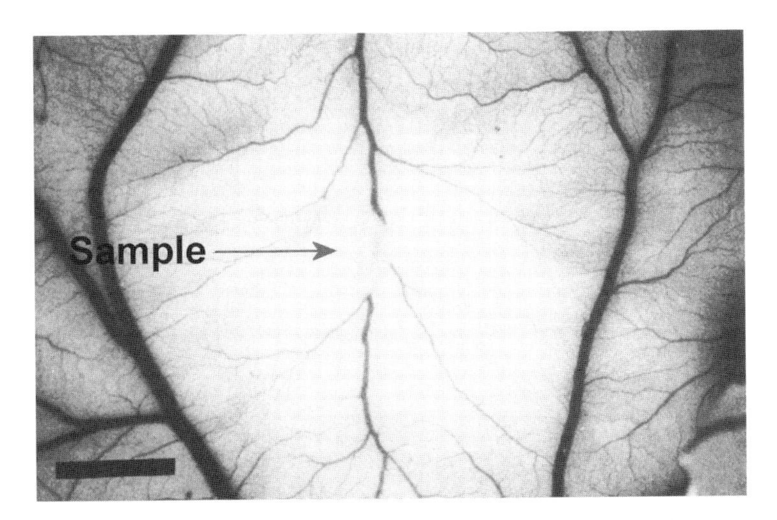

Fig. 3. *In situ* inhibition of angiogenesis in the CAM, d 8. The inhibitor was placed in the methylcellulose disc (*center*) on d 6. Note the absence of small vessels around the disc, the larger preexisting vessels being unaffected (bar, 4 mm).

2. After 20 min, remove the eggs to a radioactive type bench and rapidly inject all the eggs through the adhesive tape window, with 5 mL of ice-cold normal saline to halt metabolism (*see* **Notes 35** and **36**).
3. Remove each CAM by holding the egg in the palm of one hand, window face down, and piercing the side with pointed scissors above the level of the dropped CAM. Cut around the shell with scissors, to separate the CAM from the rest of the embryo. After cutting the umbilical vessels connecting the embryo to the CAM, decant the residual yolk and embryo into a disposal container.
4. Cut around the edge of the CAM. Generally the dropped CAM remains inflated by air above the window sealed by adhesive tape and a red ring is present at the edge of the dropped CAM where it adheres to the shell (*see* **Note 37**).
5. Rinse each CAM briefly in saline, blot dry, and place in a 20 mL Sterilin disposable vial, containing 5 mL of distilled water. These may be stored at 4°C overnight if desired.
6. Homogenize each CAM briefly in the 5 mL of H_2O using a powered homogenizer with a suitable head.
7. Add 5 mL of 10% trichloroacetic acid (TCA) to each and vortex.
8. Centrifuge the samples at approximately 1000*g* for about 8 min. Discard the supernatant and resuspend the pellet in 10 mL TCA. Vortex thoroughly and centrifuge 1000*g* for 8 min. Repeat the TCA wash to ensure adequate removal of free isotope.
9. Resuspend the final precipitate in 2.5 mL of H_2O, with vortexing, and immediately pour into scintillation vials containing 5 mL of Instagel Plus (a thixotropic

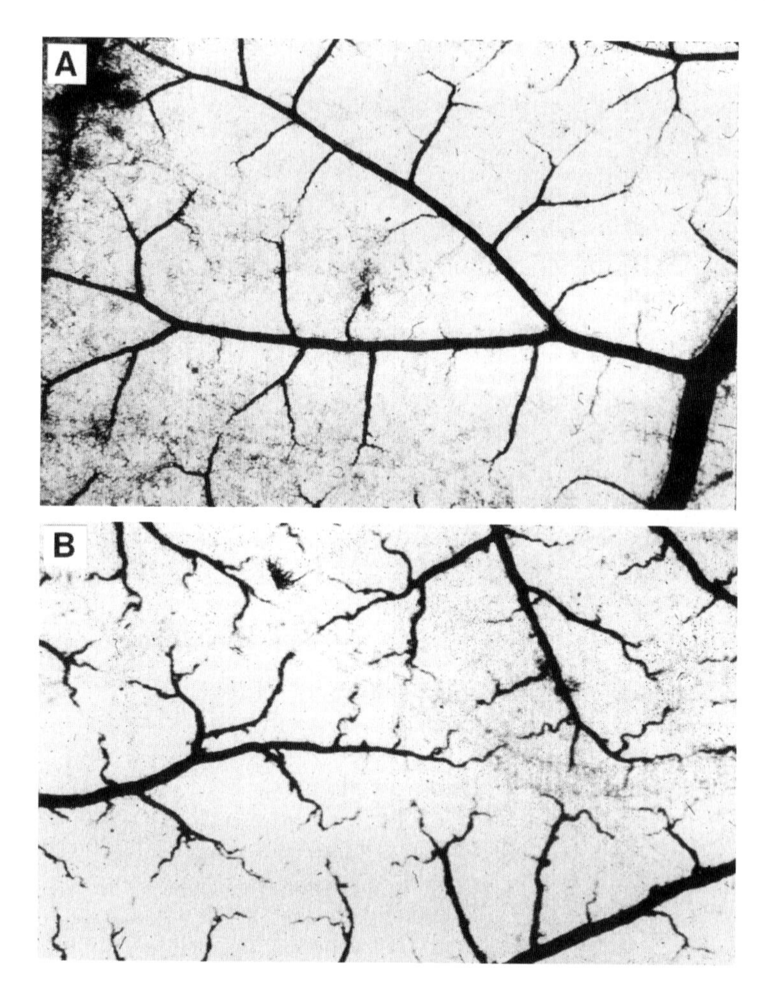

Fig. 4. Visual assessment of nonfocal application: **(A)** normal, regular branching of vessels on d 14 after application of control buffer (magnification ×30); **(B)** positive reaction showing an overall increase in vascularity and tortuosity (magnification ×30).

scintillant). Vortex individual scintillation vials until the contents become a gel (*see* **Note 38**, **39**). Count the [^{3}H] in each vial using a suitable scintillation counter (*see* **Note 40**).

3.8. Preparation of shell- less eggs

1. Prepare hen eggs as outlined in **Subheading 3.1** (*see* **Note 41**).
2. Prepare clingfilm hammocks by draping nonPVC clingfilm (clingfilm containing PVC is toxic) over a section of plastic drainpipe about 7 cm diameter and 4 cm high and securing with a rubber band.

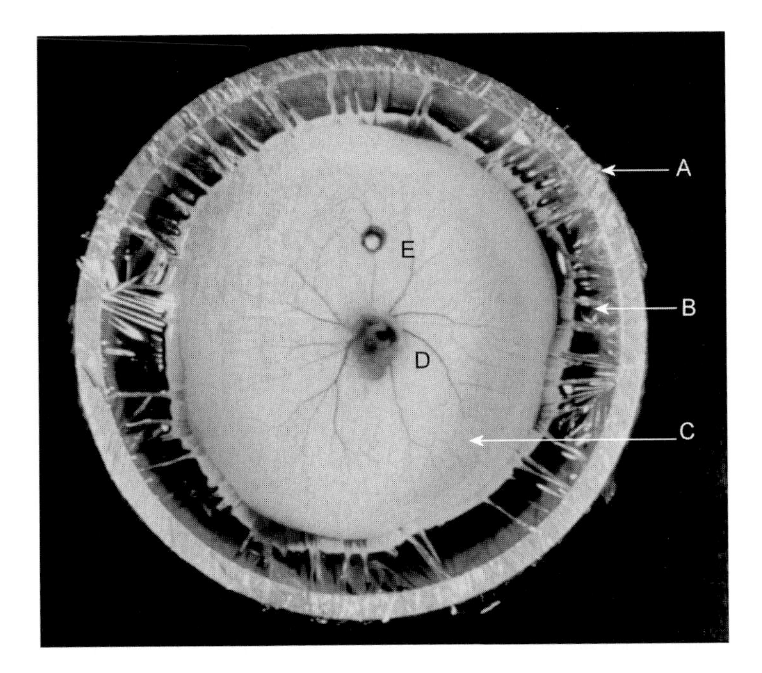

Fig. 5. Shell-less gg in clingfilm hammock. *A*, Drainpipe section as support; *B*, nonPVC clingfilm; *C*, yolk sac membrane; *D*, heart, *E*, 2 mm filter-paper sample disc.

3. On d 4, in a laminar flow cabinet, wipe the eggs with 70% alcohol and crack out of their shells into clingfilm hammocks (**Fig. 5**; *see* **Notes 42**, **43**, and **44**). Cover egg preparations with a sterile Petri dish and transfer to a humid chamber or incubator, such as a 45 × 30 cm seed propagator (accommodates 18 preparations) at 37°C.
4. On d 6, apply the sample in a 2 mm methylcellulose pellet (*see* **Subheading 3.2.**), or on 2 mm filter paper disc ***(19)***, directly over a blood vessel, preferably a major bilateral vein, on the yolk sac membrane. Veins, arteries, and blood flow (red blood cells) can be distinguished under a dissecting microscope (×30 magnification). Return the eggs to the incubator (*see* **Notes 45**, **46**, and **47**).
5. On d 6 examine the membrane for vascular effects, using a dissecting microscope and flexible oblique lighting. Preliminary vascular reactions such as hemorrhage (**Fig. 6A**; *see* **Notes 48** and **49**) and vascular narrowing, or occlusion, can be seen as early as 4 hr after sample application (**Fig. 6B**).
6. On d 8, examine the membrane for inhibition of vascularization. Inhibitors should produce a clear zone around the sample (*see* **Subheading 3.5.**; **Fig 3**). A photographic record can in most cases be made without the use of a microscope (*see* **Notes 50** and **51**).

3.8. Conclusions

We have outlined the four main variants of the CAM assay, with additional references to several others. The major feature of both the yolk sac and chorio-

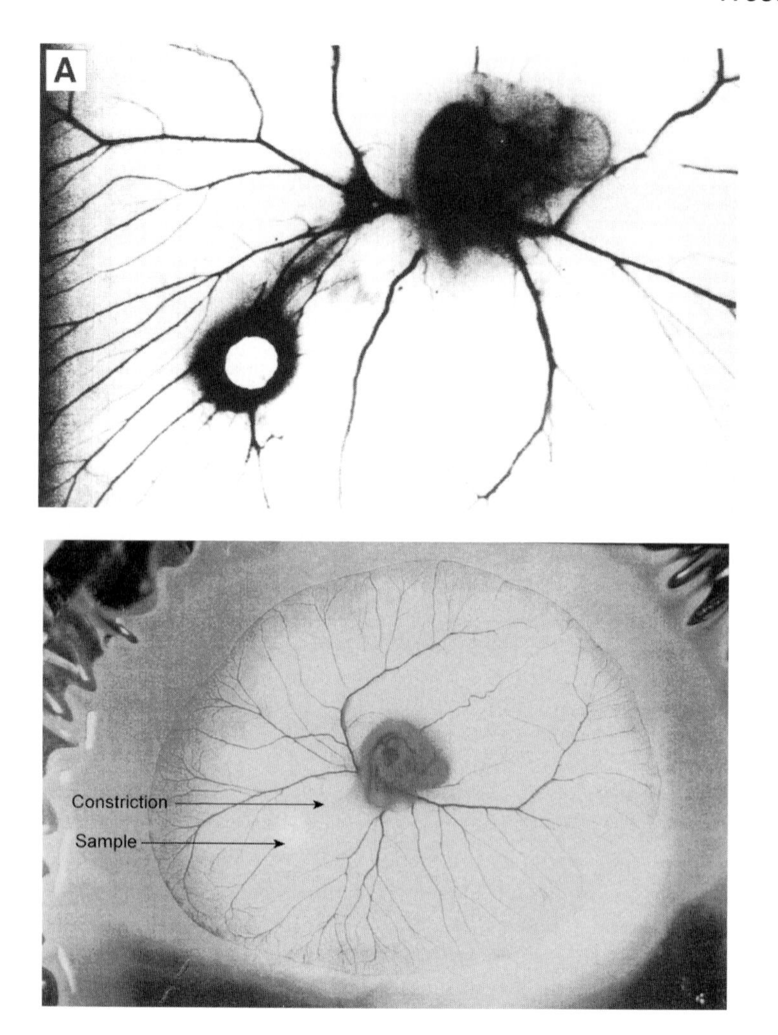

Fig. 6. Early responses (2–4 hr) after sample application: **(A)** Local hemorrhage at site of application; **(B)** vascular constriction. Note disappearance of vein proximal to the point of application.

allantoic membranes is the vascular system, which appears to be sufficiently similar to the mammalian counterpart to make this a useful in vivo model and a possible alternative to rodent models. One of the perennial problems associated with these assays is that of quantifying the response and its statistical significance. Several metabolic labeling methods have been described, as well as scoring and numerical assays. However, there is considerable variability between the eggs themselves, as commercial strains are all out-bred stock, which in the DNA assay results in a variation within a group of individual

CAMs of 30–40%, although the intrinsic assay variation is only 5%. The only answer for producing high-quality data is to use groups of 8 or 10 eggs per group. Larger groups may be necessary to obtain statistical significance.

The choice of method to use must depend on available equipment and technical backup. Focal application methods with *en face* visual assessment are quick, economical, use minimal amounts of scarce materials, and are at least semiquantitive. These methods are useful for screening many novel substances but can give inconsistent and false results in inexperienced hands, with potentially significant results requiring further exploration and confirmation. The various biochemical methods provide objective parameters of angiogenesis and the methods are generally quicker than image analysis. Although they require additional, but general, equipment (i.e., scintillation counters) and are slower than visual screening, they are valuable methods for exploring mechanisms, interactions, and blocking agents. The latter methods do depend however on a substantial work-up to obtain dose response data and time-optimization information, but this in itself can be illuminating and may expose the mechanisms of angiogenesis stimulation or inhibition.

4. Notes

1. Preferably from an egg production rather than a broiler strain. The quality of the data is dependent upon the quality and viability of the eggs. Commercial flocks reach the end of their laying lifetime after 3 mo. The shells become thin, infertility increases, and the death rate increases up to time of use at around d 10. Infection is a constant hazard with these invasive techniques and the embryo is immunodeficient until just before hatching. This means that the eggs are very susceptible to mixed bacterial infection from feces-derived shell organisms, including *Aspergillus fumigatus*, *Staphylococcus aureus*, and *Pseudomonas aeruginosa*. Nowadays, most commercial egg producers of broiler and egg-laying strains for human consumption of chickens and eggs are fanatical about checking for contaminants such as *Salmonella*. This has improved the quality considerably, but it is unrealistic to expect that eggs will have a sterile surface, so handling and incubation technique and cleanliness remain critical. Exposure of eggs to powerful general disinfectants is fatal rather than beneficial. Formalin in a hot room for eggs is intolerable for eggs and humans alike. There is, however, a useful range of products specially formulated for commercial egg production together with ample guidance from, for example, Antec International (Superhatch Chicgard and the newer product Ambicide; www.antecint.com/index.htm). A quick 1 min dip in such a diluted product before starting incubation is harmless. It may also be useful to know which antibiotics the supplier is using, as these are carried into the eggs. Some antibiotics can modulate angiogenesis.
2. The eggs may be kept for several days at 4°C (or in a cool room overnight) without harm until it is convenient to start the incubation schedule.

3. It should be noted that, in the UK, experiments on embryos after the first half of gestation come under the Animals (Scientific Procedures) Act 1986. This applies to experiments on the chick chorioallantoic membrane after d 10. The experiments are generally regarded as being of mild category.
4. Excessive humidity allows surface organisms to penetrate the shell, but trays of clean water can be placed beneath the egg trays and the egg trays should ideally have wire mesh bottoms. Addition of strong disinfectants to the water is generally too toxic, but 1% $CuSO_4 \cdot 5\ H_2O$ reduces bacterial/fungal growth and is nonvolatile.
5. When using the drill, with either the hole-making or cutting abrasive tools, it is necessary to remove the shell dust generated as this can induce false positive reactions on the CAM. It may also induce skin allergy and a mask should be worn to prevent breathing in the dust over extended periods of time, i.e., when one operator is processing large numbers of eggs.
6. Although most samples will be water soluble, small organic molecules (such as synthetic angiogenesis inhibitors) may require organic solvents. A large range of organic solvents may be used to dissolve tested compounds without adverse effects on the CAM since they will evaporate during drying of the discs. Avoid using DMSO, which does not readily evaporate.
7. Teflon is used for its nonstick properties. Teflon rods (*see* **Subheading 2.2.**) may be used rather than slabs to ensure uniform disc size, but for the screening of large numbers of compounds the slabs are more practical.
8. The eggs can be positioned horizontally (ordinary incubator), or vertically with pointed ends facing downwards (in a specialized egg incubator).
9. It may be preferable to carry out this stage in a laminar flow cabinet, but a clean bench environment is adequate. If the eggs are incubated horizontally, then the pointed end of the egg is swabbed with neat Betadine and a small hole made in the shell, using an abrasive drill bit on the electric drill, for subsequent needle aspiration. Some groups insert the needle directly through the shell, to limit shell dust. When aspirating the albumen make sure the needle opening is pointing away from the embryo. In addition, regular changes of needle and syringe limit the carryover of infection from egg to egg. For large numbers of eggs it may not be practicable to change after every egg, but every fifth egg seems adequate.
10. The diamond-tipped glass marker produces less shell dust.
11. The volume of the space above the dropped CAM is approx 7 mL.
12. One advantage of focal application is that suspensions of poorly soluble compounds and those dissolved in slow release polymers, such as Elvax, can be tested.
13. The sample may be a mixture of positive angiogenic compound (i.e., 5–10 ng vascular endothelial growth factor [VEGF]) and potential antiangiogenic compounds. Also, we have premixed tumor extracts and neutralizing antibodies to specific angiogenic cytokines to ascertain their role in the angiogenic reaction. Antibody controls are required in this instance to negate direct effects on chick angiogenesis (species cross-reactivity).

14. A high content of salts in the sample can make the methylcellulose discs hard and fragile.
15. Samples may be applied to the CAM on d 10 (score results on d 13 and/or 14), d 8 (score on d 10; *see* **Note 3**), or d 6 (score on d 8) in an inhibition assay (**Subheading 3.5.**).
16. A sparsely vascularized area between two large vessels is best. The puncture should be made 3 mm, or more, from each vessel.
17. The puncturing of the CAM is not absolutely necessary, but does reduce salt effects.
18. This assay essentially measures proline incorporation into the basement membrane collagen type IV and VIII during angiogenesis. However, if the sample induces inflammation or fibrosis, incorporation into type I collagen will also occur.
19. On d 10, if you apply sample on d 8.
20. A single operator, preferably blinded to the sample identity should grade the angiogenic response. Different test substances can produce a range of different types of angiogenic response, i.e., VEGF produces a purely microvascular response (even at grade 3), but basic fibroblast growth factor (bFGF) gives a mixed response of microvascular growth and large vessel deformation/ growth towards the point of application. Samples may also induce local bleeding (i.e., VEGF) and with some samples a local inflammatory infiltrate is evident (this is especially evident with tissue extracts and culture supernatants, desalting may help). The presence and severity of these reactions should be noted, as the response may be secondary to the bleeding or inflammation. Statistical analysis of such discrete variables may be performed using the Wilcoxon Matched-Pairs Signed-Ranks test, but >10 results per sample are required. Paired blind comparisons can be made between test and control discs on pairs of eggs. The more vascular response of each pair, viewed blinded as to substance, is recorded and permits the use of Fisher's test and tables for easy, no calculation, statistics that invoke the power of paired individual comparison. The focal spoke-wheel increase in vascularity can be quantified by point counting of the supply vessels viewed en face. This may be done on fresh or fixed and stained preparations or photographs. Objective statistical comparison is spurious if there is nothing to measure from the control applications.
21. Difficulties may be encountered owing to movement of the embryo. Focusing can be a matter of chance and a fast film of at least 200 ASA is required. In addition, the background may also be composed of vessels below the CAM itself and this can partially obscure the CAM reaction. This may be overcome by injection of paraffin oil emulsion under the CAM (*see* **Subheading 3.5.**).
22. Even after 4 hr the fixed CAM is easy to handle, but longer fixation greatly reduces its fragility.
23. A quantity of 100 nmol of compound per disc should be suitable for screening purposes. Active compounds may be subsequently tested at lower doses.
24. At least 10 eggs should be used per compound and per dose.

25. The CAM is an expanding membrane with vessels developing over its entire surface. It is not necessary to place the methylcellulose discs on its outer edges.
26. Reject eggs with an excessively humid CAM; otherwise, discs may float off during the incubation period.
27. This paraffin oil/water emulsion may be stored for several weeks.
28. The true antiangiogenic compound will inhibit the formation of new vessels, while not affecting the existing vasculature. Other effects (total absence of vessels, vessel rupture, and bleeding) should be noted.
29. Responses may be graded (e.g., +: 3–6 mm; ++, 6–10 mm; +++, >10 mm) if desired. Alternatively, images may be processed using an image analyzer (e.g., BIOCOM Visiolab™ 2000) to provide a measure of vessel density. For screening purposes, however, a simple score of positive or negative should suffice.
30. This surface is approx 3 cm diameter and 0.3 mL is sufficient to cover it. Application of volumes beyond 0.5 mL results in increased mortality, producing embryos that, to the pathologist's eye, appear to have died of congestive cardiac failure.
31. This is optimal for wound extracts and fibrin degradation products.
32. Prior treatment with osmium tetroxide or hemotoxylin staining may be necessary for good visualization of the vasculature.
33. The assay of DNA synthesis is performed, using standard precautions such as wearing disposable gloves, cleaning and monitoring of bench surfaces, and approved radioactive materials disposal for relatively minute quantities of a soft beta emitter.
34. Methyl-[^{3}H]-thymidine is used to avoid recycling of thymidine, but this is not a serious problem in any case because the exposure is short.
35. For large numbers it is best to inject eggs in sequence across the groups in order to minimize any differences in incorporation between groups.
36. The technique of removal of each CAM is simple but crucial to the rationale of achieving results expressed as total DPM per CAM. By 20 min, methyl-[^{3}H]-thymidine is incorporated only into the CAM, which has been directly exposed. Beyond 30 min, [^{3}H]-thymidine becomes detectable in adjacent CAM that is applied to the shell, via uptake from the bloodstream.
37. For difficult ones where the air space deflates and the CAM slips or detaches, it is permissible to cut wider around the presumed area of dropped CAM. The extra tissue sampled will not affect the values for incorporated [^{3}H]-thymidine.
38. This stage is completed one Sterilin at a time. To reduce chemiluminescence, the vials are stored at 4°C overnight in the dark.
39. The tissue debris becomes efficiently dispersed throughout the thixotropic scintillant as a fine emulsion suitable for counting. Nevertheless there is considerable loss of efficiency owing to color, salt, and protein quenching inherent in the nature of the tissue, and necessitates accurate quench-correction curves created by, for example, the LKB "Hat-trick" method that can be used to check a real CAM sample. Phenol extraction of DNA *(55)* gives identical results but requires more corrosive handling and entails similar quenching problems.

40. By applying this assay to the whole "dropped" area of CAM, exposed to test or control substances, the results can be expressed per CAM, and are therefore independent of changes in weight and protein content associated with, for example, edema, and of changes in cellularity due to inflammatory cell influx.
41. In this case, the eggs should be turned through 180 degrees at least once a day, to reduce adhesion to the shell.
42. Eggs not orientated so that the embryo is uppermost are discarded.
43. If careful aseptic techniques are applied, contamination is not a problem. CO_2 is not required for growth during the first 10 d *(6)*.
44. Hammocks are not essential, however. An alternative approach is to carefully remove the wide end of the shell (air pocket) with a dissecting blade just above the developing embryo and yolk sac membrane. The vascular system is clearly visible and may be prevented from drying out by covering with a piece of clingfilm.
45. By d 6 there is a 25% mortality.
46. Nutrient transfer from yolk to embryo is facilitated by a sodium-dependent potential difference (blood side positive, relative to yolk side). Absorption by diffusion of test samples from the disc may be expedited by this natural inward current flow towards the embryo *(56)*.
47. The addition of a few drops of phenol red to the methylcellulose solution produces better visualization of the sample disc. The concentration of methylcellulose may be varied according to the molecular weight of the sample. For example, for a molecular weight >70,000 Da, a methylcellulose concentration of 0.5% should facilitate dispersion.
48. This is not a specific assay for antiangiogenic reagents as nonspecific cytotoxic compounds will give the same result.
49. Hemorrhage is often seen with snake venom and this assay has also been used to screen antivenom.
50. Experimental events may be recorded either with a videorecorder or a still camera; a microscope is not normally necessary. Until d 7 or 8, the embryo is relatively still and topically applied discs containing reagent remain where they have been placed on the yolk sac membrane. At the time of writing, the resolution of video film is significantly inferior to that of transparencies or print films. Transparencies may be stored electronically on disc. Transferring the egg, and its clingfilm, from the hammock container to a Petri dish lid can further reduce depth of field, but this is not usually necessary and the change in shape will stress the membrane. Oblique lighting, using cold, flexible spotlights (Schott), is superior to flash because it eliminates reflection from the membrane. The quantification of vascular reaction to angiogenic stimuli has been attempted using laser Doppler flowmetry *(57)* with mixed results. Straightforward photographic techniques, such as the projection of transparencies so that vessels can be measured with a ruler, or the use of a microscope with eyepiece and stage calibration are probably the most successful way to measure the individual vessels of the diameter found. Color densitometry, where the proportion of yellow (yolk) background is compared with changes in the proportion of red (blood vessel) for the measurement

of vasoactive reagents, is difficult to interpret because compensating capillaries alter the yellow background reference readings. Measurement from histological samples is not recommended because of artifacts resulting from fixation, smooth muscle changes and so forth.

51. Samples may be fixed for histological examination in the following manner. First a light mineral oil (e.g., "3-in-1") is injected under the area of interest. The area is then flooded with 10% buffered formol saline or 4% paraformaldehyde for light microscopy or 2.5% phosphate buffered glutaraldehyde (pH 7.4) for electron microscopy. The area is covered with a ring of filter paper (e.g., internal diameter 8 mm, external diameter 20 mm) so that the target tissue or reaction is visible through the center of the ring. The yolk sac membrane is cut with scissors outside the filter paper, leaving sufficient margin for it to lap over onto the paper. The filter paper ring and membrane are then removed with forceps, fixed in formol saline, embedded, and sectioned for staining as appropriate.

5. Acknowledgments

DCW wishes to thank the NorthWest Cancer Research Fund for their support. PGS would like to thank Brian Getty (Medical Microbiology, Faculty of Medicine, University of Liverpool) and Peter Young (Liverpool School of Tropical Medicine) for their expert assistance. WDT acknowledges the support of the Wellcome Trust, Medical Research Council, and the Scottish Home and Health Department.

References

1. Fan, T.-P. D. and Polverini, P. J. (1997) In: *Tumor angiogenesis*, (Bicknell, R., Lewis, C. E., and Ferrara, N., eds.), Oxford University Press, Oxford, UK, pp. 5–18.
2. Jain, R. K., Schlenger, K., Hockel, M., and Yuan, F. (1997) Quantitative angiogenesis assays: progress and problems. *Nat. Med.* **3,** 1203–1208.
3. Auerbach, R., Auerbach, W., and Polakowski, I. (1991) Assays for angiogenesis: a review. *Pharmacol. Ther.* **51,** 1–11.
4. Ribatti, D., Vacca, A., Roncali, L., and Dammacco, F. (1996) The chick embryo chorioallantoic membrane as a model for in vivo research on angiogenesis. *Int. J. Dev. Biol.* **40,** 1189–1197.
5. Folkman, J. (1974) Tumor angiogenesis factor. *Cancer Res.* **34,** 2109–2113.
6. Auerbach, R., Kubai, L., Knighton, D., and Folkman, J. (1974) A simple procedure for the long-term cultivation of chick embryos. *Dev. Biol.* **41,** 391–394.
7. Ausprunk, D. H., Knighton, D. R., and Folkman, J. (1974) Differentiation of vascular endothelium in the chick chorioallantois a structural and autoradiographic study. *Dev. Biol.* **38,** 237–249.
8. Ausprunk, D. H., Knighton, D.R., and Folkman, J. (1975) Vascularization of normal and neoplastic tissues grafted to the chick chorioallantois: role of host and preexisting graft blood vessels. *Am. J. Pathol.* **79,** 597–618.

9. Folkman, J. (1975) Tumor angiogenesis. *Adv. Cancer Res.* **43,** 175–203.
10. Dunn, B., Fitzharris, T. P., and Barnett, B. D. (1981) Effects of varying chamber construction and embryo pre-incubation age on survival and growth of chick embyos in shell-less culture. *Anat. Rec.* **199,** 33–43.
11. Dugan, J. D. Jr., Lawton, M. T., Glaser, B., and Brem, H. (1981) A new technique for explantation and in vitro culture of chicken embryos. *Anat. Rec.* **229,** 125–128.
12. Jakobson, A. M., Hahnenberger, R., and Magnusson, A. (1989) A simple method for shell-less cultivation of chick embryos. *Pharmacol. Toxicol.* **64,** 193–195.
13. Vu, M. T., Smith, C. F., Berger, P. C., and Klintworth, G. K. (1985) An evaluation of methods to quantitate the chick chorioallatoic membrane assay in angiogenesis. *Lab. Invest.* **53,** 499–508.
14. Taylor, S. and Folkman, J. (1982) Protamine is an inhibitor of angiogenesis. *Nature* **297,** 307–312.
15. Folkman, J., Langer, R., Lindhardt, R., Haudenschild, C., and Taylor, S. (1983) Angiogenesis inhibition and tumor regression caused by heparin or a heparin fragment in the presence of cortisone. *Science* **221,** 719–725.
16. Rosenbruch, M. (1989) Granulation tissue in the chick embryo yolk sac blood vessel system. *J. Comp. Path.* **101,** 363–373.
17. Rosenbruch, M. (1990) Toxicity studies of the incubated chicken egg. With special reference to the extra-embryonal vascular systems. *Dermatosen Beruf. Umwelt.* **38,** 5–11.
18. Sells, P. G., Richards, A. M., Laing, G. D., and Theakston, R. D. G. (1997) The use of hens' eggs as an alternative to the conventional in vivo rodent assay for antidotes to haemorrhagic venoms. *Toxicon.* **35,** 1413–1421.
19. Sells, P. G., Ioannou, P., and Theakston, R. D. G. (1998) A humane alternative to the measurement of the lethal effects (LD_{50}) of non-neurotoxic venoms using hens' eggs. *Toxicon.* **36,** 985–991.
20. Ausprunk, D. H. (1982) Synthesis of glycoproteins by endothelial cells in embryonic blood vessels. *Dev. Biol.* **90,** 79–90.
21. Nace, G. W. and Schechtman, A. M. (1948) Development of non-vitelloid substances in the blood of the chick embryo. *J. Exp. Zool.* **108,** 217–233.
22. Sweeney, L. J., Zak, R., and Manasek, F .J. (1987) Transitions in cardiac isomyosin expression during differentiation of the embryonic chick heart. *Circ. Res.* **61,** 287–295.
23. Lillie, F. R. (1919) *The Development of the Chick*, 2nd ed. Henry Holt and Co., New York, NY.
24. Simkiss K. (1980) Water and ionic fluxes inside the egg. *Amer. Soc. Zool.* ***20,*** *385–393.*
25. Pickering, J. W. and Gladstone, R. J. (1925) The development of blood plasma. Part 1. The Genesis of the coagulable material in embryo chicks. *Proc. Royal Soc. London*, B (Ser.) **98,** 516–522.
26. Owen, J. J. and Ritter, M. A. (1969) Tissue interaction in the development of thymus lymphocytes. *J. Exp. Med.* **129,** 431–442.

27. Leene, W., Duyzings, M. J. M., and von Steeg, C. (1973) Lymphoid stem cell identification in the developing thymus and bursa of fabricus in the chick. *Z. Zellforsh.* **136,** 521–533.
28. Sugimoto, M., Yasuda, T., and Egashira, Y. (1977) Development of the embryonic chicken thymus. I. Characteristic synchronous morphogenesis of lymphocytes accompanied by the appearance of an embryonic thymus-specific antigen. *Dev. Biol.* **56,** 281–292.
29. Sugimoto, M., Yasuda, T., and Egashira, Y. (1977) Development of the embryonic chicken thymus. II. Differentiation of the epithelial cells studied by electron microscopy. *Dev. Biol.* **56,** 293–305.
30. Sugiyama, S. (1926) Origin of thrombocytes and of different types of blood cells as seen in the living chick blastoderm. *Contrib. Embryol.* **18,** 121–149.
31. Sallstrom, J. F. and Alm, G. V. (1974) Mitogen-reactive lymphocytes in the embryonic thymus in organ culture. *Int. Archs. Allergy Appl. Immun.* **47,** 388–399.
32. Wilson, D. J. (1985) Mast cells are present during angiogenesis in the chick extraembryonic vascular system. *Experientia* **41,** 269–271.
33. Jakob, W., Jentzsch, K. D., Manersberger, B., and Heider, G. (1978) The chick chorioallantoic membrane as bioassay for angiogenesis factors: reactions induced by carrier materials. *Exp. Pathol.* **15,** 241–249.
34. Spanel-Borowski, K., Schnapper, U., and Heymer, B. (1998) The chick chorioallantoic assay in the assessment of angiogenic factors. *Biomed. Res.* **9,** 253–260.
35. Wilting, J., Christ, B., and Bokeloh, M. (1991) A modified chorioallantoic membrane (CAM) assay for qualitative and quantitative study of growth factors. Studies on the effects of carriers, PBS, angiogenin and bFGF. *Anat. Embryol.* **183,** 259–271.
36. Wilting, J., Christ, B., and Weich, H. A. (1992) The effects of growth factors on the day 13 chorioallantoic membrane (CAM): a study of $VEGF_{165}$ and PDGF-BB. *Anat. Embryol.* **186,** 251–257.
37. Barnhill, R. L. and Ryan, T. J. (1983) Biochemical modulation of angiogenesis in the chorioallantoic membrane of the chick-embryo. *J. Invest. Dermatol.* **81,** 485–488.
38. Ryan, T. J. and Barnhill, R. L. (1983) Physical factors in angiogenesis. In: *Development of the Vascular System.* Ciba Symposium 100 (Nugent, J. and O'Conner, M., eds.). Pitman Books, London, pp. 80–89.
39. Thompson, W. D. and Kazmi, M. A. (1989) Angiogenic stimulation compared with angiogenic reaction to injury: distinction by focal and general application of trypsin to the chick chorioallantoic membrane. *Brit. J. Exp. Pathol.* **70,** 627–635.
40. Thompson, W. D., Evans, A. T., and Campbell, R. (1986) The control of fibrogenesis: stimulation and suppression of collagen synthesis in the chick chorioallantoic membrane with fibrin degradation products, wound extracts and proteases. *J. Pathol.* **148,** 207–215.
41. Thompson, W. D., McGuigan, C. J., Snyder, C., Keen, G. A., and Smith, E. B. (1987) Mitogenic activity in human atherosclerotic lesions. *Atherosclerosis* **66,** 85–93.

42. Thompson, W. D. and Brown, F. I. (1987) Measurement of angiogenesis: mode of action of histamine in the chick chorioallantoic membrane is indirect. *Int. J. Microcirc.* **6,** 343–357.
43. Ribatti, D., Roncalj, L., Nico, B., and Bertossi, M. (1987) Effects of exogenous heparin on the vasculogenesis of the chorioallatoic membrane. *Acta Anat.* **130,** 257–263.
44. Folkman, J. and Cotran, R. (1976) Relation of vascular proliferation to tumor growth. *Int. Rev. Exp. Pathol.* **16,** 207–248.
45. Form, D. M. and Auerbach, R. (1983) PGE_2 and angiogenesis. *Proc. Soc. Exp. Biol.* **172,** 214–218.
46. Phillips, P. and Kumar, S. (1979) Tumor angiogenesis factor (TAF) and its neutralization by a xenogeneic antiserum. *Intl. J. Cancer* **23,** 82–.
47. West, D. C., Hampson, I. N., Arnold, F., and Kumar, S. (1985) Angiogenesis induced by degradation produces of hyaluronic acid. *Science* **228,** 1324–1326.
48. Dusseau, J. W., Hutchins, P. M., and Malbasa, D. S. (1986) Stimulation of angiogenesis by adenosine on the chick chorioallantoic membrane. *Circ. Res.* **71,** 33–44.
49. Steiner, R. (1992) Angiostatic activity of anticancer agents in the chick embryo chorioallantoic membrane (CHE-CAM) assay. In: *Angiogenesis: key principles—technology—medicine* (Steiner, R., Weisz, P. B., and Langer, R., eds). Birkhauser Verlag, Basel, Switzerland, pp. 449–454.
50. Harris-Hooker, S. A., Gajdusek, C. M., Wright, T. N., and Schwartz, S. M. (1983) Neovascular response induced by cultured aortic endothelial cells. *J. Cell Physiol.* **114,** 302–310.
51. Maragoudakis, M. E., Haralabopoulos, G. C., Tsopanoglou, N. E., and Pipili-Synetos, E. (1995) Validation of collagenous protein synthesis as an index for angiogenesis with the use of morphological methods. *Microvasc. Res.* **50,** 215–222.
52. Voss, K., Jakob, W., and Roth, K. (1984) New image analysis method for the quantification of neovascularization. *Exp. Pathol.* **26,** 155–161.
53. Strick, D. M., Waycaster, R. L., Montani, J., Gay, W. J., and Adair, T. H. (1991) Morphometric measurements of the chorioallantoic membrane vascularity: effects of hypoxia and hyperoxia. *Am. J. Physiol.* **29,** H1385–H1389.
54. Nguyen, M., Shing, Y., and Folkman, J. (1994) Quantitation of angiogenesis and antiangiogenesis in the chick embryo chorioallatoic membrane. *Microvasc. Res.* **47,** 31–40.
55. Maniatis, T., Fritsch, E. F., Sambrook, J. (1982) *Molecular Cloning—a Laboratory Manual.* Cold Spring Harbor Laboratories, Cold Spring Harbor, NY.
56. Takada, M. and Clark, N. B. (1992) Sodium-dependent short-circuit current across the yolk sac membrane during embryonic development in normal and shell-less cultured chicks. *J. Comp. Physiol. [B]* **162,** 496–501.
57. Gush, R. J., Thompson, J. M., and Weiss, J. B. (1990) Measurement of blood flow in the chick egg yolk sac membrane. *J. Med. Eng. Technol.* **14,** 205–209.

10

Corneal Assay for Angiogenesis

Marina Ziche

1. Introduction

In order to develop angiogenic and antiangiogenic strategies, concerted efforts have been made to provide animal models for more quantitative analysis of in vivo angiogenesis. In vivo techniques consist of the cornea pocket and iris implant in the eye, the rabbit ear chamber, the dorsal skinfold chamber, the cranial window, the hamster cheek pouch window, the sponge implant assay, the fibrin clots, the sodium alginate beads and the Matrigel plugs, the rat mesenteric window, the chick embryo chorioallantoic membrane and the air sac in mice and rats *(1)*. In this chapter we will discuss the avascular cornea assay, and the advantages and disadvantages of using this assay in different species. The cornea assay is based on the placement of an angiogenic inducer (tumor tissue, cell suspension, growth factor) into a corneal pocket in order to evoke vascular outgrowth from the peripherally located limbal vasculature. In comparison to other in vivo assays, this assay has the advantage of measuring only new blood vessels, because the cornea is initially avascular.

2. Materials

1. Sodium pentothal or sodium pentobarbital.
2. Pliable iris spatula (1.5 mm wide).
3. Ethylene-vinyl-acetate copolymer (Elvax-40, DuPont de Nemours, Wilmington, DE).
4. Absolute alcohol.
5. Methylene chloride.
6. Hydroxyethyl-methacrylate polymer (Hydron, Interferon Science, New Brunswick, NJ).
7. Sucrose aluminum sulfate (Sucralfate, BukhMeditec, Copenhagen, Denmark).
8. Slit-lamp stereo-microscope with ocular grid

From: *Methods in Molecular Medicine, Vol. 46: Angiogenesis Protocols*
Edited by: J. C. Murray © Humana Press Inc., Totowa, NJ

3. Methods

3.1. Rabbit Cornea Assay

This assay was first described by Gimbrone et al. *(2)* for use in New Zealand white rabbits (*see* **Note 1**), owing partly to the absence of a vascular pattern, and for the ease of manipulation and monitoring of neovascular growth. We have used this technique extensively for several years and have substantially modified it to fulfil different experimental requirements.

3.1.1. Sample Preparation

The material to be tested can be in the form of slow-release pellets incorporating recombinant growth factors, cell suspensions, or tissue samples *(3,4)*. Recombinant growth factors are prepared as slow-release pellets by incorporating the protein under test into an ethylene-vinyl-acetate copolymer (Elvax-40; *see* **Note 2**). In order to avoid nonspecific reactions, Elvax-40 has to be carefully prepared as follows:

1. Elvax-40 beads are extensively washed in absolute alcohol at 37°C (*see* **Note 3**).
2. Prepare a 10% casting stock solution in methylene chloride and test for biocompatibility *(5)*. (The casting solution can be used if none of the implants performed with this preparation induces a histological reaction in the rabbit cornea.)
3. A predetermined volume of Elvax-40 casting solution is mixed with a given amount of the compound to be tested on a flat surface and the polymer is allowed to dry under a laminar flow hood.
4. After drying, cut the film sequestering the test compound into 1 mm × 1 mm × 0.5 mm pieces.

3.1.2. Cell and Tissue Implants

Cell suspensions are obtained by trypsinization of confluent cell monolayers. Five microliters containing ~2×10^5 cells in medium supplemented with 10% serum are introduced into the corneal micropocket. If the overexpression of growth factors by stable transfection of specific cDNA is being studied, one eye is implanted with transfected cells and the other with the wild-type cell line. When tissue samples are tested, samples of 2–3 mg are obtained by cutting the original fragments under sterile conditions. The angiogenic activity of tumor samples should be compared with macroscopically normal tissue.

3.1.3. Surgical Technique

1. Anesthetize rabbit with sodium pentothal (30 mg/kg intravenously) (*see* **Note 4**).
2. Under aseptic conditions, surgically produce a micropocket (1.5 mm × 3 mm) in the lower half of the cornea using a pliable iris spatula (*see* **Notes 5**, **6**, **7**).
3. Drain a small amount of the aqueous humor from the anterior chamber to reduce corneal tension (*see* **Note 8**).

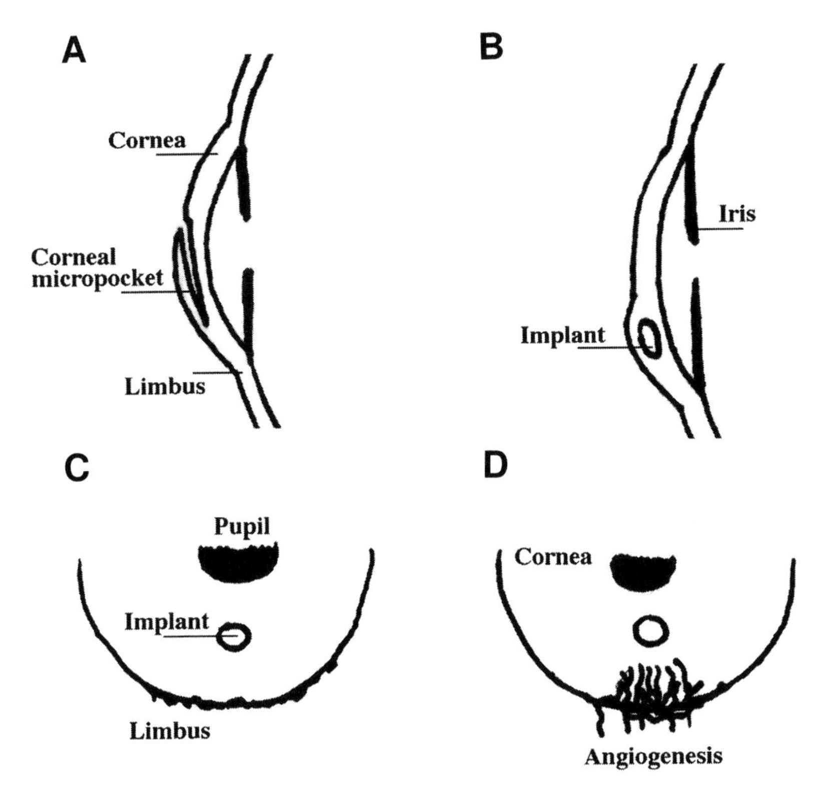

Fig. 1. Schematic representation of the corneal micropocket assay. A micropocket is surgically produced in the corneal stroma with a pliable spatula (**A**) and the test substance is inserted into the micropocket (**B**). The newly formed vessels start from the limbal vasculature (**C**, t = 0) and progress toward the implanted stimulus (**D**, t = 7 d).

4. Position the implant 2.5–3 mm from the limbus to avoid false positives owing to mechanical procedure. This also allows diffusion of test substances and the subsequent formation of a gradient for the endothelial cells of the limbal vessels (**Fig. 1 A, B**).
5. Implants sequestering test materials, and appropriate controls (empty pellets of Elvax-40 can be used as controls), should be coded and implanted in a double-blinded manner.

3.1.4. Quantification

Subsequent daily observation of the implants is made with a slit lamp stereomicroscope without anesthesia. Angiogenesis, edema, and cellular infiltrate are recorded daily with the aid of an ocular grid by an independent operator who does not perform the surgery. An angiogenic response is scored positive when budding of vessels from the limbal plexus occurs after 3–4 d and capillaries reach the implanted pellet in 7–10 d (**Fig. 1 C,D**). Implants that fail to produce neovascular growth within 10 d are considered negative, while implants showing an inflammatory reaction are discarded *(6)*. The ratio of positive implants to total implants performed is determined during each observation. The potency of angiogenic activity is evaluated on the basis of the number and growth rate of newly formed capillaries, and an angiogenic score is calculated by the formula (vessel density) × (distance from limbus) *(7–9)*. A density value of 1 corresponds to 0–25 vessels per cornea, 2 from 25–50, 3 from 50–75, 4 from 75–100, and 5 for more than 100 vessels. The distance from the limbus (in mm) is graded with the aid of an ocular grid.

3.1.5. Histological Examination

Corneas are removed at the end of the experiment as well as at defined intervals after surgery and/or treatment and fixed in formalin for histological examination. Newly formed vessels and the presence of inflammatory cells are detected by hematoxylin & eosin staining or specific immunohistochemical procedures (e.g., antirabbit macrophage [RAM11], antiCD-31 for endothelium) *(8)*. Double staining (i.e., antiCD-31 for vascular endothelium and specific markers for tumor cells) could be useful to label newly formed vessels of the host and proliferating tumor cells implanted in the cornea.

3.2. Mouse Cornea Micropocket Assay

The mouse cornea micropocket assay was first described by Muthukkaruppan and Auerbach *(10)*. After the animals are anesthetized with methoxyflurane, corneal micropockets are made in both eyes, reaching within 1 mm of the limbus, and pellets containing substances to be tested coated with Hydron are implanted. Hydron is used as a casting solution of 12% (w/v) prepared by dissolving the polymer in absolute alcohol at 37°C *(5)*. When peptides are tested, sucralfate is added to stabilize the molecule and to slow its release from Hydron *(11,12)*. The vascular response, measured as the maximal vessel length and number of clock hours of neovascularization, is scored at fixed time (usually at d 5 and 7 postoperation). For this purpose, the eyes are examined by slit-lamp biomicroscopy and photographed. To quantify the section of the cornea in which new vessels are sprouting from the preexisting limbal vessels, the circumference of the cornea is divided into the equivalent of 12 clock hours. The measure of the number of clock hours of neovascularization for each eye is performed during each observation.

3.3. Rat Corneal Assay

Purified growth factors are combined 1/1 with Hydron, as described by Polverini and Leibovich *(13)*. Pellets are implanted 1–1.5 mm from the limbus of the cornea of anesthetized rats (sodium pentobarbital, 30 mg/kg, ip). Neovascularization is assessed at fixed days (usually d 3, 5, and 7). To photograph responses, animals are perfused with colloidal carbon solution to label vessels, and eyes are enucleated and fixed in 10% neutral-buffered formalin overnight. The following day, corneas are excised, flattened, and photographed. A positive neovascularization response is recorded only if sustained directional ingrowth of capillary sprouts and hairpin loops toward the implant are observed. A negative response is recorded when either no growth is observed or when only an occasional sprout or hairpin loop showing no evidence of sustained growth is detected.

3.4. Features of the Corneal Assay for Angiogenesis

Species: Rabbit cornea has been found to be avascular in all strains examined so far. In some strains of rat the presence of preexisting vessels within the cornea and the development of keratitis are serious disadvantages.

Animal size: Rabbits are more docile and amenable to handling and experimentation than mice or rats. The size of a rabbit (2–3 kg) allows easy manipulation of the whole animal, and the eye to be easily extruded from its location and surgically manipulated. General anesthesia is required only for surgery, while daily examinations occasionally need local anesthesia. Conversely, the small size of rats and mice allows the use of small amounts of drugs given systemically.

Measurements: In mice and rats it is possible to obtain time-point results. Following the evolution of the angiogenic response in only one animal is not recommended because each time the cornea is observed the animal has to be anesthetized. Experiments are made with a large number of animals, and vessel growth over time can be visualized by perfusion of individual animals with colloidal carbon. Multiple observations are however possible with rabbits. The use of a slit lamp stereomicroscope to observe unanesthetized animals allows the observation of newly formed vessels for periods up to 1–2 mo.

Monitoring of inflammation: Inflammatory reactions are easily detectable by stereomicroscopic examination of rabbits as corneal opacity.

Different experimental procedures: In the wide area of rabbit eye, stimuli can be placed in different forms. In particular, the activity of specific growth factors can be studied in the form of slow-release pellets *(3,14–17)* (**Fig. 2**), and of tumor or nontumor cell lines stably transfected for the overexpression of angiogenic factors *(8,18–22)* (**Fig. 3**). The modulation of the angiogenic responses by different stimuli can be assessed in the rabbit cornea assay through the implantation and/or removal of multiple pellets placed in parallel micropockets produced in the same cornea *(6,7)* (**Fig. 4**). Implanting tumor samples from different locations can be performed both in corneal micropockets and in the anterior chamber of the eye to monitor angiogenesis

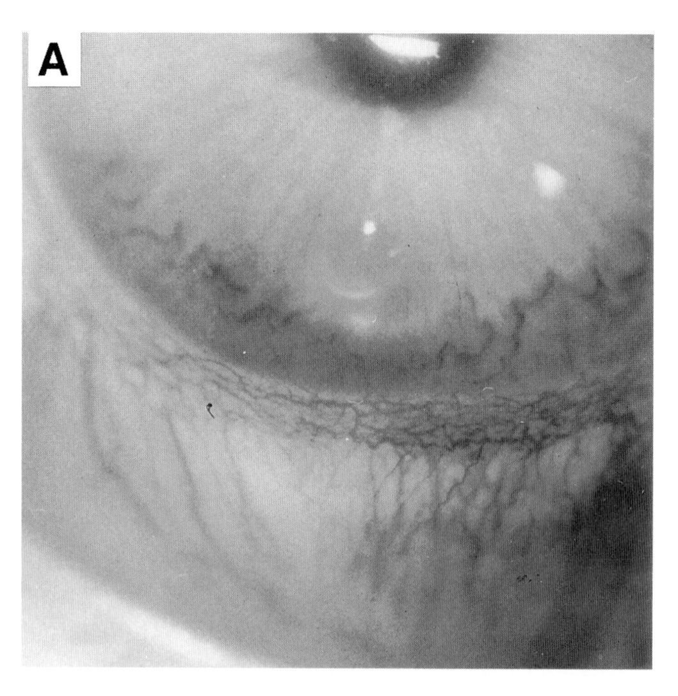

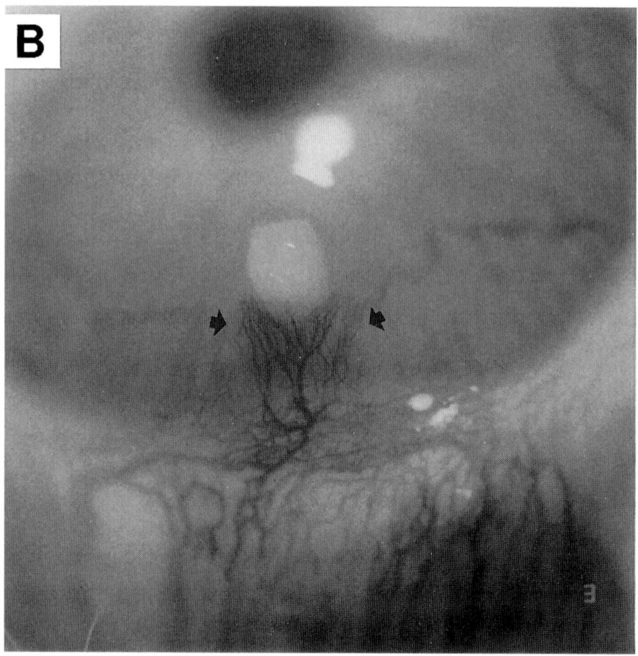

Fig. 2. Representative pictures of a negative implant (**A**) and an implant producing neovascular growth (**B**). In A, a control pellet of Elvax alone has been inserted into the micropocket. In B arrows indicate the newly formed vessels that progress into the corneal stroma toward a pellet of FGF-2 (200 ng). Photographs were taken at d 10 after surgical implantation, through a slit stereomicroscope (×18).

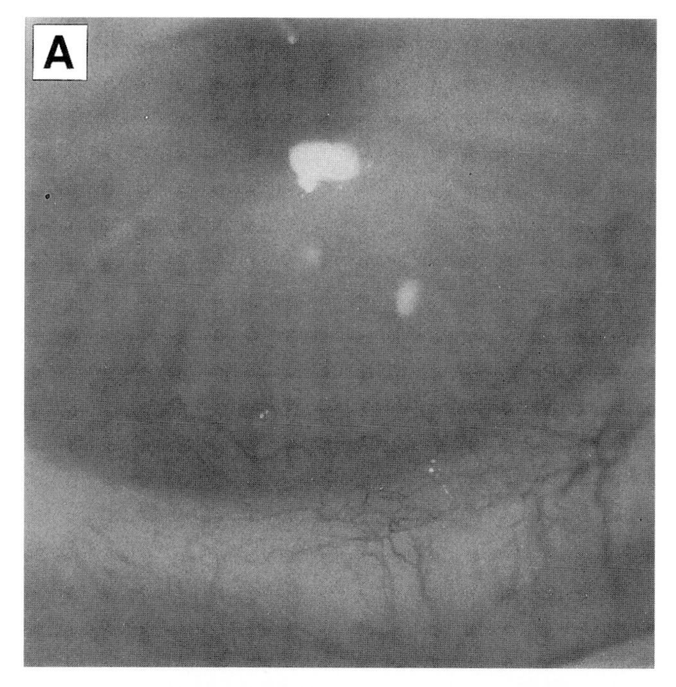

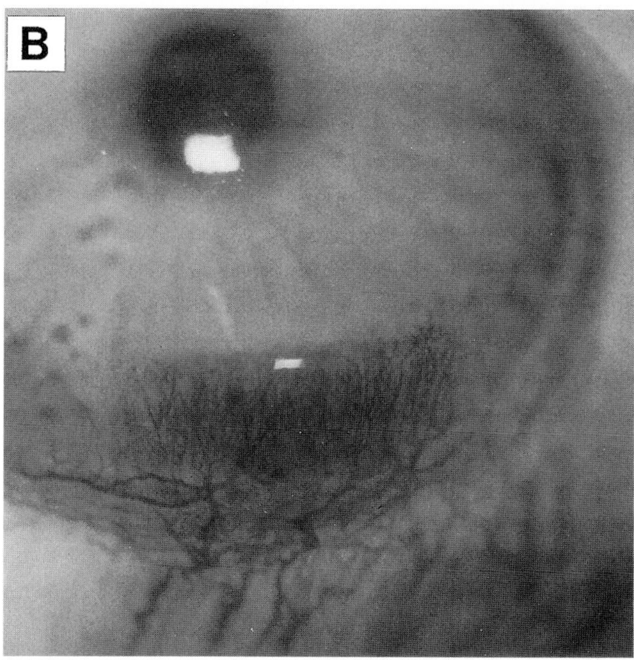

Fig. 3. Representative pictures of corneal angiogenesis induced by wild type MCF-7 cells (**A**) and by MCF-7 stably transfected with $VEGF_{121}$ (**B**). Wild-type MCF-7 cells induce the growth of few slowly progressing capillaries from preexisting limbal vessels, while VEGF transfectants induce a strong angiogenic response. Photographs were taken at d 18 after surgical implantation, through a slit stereomicroscope (×18).

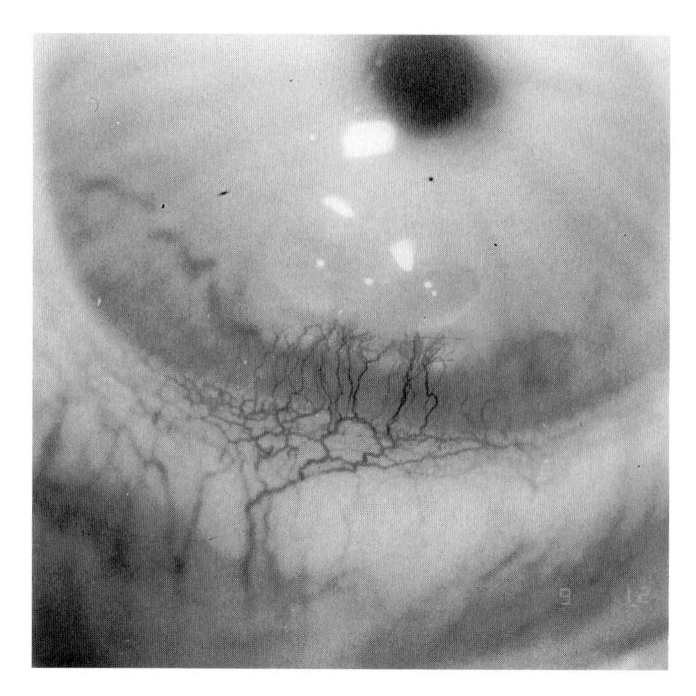

Fig. 4. Representative pictures of corneal angiogenesis induced by implanting in the same eye two pellets containing two different angiogenic triggers at suboptimal doses. Photograph taken at d 10 after surgical implantation, through a slit stereomicroscope (×18).

produced by hormone-dependent tissues or tumors (e.g., human breast or ovarian carcinoma in female rabbits). This also allows the detection of neovascular growth in both the iris and the cornea *(4)*.

The use of nude mice allows the study of angiogenesis in response to effectors produced and released by tumors, or tumor cell lines, of human origin growing subcutaneously *(11)*. Treatment of mice with antiangiogenic or antitumor drugs allows the simultaneous measurement of tumor growth and metastasis, and corneal angiogenesis.

3.5. Morphological Analysis and Quantification of Neovascularization

In the past angiogenesis has been described qualitatively (strong, weak, absent) or by measuring the length of the most advanced vascular sprout at daily intervals. Moreover, flat-mount preparations of cornea have been used to obtain more quantitative assessments using imaging techniques and vector analysis to determine the directional properties of newly penetrating blood vessels *(23)*. The use of fluorescence to provide better imaging properties has led to more precise measurements of neovascularization. The availability of sensitive video cameras and image recording now permit sequential recording of the neovascular process in individual animals without the need to resort to dissection and fixation of tissue previously needed for obtaining imaging data from flat-mount preparations *(24)*.

Inherent in the in vivo approach is the capacity to establish, quantify, and characterize the spatial and temporal pattern of corneal angiogenesis elicited by distinct angiogenesis effectors in each animal over a long period of time. Using image analysis, the effect of intervention on the process can be shown and quantified. This information can provide an accurate and reproducible method for quantifying the "angiogenic score" obtained under different conditions in the same model, as well as providing a reference for the efficacy of treatment *(11)*. Computer-assisted image analysis of corneal vascularization partially overcomes the problem of subjective evaluation by the operator, and provides multiple data points from each animal. After collection and analysis of data obtained from different stored images, the characteristics of distinct angiogenic phenotypes can be drawn to provide a full characterization of angiogenesis and to complete the partial data obtained by post-mortem techniques. Furthermore, it provides a permanent record for further review and/or additional studies.

3.6. Concluding Remarks

Continuous monitoring of angiogenesis in vivo is required for the development and evaluation of drugs acting as suppressors or stimulators of angiogenesis. In this respect, the avascular cornea of New Zealand albino rabbits offers a unique model, allowing the progression of neovascularization to be monitored for extended periods of time with a noninvasive approach, and the comparison in the same animal of distinct effectors. Measurement of corneal angiogenesis is useful for quantitating the effects of angiogenic stimuli and for evaluating the efficacy of potential inhibitors of neovascularization. Because accurate methods suitable for recording the entire pattern of corneal neovascularization over time and for obtaining a quantitative evaluation of the process in individual living animals do not exist, we are developing a noninvasive method to achieve this goal through the use of computerized image analysis.

4. Notes

1. Body weight of rabbits should be in the range 1.8–2.5 kg for ease of handling and prompt recovery from anesthesia.

2. Polymers other than Elvax-40 can also be used. In our experience, polyvinyl alcohol and Hydron, a polymer of hydroxyethyl-methacrylate *(5)*, gave less satisfactory results than Elvax-40.
3. Elvax-40 beads should be carefully washed in absolute alcohol as indicated to avoid inflammatory reactions.
4. Immobilization during anesthetic procedure and observation is important, to avoid self-induced injury.
5. Sterility of materials and procedures is crucial to avoid nonspecific responses.
6. Make the cut in the cornea in correspondence to the pupil, and orient the micropocket toward the lower eyelid.
7. When two factors are tested make two independent micropockets.
8. Drain a small amount of aqueous humor when implanting cells or tissue fragments to reduce corneal tension.

Acknowledgments

I wish to thank Dr. Lucia Morbidelli for her excellent contribution to the preparation of this chapter. This work was supported by funds from the Italian Association for Cancer Research, the National Council of Research (Progetto Finalizzato Biotechnologie), and the Italian Ministry of University and Scientific and Technological Research (MURST).

References

1. Jain, R. K., Schlenger, K., Hockel, M., and Yuan, F. (1997) Quantitative angiogenesis assays: Progress and problems. *Nature Med.* **3,** 1203–1208.
2. Gimbrone, M. Jr., Cotran, R., Leapman, S. B., and Folkman, J. (1974) Tumor growth and neovascularization: An experimental model using the rabbit cornea. *J. Natl. Cancer Inst.* **52,** 413–427.
3. Ziche, M., Jones, J., and Gullino, P. M. (1982) Role of prostaglandin E1 and copper in angiogenesis. *J. Natl. Cancer Inst.* **69,** 475–482.
4. Gallo, O., Masini, E., Morbidelli, L., Franchi, A., Fini-Storchi, I., Vergari, W. A., and Ziche, M. (1998) Role of nitric oxide in angiogenesis and tumor progression in head and neck cancer. *J. Natl. Cancer Inst.* **90,** 587–596.
5. Langer, R. and Folkman, J. (1976) Polymers for the sustained release of proteins and other macromolecules. *Nature* **363,** 797–800.
6. Ziche, M., Alessandri, G., and Gullino, P. M. (1989) Gangliosides promote the angiogenic response. *Lab. Invest.* **61,** 629–634.
7. Ziche, M., Morbidelli, L., Masini, E., Amerini, S., Granger, H. J., Maggi, C. A., et al. (1994) Nitric oxide mediates angiogenesis in vivo and endothelial cell growth and migration in vitro promoted by substance P. *J. Clin. Invest.* **94,** 2036–2044.
8. Ziche, M., Morbidelli, L., Choudhuri, R., Zhang, H. T., Donnini, S., Granger, H. J., and Bicknell, R. (1997) Nitric oxide-synthase lies downstream of vascular endothelial growth factor but not basic fibroblast growth factor induced angiogenesis. *J. Clin. Invest.* **99,** 2625–2634.

9. Ziche, M., Maglione, D., Ribatti, D., Morbidelli, L., Lago, C. T., Battisti, M., et al. (1997) Placenta growth factor-1 (PlGF-1) is chemotactic, mitogenic and angiogenic. *Lab. Invest.* **76,** 517–531.
10. Muthukkaruppan, V. and Auerbach, R. (1979) Angiogenesis in the mouse cornea. *Science* **206,** 1416–1418.
11. Chen, C., Parangi, S., Tolentino, M. T., and Folkman, J. (1995) A strategy to discover circulating angiogenesis inhibitors generated by human tumors. *Cancer Res.* **55,** 4230–4233.
12. Voest, E. E., Kenyon, B. M., O'Reilly, M. S., Truitt, G., D'Amato, R. J., and Folkman, J. (1995) Inhibition of angiogenesis in vivo by interleukin 12. *J. Natl. Cancer Inst.* **87,** 581–586.
13. Polverini, P. J. and Leibovich, S. J. (1984) Induction of neovascularization in vivo and endothelial cell proliferation in vitro by tumor associated macrophages. *Lab. Invest.* **51,** 635–642.
14. Ziche, M., Morbidelli, L., Pacini, M., Geppetti, P., Alessandri, G., and Maggi, C. A. (1990) Substance P stimulates neovascularization in vivo and proliferation of cultured endothelial cells. *Microvasc. Res.* **40,** 264–278.
15. Bussolino, F., Ziche, M., Wang, J.M., Alessi, D., Morbidelli, L., Cremona, O., et al. (1991) In vitro and in vivo activation of endothelial cells by colony-stimulating factors. *J. Clin. Invest.* **87,** 986–995.
16. Bussolino, F., Di Renzo, M. F., Ziche, M., Bocchietto, E., Oliviero, M., Naldini, L., et al. (1992) Hepatocyte growth factor is a potent angiogenic factor which stimulates endothelial cell motility and growth. *J. Cell Biol.* **119,** 629–641.
17. Albini, A., Benelli, R., Presta, M., Rusnati, M., Ziche, M., Rubartelli, A., et al. (1996) HIV-tat protein is a heparin-binding angiogenic growth factor. *Oncogene* **12,** 289–297.
18. Zhang, H. T., Kraft, P., Scott, P. A. E., Ziche, M., Weich, H. A., Harris, A. L., and Bicknell R. (1995) Enhancement of tumour growth and vascular density by transfection of vascular endothelial cell growth factor (VEGF) into MCF7 human breast carcinoma cells. *J. Natl. Cancer Inst.* **87,** 213–219.
19. Coltrini, D., Gualandris, A., Nelli, E.M., Parolini, S., Molinari-Tosatti, M.P., Quarto, N., et al. (1995) Growth advantage and vascularization induced by basic fibroblast growth factor over-expression in endometrial HEC-1-B cells: an export-dependent mechanism of action. *Cancer Res.* **55,** 4729–4738.
20. Gualandris, A., Rusnati, M., Belleri, M., Nelli, E. M., Bastaki, M., Molinari-Tosatti, M. P., et al. (1996) Basic fibroblast growth factor over-expression in mouse endothelial cells: an autocrine model of angiogenesis and angioproliferative diseases. *Cell Growth Differentiation* **7,** 147–160.
21. Ziche, M., Donnini, S., Morbidelli, L., Parenti, A., Gasparini, G., and Ledda, F. (1998) Linomide blocks angiogenesis by breast carcinoma vascular endothelial growth factor transfectants. *Br. J. Cancer* **77,** 1123–1129.
22. Choudhuri, R., Zhang, H. T., Donnini, S., Ziche, M., and Bicknell, R. (1997) An angiogenic role for the neurotrophins midkine and pleiotrophin in tumorigenesis. *Cancer Res.* **57,** 1814–1819.

23. Proia, A. D., Chadler, D. B., Haynes, W. L., Smith, C. F., Suvarenamani, C., Erkel, F. H., and Klintworth, G. K. (1988) Quantification of corneal neovascularization using computerized image analysis. *Lab. Invest.* **58,** 473–479.
24. Conrad, T. J., Chandler, D. B., Corless, J. M., and Klintworth, G. K. (1994) In vivo measurement of corneal angiogenesis with video data acquisition and computerized image analysis. *Lab. Invest.* **70,** 426–434.

III

Angiogenesis Protocols in Vitro

11

Collagen Gel Assay for Angiogenesis

Induction of Endothelial Cell Sprouting

Ana M. Schor, Ian Ellis, and Seth L. Schor

1. Introduction

The inner lining of blood vessels, the endothelium, consists of a monolayer of endothelial cells (ECs), that present a free luminal surface and attach on their abluminal side to the underlying basement membrane (apart from a minimal amount of cell–cell overlap). A great deal of heterogeneity exists in the morphology of the endothelium and in the phenotype displayed by individual ECs. In spite of this, all ECs may be defined by two general criteria: anatomical location (i.e., luminal wall of blood vessels) and functionality (e.g., provision of a nonthrombogenic surface). In a mature resting vessel, the functionality and integrity of the endothelium is maintained under steady state conditions by the biosynthetic activity of the ECs, in conjunction with low levels of cell proliferation and motility. Significant changes in the motility of the endothelial cells, often accompanied by cell proliferation, occur during angiogenesis and in response to vessel injury.

Angiogenesis is a complex morphogenetic process, which involves both cellular and extracellular components *(1–5)*. Key events involving the endothelium include:

1. The initial activation of the resting endothelium in the target (preexisting) vessel. Activation is commonly defined by morphological criteria (including cell enlargement, blebbing, and loosening of cell junctions) which herald a switch from a resting to a migratory EC phenotype.
2. Activated ECs migrate through their basement membrane into the surrounding tissue stroma, thus moving from a 2D to a 3D macromolecular environment,

From: *Methods in Molecular Medicine, Vol. 46: Angiogenesis Protocols*
Edited by: J. C. Murray © Humana Press Inc., Totowa, NJ

where they form solid sprouts. Sprouting can occur simultaneously from several sites along the parent vessel.
3. The newly formed endothelial sprouts move towards the source of angiogenic stimulus by a combination of cell migration and cell proliferation, although the latter is not essential for angiogenesis to occur.
4. The solid sprouts become patent vessels in the final stages of angiogenesis, thus completing the morphogenetic process.

During angiogenesis the phenotype of the ECs changes in a reversible manner, first from resting to sprouting, and then back to resting. Apart from providing a substratum for cell attachment and migration, the precise nature of the extracellular matrix actively contributes to the control of EC behavior during angiogenesis by a number of mechanisms ***(2,4,6–8)***, including:

1. Acting as a reservoir of soluble angiogenic factors that may be either bound to the matrix macromolecule or cryptic within it. In the latter case, the cryptic soluble factor(s) may be released by proteolytic degradation or by conformational change.
2. Altering the pattern of gene expression in adherent cells and their response to soluble effector molecules.
3. Providing a substratum that may be either permissive or nonpermissive (i.e., inhibitory) for angiogenesis and therefore modulate the activity of soluble angiogenic factors.

Injury to the endothelium (e.g., a break in its continuity) also elicits a rapid response of ECs peripheral to the damage. As is the case with angiogenesis, these cells become activated and migrate into the denuded area in order to restore endothelium continuity. In spite of these similarities, EC migration during angiogenesis and vessel repair differ with respect to the substratum upon which cells move: during angiogenesis cell migration occurs within a 3D macromolecular matrix, whilst during the repair of a damaged endothelium the mobility of endothelial cells is restricted to a 2D surface. ***(4)***. In this regard, it should be noted that cells may respond quite differently to the same motogenic factor when assayed within 3D macromolecular matrices as compared to commonly used experimental systems employing 2D substrata, such as the "Boyden chamber" or "wounded" monolayer assays ***(6)***.

We have developed a simple in vitro culture system in which it is possible to induce ECs to express either a resting or sprouting phenotype ***(2,4,8–13)***. An essential feature of this system is that cell migration occurs *within a 3D macromolecular matrix*, thereby modelling angiogenesis as opposed to vessel repair. Key characteristics of this culture system and its relevance to angiogenesis in vivo can be summarised as follows:

1. The appearance of the ECs expressing both resting and sprouting phenotypes in vitro is reminiscent of their in vivo counterparts. The resting cells present a free apical surface, are attached on their basal surface to a subendothelial matrix, and form a cobblestone monolayer. The sprouting cells, on the other hand, are found within a 3D matrix and therefore able to establish adhesive interactions along their entire surface. Their morphology is elongated and fusiform, and they tend to associate in a head-to-tail fashion, forming complex 3D tubule-like structures.
2. Like their in vivo counterparts, expression of a resting or sprouting phenotype is completely reversible by appropriate manipulation of the extracellular matrix.
3. The resting phenotype may be maintained indefinitely under steady state conditions. These cells display a low level of proliferation and characteristic biosynthetic profile.
4. Resting cells can be induced to become sprouting cells by the addition of exogenous angiogenic factors. Morphological changes in the resting monolayer (consistent with their "activation") precede overt sprouting.
5. Sprouting cells migrate underneath the cobblestone monolayer, through the matrix deposited by the latter and into the underlying 3D matrix provided in the assay.

These various observations form the basis of several in vitro assays for angiogenesis that mimic one or more of the cellular events of angiogenesis in vivo, but remain relatively simple and easy to score. The particular angiogenesis assay described in this chapter is based on the ability of cultured resting ECs to adopt a sprouting phenotype in response to an applied angiogenic factor. The migration of homogeneous cultures of sprouting cells within 3D matrices may be studied as described in Chapter 12 ***(14)***.

The induction of endothelial sprouting assay entails: (1) preparation of a 3D matrix of type I collagen fibres (to be referred to as the "collagen gel"), (2) plating ECs on the surface of the gel and allowing these to form a stable resting monolayer under steady state conditions, and (3) activating the resting cells by the addition of the factor to be tested, and monitoring (or quantifying, as required) changes in the morphology and the appearance of the resultant sprouting cells.

Considering the well-documented heterogeneity of ECs and the role of the extracellular matrix in the control of cell behavior, it is clear that the effects of a particular soluble angiogenic factor or inhibitor should be defined in the context of the specific target cell population and matrix substratum. We will describe a basic protocol using bovine aortic ECs and native type I collagen matrix, although it must be emphasized that other target ECs and macromolecular substrata may be employed. Variations of the basic assay described here include the use of different substrata as explained in **Fig. 1** and **Note 1**.

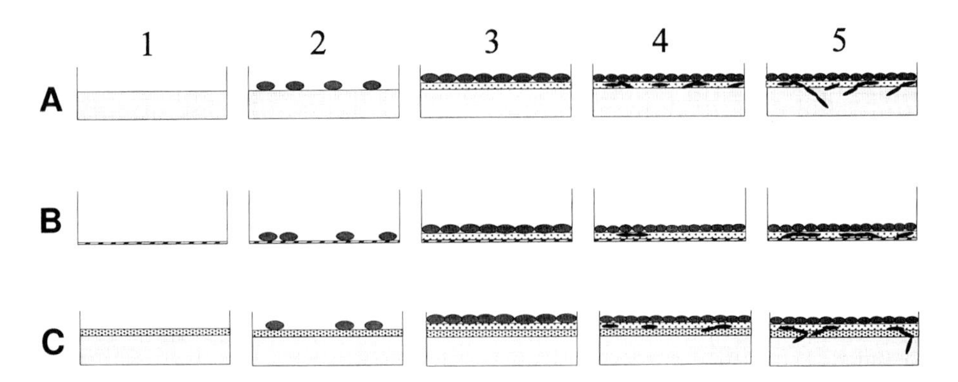

Fig. 1. Induction of endothelial sprouting: Variations of the assay. **(A)** Standard assay using a 3D macromolecular matrix (such as type I collagen gel). **(B)** Simplified assay using tissue culture dish coated with thin film of matrix molecule (such as gelatin). **(C)** More complex variation of assay using 3D macromolecular matrix covered with cell-produced extracellular matrix (ECM). *1*, Starting substratum. *2*, Endothelial cells plated at subconfluent density. *3*, After several days of incubation, the endothelial cells have produced a confluent "cobblestone" monolayer of resting cells adherent to a cell-produced ECM. These resting cultures are stable for several months when kept under maintenance conditions. *4*, Following the addition of an angiogenic factor, a proportion of the resting cells become activated, migrate down into the cell-produced extracellular matrix and adopt a sprouting phenotype within this 3D environment. *5*, After 7–14 d, there may be an extensive network of sprouting cell aggregates (depending on the culture conditions); these will have migrated down into the 3D macromolecular matrix, in assay variants where this has been provided (**A,C**).

2. Materials

2.1. Preparation of Native Type I Collagen

1. Rat tails from healthy young rats: cut off the tail approximately 1 cm from the base using a single edge razor blade (SLS/BDH); wipe thoroughly with 70% ethanol and freeze at –20° C in batches of 10 or as required (NB: 10 tails yield approximately 2 g of collagen). The tails may be stored at –20° C for 1 yr (*see* **Notes 2** and **3**).
2. Forceps (straight artery clamp, such as Spencer Wells, 125–150 mm long), scissors, a magnetic stirrer bar, one 500 mL beaker, two 1000 mL centrifuge buckets and filter funnels. All sterile.
3. 70% ethanol.
4. 3% acetic acid Analar (or Aristar). Make fresh and pass through 0.2 μm filter to sterilize. Five hundred mL will be required for 10 rat tails (*see* **Note 4**).
5. 50 m*M* ethylenediaminotetraacetic acid (EDTA) with 50 m*M* sodium bicarbonate.
6. Sterile and treated dialysis tubing. Boil approximately 5 m of dialysis tubing (size 8, internal diameter 32/32 from Medicell International) for 1–2 hr in a

solution of 50 m*M* EDTA and 50 m*M* sodium bicarbonate. The tubing is then washed outside and inside by filling it up with water through one end while keeping the other end closed and then releasing the closed end to pour out the water. This operation is repeated at least three times with tap water and then three times with distilled water. Cut into 25-cm pieces, tie a knot at one end of each piece, and autoclave in distilled water. It may then be stored at 4°C for 2–3 mo.

2.2. Preparation of 3D Collagen Gels

1. Collagen working solution at 2.2 mg/mL in distilled water (pH 4.1). May be stored at 4°C for 5–6 mo.
2. Medium: 10X concentrated Eagle's Minimum Essential Medium, with Eagle's salts and sodium bicarbonate (10X MEM) (Sigma). Stored at 4°C.
3. 7.5% Sodium bicarbonate solution (Sigma). Stored at 4°C.

2.3. Maintenance and Plating of ECs

1. Medium: Eagle's Minimum Essential Medium, with Eagle's salts and sodium bicarbonate (MEM) (Sigma). Stored at 4°C.
2. Medium additives: nonessential amino acids, sodium pyruvate, penicillin/streptomycin (optional), glutamine (added fresh) (Sigma). Aliquot and store at –20°C. All added at 1% (v/v) final dilution.
3. Donor calf serum: Aliquot and store at –20°C (*see* **Note 5**). Thaw and heat at 56°C for 20 min to inactivate complement before adding to medium.
4. Ascorbic acid (Analar, BDH). Make up a stock solution at 50 mg/mL in PBS. Sterilize by filtering. Aliquot and store at –20° C for up to 1 mo and thaw only once.
5. Hanks' Balanced Salts solution without Ca^{2+}/Mg^{2+} (Sigma). Store at room temperature.
6. Trypsin. 10X solution (2.5 g porcine trypsin powder/100mL) (Sigma). Aliquot and store at –20° C.
7. Soya bean trypsin inhibitor (Sigma). Aliquot and store at –20° C.
8. EGTA (ethylene glycol-bis (β-aminoethyl ether)-N,N,N′,N′-tetra acetic acid): 0.76 g/L in PBS (Sigma). Aliquot and store at –20° C.
9. Trypsin-EGTA: 0.05% (w/v) trypsin, 2 m*M* EGTA in phosphate buffered saline (PBS). Stored at –20°C for 6 mo. Thaw only once; once thawed may be stored at 4°C for 2–3 wk.
10. Tissue culture dishes, 90 mm and 30 mm diameter (Life Technologies).
11. Gelatin: 100 mg/mL in distilled water, autoclaved (BDH). Stored at 4°C for 6 mo.
12. Native type I collagen: 100 mg/mL in distilled water, diluted from stock collagen solution. Stored at 4°C for 6 mo.
13. Collagenase (type I, CLS1, Worthington Enzymes): 0.5 mg/mL in MEM. Filter sterilized, may be aliquoted and stored at –20°C for 6 mo. Thaw only once.

2.4. Quantification of Endothelial Sprouting

1. PBS (phosphate buffer saline). May be made with tablets (Sigma) dissolved in distilled water.

2. PBS with 0.01% sodium azide.
3. 10% formaldehyde in PBS.
4. Graticule for microscope eyepiece (Graticules Limited, Tonbridge, Kent)

3. Methods

3.1. Preparation of Native Type I Collagen Gels

3.1.1. Extraction of Collagen from Rat Tails

1. Thaw one batch of 10 tails directly into a beaker containing enough 70% ethanol to cover the tails (about 200 mL). Place the beaker in a laminar flow cabinet and do all subsequent work under sterile conditions. Have ready in the cabinet a sterile 1000 mL centrifuge bucket (with screw cap) containing approximately 100 mL 70% ethanol. When the tails have thawed (20–30 min), take one tail out of the beaker. Using two pairs of forceps, one in each hand, hold the tail at two points: one 2–3 cm from the tip end of the tail, and the second about 2.5–3.0 cm from the first, closer to the base of the tail. The tail should be held firmly but without exerting pressure. Break the tail with a quick flick of the wrists (as if to bring the two forceps together under the tail) and pull to separate the two parts. This operation will leave a short length of tail (on the forceps closest to the tip) with tendons protruding from it. Place this piece on top of the centrifuge bucket containing ethanol and cut the tendons cleanly with sharp scissors so that they fall into the bucket (*see* **Note 6**). Discard that piece and repeat the operation moving along the tail.
2. When all the tails have been processed, decant the ethanol taking care not to lose the tendons (this is quite straightforward, as they tend to adhere to the sides of the bucket). Add 500 mL of sterile 3% acetic acid (approximately 100 mL per gram of tendons), making sure that all the tendons are submerged. Insert the sterile magnetic stirrer bar, put the cap on the bucket, and stir overnight at 4°C.
3. Centrifuge at 2500*g* at 4°C for 2.5 hours (*see* **Note 7**).
4. Decant supernatant, using a sterile funnel, into sterilized dialysis tubing pretied at the bottom. Tie the top securely.
5. Dialyse for 2 d against distilled water, changing the water 2 times per day (*see* **Note 8**).
6. Wipe dialysis tubing with 70% ethanol and drain into a sterile centrifuge bucket.
7. Centrifuge at 2500*g* at 4°C for 2.5 hr.
8. Decant fluid into a sterile 500 mL bottle.
9. Add penicillin/streptomycin solution to a final concentration of 1% (*see* **Note 9**).
10. Store at 4°C. Check the pH of the collagen solution; this should be between pH 3.9 and 4.3 (*see* **Note 10**). Determine collagen concentration (*see* **Note 11**) and dilute to 2.2 mg/ml. Check pH again and adjust if necessary. Incubate approximately 5mL in a tissue culture incubator for 48 hr to check sterility.

3.1.2. Preparation of Gels

1. Working under sterile conditions in a laminar flow cabinet, pipette 1 mL of 10X MEM and 0.5 mL of 7.5% sodium bicarbonate solution into a 20 mL Universal tube.

2. Add 8.5 mL of collagen working solution (2.2 mg/mL) to the above mixture at 4°C, thus making a total of 10 mL gelling solution (*see* **Notes 12** and **13**).
3. Mix by pipetting up and down with a 10 mL pipet, taking care to avoid air bubbles. The mixing has to be done thoroughly as well as rapidly.
4. Transfer all the gelling solution into the pipette and plate 2 mL onto each of five 35-mm dishes (*see* **Note 13**). The tip of the pipet may rest on the base of the dish while pipeting, to facilitate accurate delivery.
5. Gently tilt and swirl the dish to spread the collagen evenly.
6. Without moving the dishes further, allow the collagen to set in the laminar flow cabinet. This will take approximately 5 min, but it is better not to move them for another 5–10 min. Make sure that the surface of the cabinet is level!
7. Repeat steps 1–6 to make the total number of gels required.
8. Once the gels are set, transfer to a humidified tissue culture incubator at 37°C with 5% CO_2 and incubate for 1–2 hr before adding medium.
9. Add 1 mL appropriate medium per gel (e.g., SF-MEM). Gels may be made 1–3 d in advance and stored in the incubator.

3.2. Maintenance and Plating of ECs on Collagen Gels

Stock cultures of bovine aortic ECs are maintained on gelatinized dishes in MEM supplemented with 20% donor calf serum (DCS), 1 m*M* sodium pyruvate, nonessential amino acids (81.4 mg/L), 2 m*M* glutamine, and 50 µg/mL ascorbic acid. Glutamine and ascorbic acid are stored at –20°C in aliquots of convenient volume; these are thawed and added to the volume of medium required for either 7–10 d (glutamine) or for 1 d (ascorbic acid). Penicillin (100 U/mL) and streptomycin (100 µg/mL) may also be added. The complete growth medium is referred to as 20% DCS-MEM (*see* **Note 5**). When serum is omitted, this complete medium is referred to as SF-MEM.

3.2.1. Routine Passage of ECs

Stock cultures of ECs are split at a 1/3 ratio when confluent. This is accomplished as follows:

1. Gelatinize the number of new dishes required by adding sufficient gelatin solution to cover the surface (e.g., 2 mL/90-mm dish) (*see* **Note 14**). Incubate for 1 hr at 37°C, remove the gelatin solution, and wash the dish twice with Hanks' (*see* **Note 15**).
2. Remove the medium from the stock culture to be passaged and wash the attached cells twice with 4–5 mL Hanks'.
3. Add a small volume of trypsin EGTA to cover the dish (e.g., 1.5 mL/90-mm dish). Incubate at 37°C until the cells begin to round and detach from the dish (about 5 min) (*see* **Note 16**).
4. Pipet up and down to detach all cells. Transfer trypsin and floating cells to a Universal containing 7.5 mL of 40% DCS-MEM (soya bean trypsin inhibitor

may be used if the experiment requires serum-free conditions). Wash the dish twice with 3 mL SF-MEM, transferring the washes to the same Universal.

5. Distribute the total volume (e.g., 15 mL of 20% DCS-MEM containing the cells harvested from one 90-mm dish) into 3 gelatinized dishes.
6. Cell attachment occurs rapidly, with 90% of the cells adherent to the dish within 15–30 min. Check microscopically that this is the case. After 1–2 hr remove medium with nonattached cells and replace with fresh medium. The medium is normally changed on Mondays, Wednesdays, and Fridays. The volume added per 90-mm dish is increased from 5 mL when the cells are sparse to 10 mL when confluent or over the weekend.

3.2.2. Plating ECs on Collagen Gels

Gels of different volumes and surface area may be prepared (*see* **Note 13**). The method described here is for 2 mL gels cast in 30-mm dishes:

1. The gels contain MEM without additives; before plating the cells, they must be equilibrated so that the interstitial fluid within the gel has the same composition as the medium in which the cells are to be plated. For example, when the cells are to be plated and grown in 20% DCS-MEM, the gels are first incubated for 1–2 hr with 1 mL 60% DCS-MEM and threefold concentration of glutamine, sodium pyruvate, ascorbic acid, and nonessential amino acids (*see* **Note 17**). The equilibrating medium may be removed before plating the cells if required.
2. Bring the cells to be plated into a stock cell suspension as described in **Subheading 3.2.1., steps 2, 3**, and **4**.
3. Take a small aliquot (e.g., 0.1–0.5 mL) to count the number of cells present (e.g., by Coulter counter or hematocytometer).
4. Take a volume of cell suspension that contains the number of cells required plus approximately 10–20%. For example to plate 20 gels at 2×10^5 cells per gel take 5×10^6 cells.
5. Centrifuge the cells at 100*g* for 5 min.
6. Remove the supernatant and resuspend the cell pellet in the appropriate volume of 20%DCS-MEM to achieve the desired concentration of cells to be plated per gel in 1 mL of medium (e.g., 2×10^5 cells per mL) (*see* **Note 18).**
7. Plate 1 mL of final cell suspension per gel, taking care of distributing it homogeneously throughout the gel and not to touch or damage the gel (e.g., by a fast stream of medium). Shake the dish horizontally in various orientations to achieve a homogeneous distribution of the cells.
8. Transfer the gels to the tissue culture incubator and incubate for 24 hr.
9. Remove the medium and replace with 1 mL 20%DCS-MEM. Change medium every 2–3 d (*see* **Note 19**).
10. When the cells reach saturation density (**Note 20**) remove the medium and wash the gels 4 times with serum-free MEM, incubating the cultures for 1 hr after each wash.
11. Add 1.5 mL of "maintenance medium" (e.g., 1% DCS-MEM) per gel (*see* **Note 21**) and change the medium every 2 d. These cultures are referred to as "resting".

3.3. Addition of Test Substances

The required number of resting cell cultures is prepared as described above. The resting cells may be maintained for several months. These cells remain metabolically active, and continue to synthesize and deposit a subendothelial matrix. It is therefore important that care be taken to use cultures of the same age in order to ensure an optimal level of experimental reproducibility (*see* **notes 1** and **16**).

The factor to be tested is dissolved in maintenance medium and added to duplicate cultures at a wide range of concentrations. If it is necessary to dissolve the factor in an organic solvent first (e.g., the phorbol ester PMA is dissolved in DMSO or acetone), the concentration of the organic solvent should be kept constant for the different concentrations of the active compound and for the controls. Alternatively, various solvent concentrations have to be used for both the active compound and the controls.

In initial experiments, positive controls should be included at different concentrations, for example bFGF or vascular endothelial growth factor (VEGF), at concentrations from 1–200 ng/mL. The various factors can be added in 1–2 mL volumes per gel and the medium changed every 2 d (*see* **Note 22**). Sprouts normally appear in the positive controls after 24–48 hr and the experiment is usually terminated in 4–6 d. At this point a small percentage of sprouting cells have usually migrated measurable distances into the collagen gel. If the cultures are to be maintained for longer periods, it is advisable to change the medium every day. The morphology of the cells at different stages of the assay is shown in **Fig 2**.

3.4. Quantification of Sprouting

The initial assessment of factor activity may be done using duplicate cultures. Once activity has been detected and the active concentration approximated, it may be necessary to quantify the sprouting-inducing ability of the test factor (*see* **Note 23**). In this case triplicate cultures are used for every concentration and time point tested, as well as positive and negative controls.

The cultures are fixed by removing the medium, washing once with PBS, and adding 1.5 mL of 10% formaldehyde. These cultures may be left at room temperature for 1–2 hr or stored at 4°C for 2–3 d. Before quantitation, the cultures are washed twice with PBS and then 1.5 ml of PBS containing sodium azide is added. These cultures may be stored for months at 4°C provided that the PBS is not allowed to dry out (*see* **Note 24**).

The extent of sprouting may be quantified by various means. Here we describe two methods that we have found to be rapid and reproducible and only require standard phase-contrast microscopy. Both methods may be scored at the same time, as the information they provide is complementary.

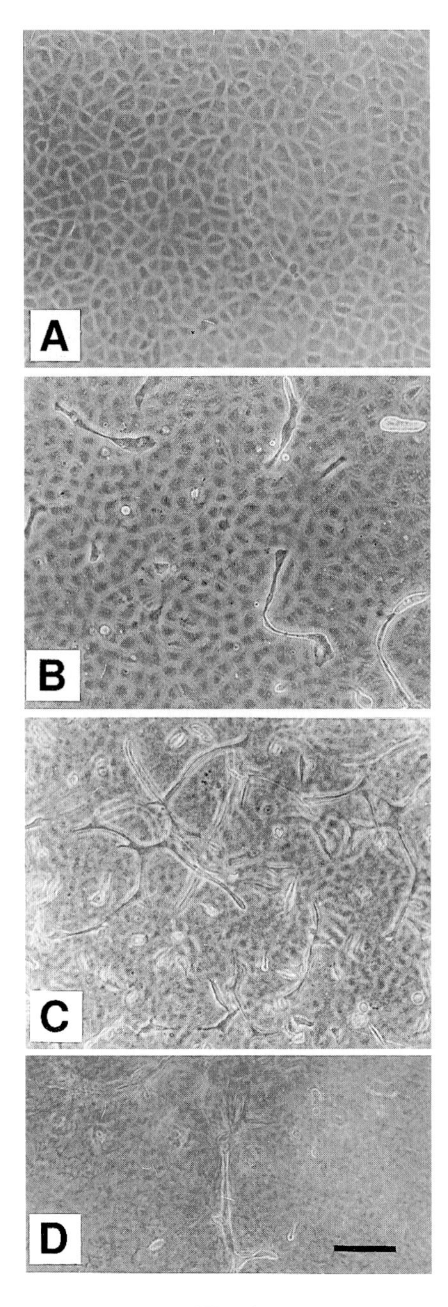

Fig. 2

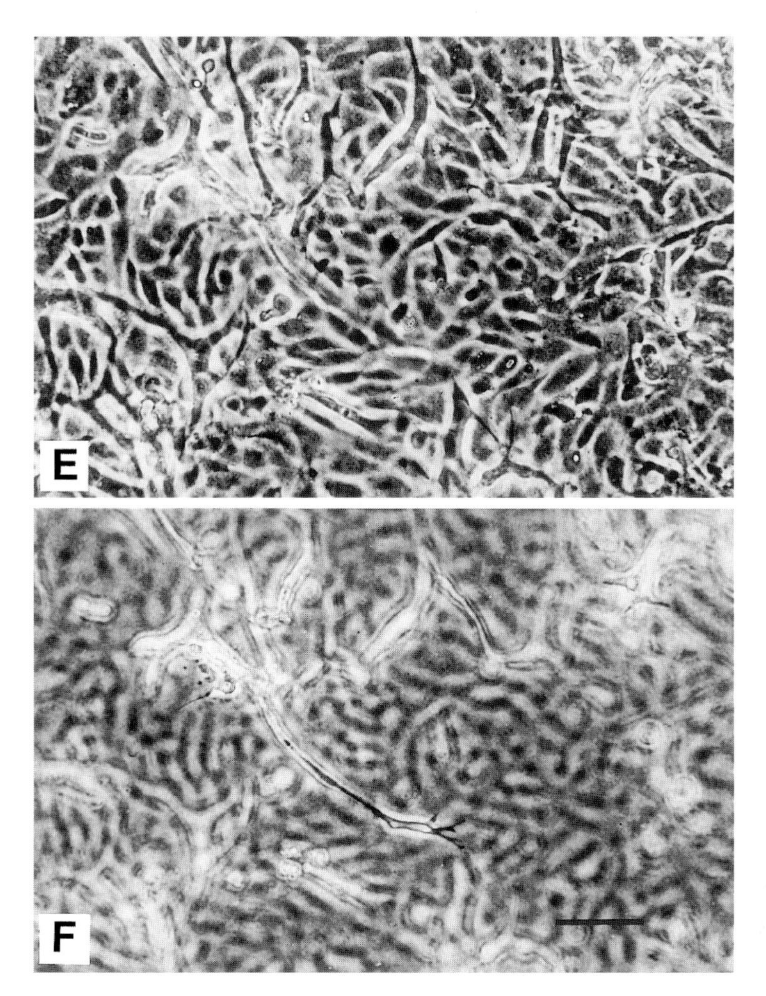

Fig. 2. Morphology of the cells at different stages of the assay. **(A)** Confluent cobblestone monolayer of resting cells under maintenance conditions. **(B–F)** Induction of sprouting cells by addition of angiogenic factors. **(B)** Early phase. Single sprouting cells directly underneath the cobblestone monolayer, which is seen slightly out of focus. **(C)** Late phase. As in **A**, but showing also sprouting cells connected into complex networks or "aggregates". **(D)** Sprouting cells deeper into the 3D gel. **(E)** Culture showing extensive network of sprouting cells and cobblestone monolayer of "activated" appearance (compare with resting cobblestone monolayer in **A** and **B**). **(F)** Same field as shown in **E**, but focusing down shows sprouting cells deeper into the gel and connected to cells underneath to the monolayer. Bar, 100 µm

3.4.1. Method 1: Percentage of Microscopic Fields Containing Sprouts

An arbitrary field is defined by a graticule placed in the eyepiece of the microscope. The cultures are viewed using phase-contrast optics set at the lowest magnification that allows a clear distinction between resting cells of the monolayer (cobblestone) and sprouting cells (e.g., 10× or 20× objective, and 10× eyepiece). Between 10 and 15 fields are randomly selected per culture, using the same procedure for each culture. The presence of single sprouting cells and sprouting cell aggregates is recorded as yes or no in each field. The aggregates are defined by the presence of at least three connected sprouting cells. This scoring is carried out first focussing on (and directly under) the surface monolayer and then focussing down into the gel (**Fig. 2**). Four variables are therefore recorded per culture, and the results are expressed as percentage of fields containing (a) single sprouting cells on the surface, (b) single sprouting cells in the gel, (c) aggregates on the surface, and (d) aggregates in the gel.

3.4.2. Method 2: Percentage of Sprouting Cells

A graticule that contains a fixed number of points (e.g., a Chalkley grid with 100 points) is placed in the eyepiece of the microscope. Using the same procedure as for Method 1 (**Subheading 3.4.1.**), 10–15 fields are randomly selected per culture (the same fields and graticule may be used to quantify by both methods). Each point that coincides with a sprouting cell is recorded, making it possible that a single sprouting cell be recorded more than once (**Fig 3**). This is a classical stereological method that reflects the volume occupied by the sprouting cells. For each culture two variables are recorded: (a) percentage of sprouting cells on the surface, and (b) percentage of sprouting cells in the gel. When scoring points counted within the 3D gel, a fixed number of focussing planes are predefined (e.g., four planes at fixed distances under the resting monolayer) and only those sprouting cells that coincide with the points of the grid and are in focus are recorded (note that a single cell may be recorded more than once).

4. Notes

1. In this assay the 3D extracellular matrix is formed by two components (i) a 3D substratum (type I collagen matrix or "collagen gel") on which the endothelial cells are cultured, and (ii) a cell-produced subendothelial matrix which is synthesised and deposited by the cells on top of the collagen gel. The basic assay described here may be adapted in several ways, as follows:
   a. The collagen gel used as substratum may be modified by the addition of other matrix macromolecules in the medium or within the gel *(**14,15**)*.

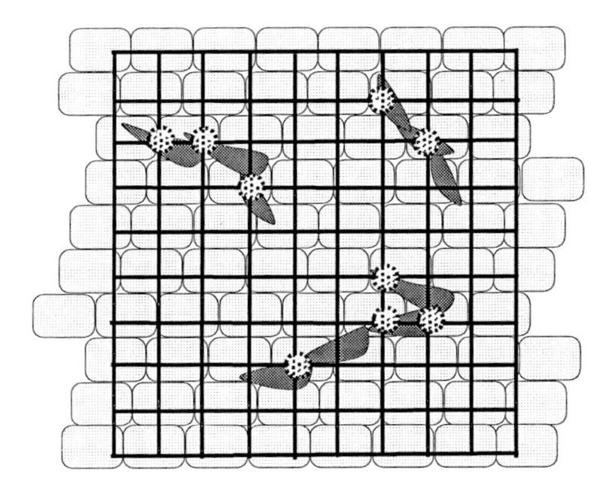

Fig. 3. Quantification of sprouting cells by point counting. Diagram representing a cobblestone monolayer, sprouting cells and a graticule containing 100 points. Each point that coincides with a sprouting cell is recorded as positive (*marked in the diagram*).

b. The collagen gel covered by a cell-produced matrix may be used as substratum following lysis of the producing cell monolayer ***(8,10,16)***.
c. A simpler version of this assay is achieved by using tissue culture dishes coated with native collagen (or other extracellular matrix component) instead of the collagen gel. In this case the 3D matrix available for sprouting cell migration is very thin, being provided exclusively by the cell-produced subendothelial matrix ***(9,10)***.

In all cases it is important to bear in mind that the composition of the cell-produced subendothelial matrix can be modified by the composition and structure of the substratum in contact with the cells, exogenous factors that may be added to the culture medium, and by other experimental conditions, such as cell density of the cells and time in culture. It is therefore essential that experimental variables are kept constant. We have used this assay with ECs derived from bovine aorta, retina and brain, human subcutaneous fat, and umbilical vein. The latter was the only unsatisfactory cell type. Interestingly, new blood vessels may originate from all types of microvessels and from luminal endothelium of large vessels ***(4)***, but, to our knowledge, it is not known whether umbilical endothelium is angiogenic.

2. Type I collagen preparations (bovine dermal and rat tail) suitable to make gels for tissue culture are commercially available (Collagen Bio-Materials, Becton & Dickinson, Sigma). We cannot comment on these, because we find it more efficient to make our own collagen preparation. In spite of carefully controlling the protocol, not all collagen batches are suitable to use with ECs. Occasionally a batch appears to prevent proliferation or even be "toxic" for these cells. Interest-

ingly, such toxic batches are only so for ECs and have no detectable deleterious effect on other cell types, such as fibroblasts. We have also observed that a suitable batch of collagen may become toxic upon storage. It is therefore important to test in a pilot experiment the collagen batch to be used. Ideally, several batches should be tested before carrying out the assay and the chosen batch retested if it has been stored for more than three months. The reason for differences amongst batches prepared in the same way is not clear. It may be due to variations in collagen crosslinking or even to the presence of active TGFβ-1 (*see* **Note 5**).

3. To avoid problems due to variations amongst collagen batches it is important to use the same type of rats. Young rats (e.g., 6 mo-old) are preferable to old ones, possibly owing to the increase in collagen crosslinking that occurs with aging. Variations also occur amongst strains. The tails are frozen for convenience and to facilitate excision of the tendons.
4. Analar or Aristar grade acetic acid is essential for the preparation of collagen; if inferior grades are used, the collagen may set erratically or precipitate during dialysis.
5. Either donor or fetal calf sera may be used, but batches should be tested for growth-promoting properties on the ECs. Great variations occur among batches, irrespective of their origin. We have found that some serum batches contain active TGFβ-1; these should not be used. However, TGFβ-1 becomes activated by prolonged storage (e.g., 6 mo at –20°C). Therefore, batches should be retested when used for more than 3–4 mo. Serum-free medium with added growth (and angiogenic) factors to promote endothelial proliferation is commercially available (e.g., TCS Biologicals). It should be possible to use this "serum-free endothelial growth medium" instead of serum-supplemented medium (*see also* **Note 19**). This, however, has not been tested by us.
6. Cutting the tendons should be done by a second operator. Although this is not essential, it makes an enormous difference in efficiency. With two operators working together, the whole procedure takes about 1 min per tail.
7. Following this step, the supernatant may be mixed with an equal volume of 20% NaCl to precipitate the collagen. The precipitate is recovered by another centrifugation (as for **step 3**), redissolved in 0.3% acetic acid and dialysed (**step 4**). We normally omit this NaCl precipitation, since tests show that it did not appear to affect the final product or results obtained.
8. The collagen tends to thicken during dialysis and occasionally it may aggregate and precipitate in the tubing. This is most likely caused by an elevated pH and can be reversed when the water used for dialysis is brought to pH 4.1 by the addition of acetic acid.
9. The addition of antibiotics is optional. Products to control fungal contamination may also be added, but Fungizone should not be used when working with ECs, for which it appears to be toxic.
10. The collagen solution should only be stored in the refridgerator. Freezing and thawing alters the physical properties of the resultant gels. The pH of the collagen solution needs to be kept as close to 4.1 as possible (NB: pH 3.9–4.4 is

acceptable). If it is lower than pH 3.9 the collagen will not set, if higher than pH 4.4, it will set too quickly for proper gel formation. *See* **Note 2** regarding time of storage.

11. A collagen standard curve is made using a known concentration of collagen diluted in distilled water acidified to pH 4.1 with acetic acid. In the first instance, the concentration of collagen is assessed by freeze drying and weighing known volumes of the stock collagen solution, or using commercially available type I collagen (such as Vitrogen, 2 mg/mL). At least six dilutions are made so that the standard curve should include from 0.1–0.6 mg/mL. The absorbance is read at optical density 230 nm. The concentration of collagen in new batches is determined by diluting aliquots with distilled water (1/5, 1/10. and 1/15 is usually sufficient), measuring the optical density at 230 nm and comparing the results with the standard curve. The values obtained for each dilution that fall within the curve are used to calculate the concentration of collagen (mean ± SD). The stock collagen solution is then diluted to a working dilution of 2.2 mg/mL.
12. According to this protocol, the concentration of collagen in the gelling solution is 1.87 mg/mL. The concentration of the collagen working solution may vary (e.g., 2.0–2.6 mg/mL), as long as the final concentration in the gels used for the assay is constant and close to 2.0 mg/mL. This is the concentration that allows maximal migration to occur for various cell types tested ***(17)***.
13. The temperature is very important, as the collagen solution is induced to set by raising the pH and temperature. It is usually convenient to keep the collagen solution on ice. The total volume of gelling solution (10 mL) may be adjusted to smaller volumes if required. However, we do not recommend adjusting to volumes larger than 12 mL, because it is crucial to cast the gelling solution as rapidly as possible. In practice, we mix 10.5 mL of gelling solution in order to cast 2.0 mL per 5 × 30 mm dishes. Alternatively, one can make 10 mL and cast 1.9 mL per dish. The volume per dish may be reduced, for example 1 mL per 30-mm dish. However, thin gels have a more pronounced meniscus and this produces problems regarding heterogeneous distribution of cells and their visualization. The volume can be adapted for use on smaller dishes (e.g., 24-well trays). It is convenient to make more gels than required for the experiment, as some may have to be discarded.
14. To speed up this step, we first wet the surface of the dishes by "washing" them with the gelatin solution, e.g., 5–10 mL of the solution is transferred from one dish to the next.
15. Gelatinized dishes may be prepared 1–3 d in advance for convenience and also to check sterility. In this case, add 3 mL of 20% DCS-MEM per dish and store them in the tissue culture incubator.
16. If the cells are postconfluent it is difficult to obtain a single cell suspension by trypsinization. This may be remedied by incubation with a solution of collagenase (0.5 mg/mL in MEM) for 15–20 min prior to trypsinization. For optimal experimental reproducibility, the stock cultures used to set up experiments should be of similar density, just confluent.

17. It is advisable to make more gels than required for the experiment. The gels may be made in advance and incubated with equilibrating medium for 1–2 d before plating the cells. This is for convenience, to recheck sterility, and to discard any defective gel (e.g., those containing air bubbles).
18. In order to obtain a single cell suspension, the cell pellet is first resuspended in 0.5–1.0 mL of medium and pipeted up and down with a fine bore pipet. More medium is next added to bring the volume to approximately half of the final volume required. This intermediate cell suspension is mixed well and a 0.5 mL aliquot taken to count cell number and estimate viability. Based on these results the intermediate cell suspension is diluted to the final concentration to be plated. When the volume of the final cell suspension is more than 25–30 mL, we aliquot it into plastic universals (15–20 mL per universal). Each is kept closed and mixed well before plating the cells. The number of cells plated per gel may be suspended in any volume of medium, from 0.2–2.0 mL as long as cell numbers are accurate and the total volume on the gel is between 1.0 and 2.0 mL (e.g., the gel may contain 1 mL of equilibrating medium and the cells plated in 0.5 mL). It is advisable to practice with pilot experiments of varying volumes, and using techniques to plate the cells in order to obtain a homogeneous distribution of the cells on the gel surface when they attach.
19. At all times take care not to touch the gel. Depending on the type of experiment, the volume of growth medium added may vary from 1–2 mL/gel, as long as it is kept constant for every gel in the experiment.
20. The time taken to reach confluence will vary depending on the culture conditions; it is usually 3 d with the conditions given in this example. The cells reach saturation density 1–2 d after confluence. To gain experience, cell numbers may be counted and compared with the morphology of the cells. It is, nevertheless, safe to change from 20% to 1% serum 1–2 d after confluence. Confluence is defined as the first instance when the cells cover 100% of the surface area. The cells still divide after that stage, still occupying 100% of the surface area, but less area per cell.
21. The concentration of serum used at this stage may vary from 0.1–5%, depending on the growth-promoting properties of the particular batch in use. The aim is to provide the minimum amount necessary to maintain the cells in a healthy, confluent, and metabolically active state. As an alternative, a commercially available serum-free growth medium may be diluted to achieve the same effect (*see* **Note 5**). In either case the right serum (or "serum-free endothelial growth medium") concentration to be used for the "maintenance medium" can be tested in preliminary experiments as follows:
    a. *Experiment 1:* Plate the cells on gelatinized 30-mm dishes at 10^5 cells/dish in 1 mL 20% DCS-MEM. Two days later (cells should be semiconfluent) wash the cultures three times with SF-MEM, count the number of cells present (baseline), and add MEM containing from 0.1–10%DCS to replicate dishes. Change medium every 2 d. Count the number of cells present and viable after 2, 4, and 6 d.

b. *Experiment 2:* (This may be planned as the second part of *Experiment 1* or as a separate experiment.) Plate the cells on gelatinized 30-mm dishes at 1–2 × 10^5 cells/dish in 1 mL 20% DCS-MEM. Change medium every 2 d. When the cells reach saturation density, wash the cultures with SF-MEM, count the number of cells present, and add MEM as for *Experiment 1*. Count the number of cells present and viable after 2, 4, and 6 d. The maintenance medium should promote the maintenance of cell numbers in both experiments; a minimum degree of cell growth in *Experiment 1* and the appearance of very few sprouting cells in *Experiment 2* are also acceptable. It should be noted that serum may contain angiogenic factors and sprouting is commonly induced by high serum concentrations.

22. In this assay the factors to be tested are added in the medium, therefore directly onto the apical (luminal) surface of the cells. It is also possible to add the factors in a slow release pellet placed under the gel, so that they diffuse as if from the perivascular tissue in vivo. In this case the cells are plated at high density (1–2 × 10^6 cells/gel) in order to achieve a confluent resting monolayer in as short time as possible. It should be noted, however, that the nature of the subendothelial matrix deposited by the cells is affected by variations in experimental conditions. Therefore, results obtained using this variation of the assay may not be directly comparable to those obtained using the standard assay. For this variation of the assay a methyl-cellulose pellet is anchored under a 2 mL collagen gel as described in Chapter 12, Note 25, and the cells are plated in maintenance medium. The cells should be confluent within the next 24 hr and the medium is changed only once.
23. This assay may be equally used to quantify the inhibition of sprouting induced by a compound in the presence of a known stimulator.
24. It is feasible to quantify live cultures. In this case the cultures have to be taken out of the incubator one (or two) at a time, and the time recorded to ensure that: (a) every culture is out of the incubator for the same amount of time and (b) such amount of time does not exceed a certain limit (e.g., 15 min every 2 hr).

Acknowledgments

This work was funded by the Cancer Research Campaign (UK) and the Biotechnology and Biological Sciences Research Council.

References

1. Ausprunk, D. H. and Folkman, J. (1977) Migration and proliferation of endothelial cells in preformed and newly formed blood vessels during tumor angiogenesis. *Microvascular Res.* **14,** 53–65.
2. Schor, A. M. and Schor, S. L. (1983) Tumour angiogenesis: a review. *J. Exp. Path.* **141,** 385–413.
3. Ishibashi, T., Miller, H., Orr, G., Sorgente, N., and Ryan, S. J. (1987) Morphologic observation on experimental subretinal neovascularization in the monkey. *Invest. Ophthalmol. Vis. Sci.* **28,** 1116–1130.

4. Schor, A. M., Schor, S. L., and Arciniegas, E. (1997) Phenotypic diversity and lineage relationships in vascular endothelial cells. In: *Stem Cells* (Potten, C. S., ed.). Academic Press Limited, London, pp. 119–146.
5. Iruela-Arispe, M. L. and Dvorak, H. F. (1997) Angiogenesis: a dynamic balance of stimulators and inhibitors. *Thromb. Haemost.* **78,** 672–677.
6. Schor, S. L. (1994) Cytokine control of cell motility: modulation and mediation by the extracellular matrix. *Prog. Growth Factor Res.* **5,** 223–248.
7. Boudreau, N. and Bissel, M. J. (1996) Regulation of gene expression by the extracellular matrix. In: *Extracellular Matrix*, vol. 2, *Molecular Components and Interactions* (Comper, W. D., ed.) *Overseas Publishers Association*, Amsterdam.
8. Canfield, A. E. and Schor, A.M. (1995) Evidence that tenascin and thrombospondin-1 modulate sprouting of endothelial cells. *J. Cell Sci.* **108,** 797–809.
9. Schor, A. M., Schor, S. L., and Allen, T. D. (1983) The effects of culture conditions on the proliferation and morphology of bovine aortic endothelial cells in vitro: reversible expression of the sprouting cell phenotype. *J. Cell Sci.* **62,** 267–285.
10. Schor, A. M., Schor, S. L., and Allen, T. D. (1984) The synthesis of subendothelial matrix by bovine aortic endothelial cells in culture. *Tissue and Cell* **16,** 677–691.
11. Schor, A. M. and Schor, S. L. (1986) The isolation and culture of endothelial cells and pericytes from the bovine retinal microvasculature: a comparative study with large vessel vascular cells. *Microvascular Res.* **32,** 21–38.;
12. Schor, A. M. and Schor, S. L. (1988) Inhibition of endothelial cell morphogenetic interactions in vitro by alpha- and beta-xylosides. *In Vitro* **24,** 659–668.
13. Sutton, A. M., Canfield, A. E., Schor, S. L., Grant, M. E., and Schor, A. M. (1991) The response of endothelial cells to TGF-b1 is dependent upon cell shape, proliferative state and the nature of the substratum. *J. Cell Sci.* **99,** 777–787.
14. Schor, A. M., Ellis, I., and Schor, S. L. (2001) Chemotaxis and chemokinesis in 3D macromolecular matrices. Relevance to angiogenesis. In: *Methods in Molecular Medicine, Angiogenesis Protocols* (Murray J. C., ed.). Humana Press, Totowa, NJ, pp. 163–183.
15. Schor, S. L., Schor, A. M., Grey, A. M., Chen, J., Rushton, G., Grant, M. E., and Ellis, I. (1989) Mechanism of action of the migration stimulating factor (MSF) produced by fetal and cancer patient fibroblasts: effect on hyaluronic acid synthesis. *In Vitro* **25,** 737–746.
16. Schor, S. L., Schor, A. M., Allen, T. D., and Winn, B. (1985) The interaction of melanoma cells with fibroblasts and endothelial cells in three-dimensional macromolecular matrices: a model of tumour cell invasion. *Int. J. Cancer* **36,** 93–102.
17. Schor, S. L., Schor, A. M., Winn, B., and Rushton, G. (1982) The use of 3D collagen gels for the study of tumour cell invasion in vitro: experimental parameters influencing cell migration into the gel matrix. *Int. J. Cancer* **29,** 57–62.

12

Chemotaxis and Chemokinesis in 3D Macromolecular Matrices

Relevance to Angiogenesis

Ana M. Schor, Ian Ellis, Seth L. Schor

1. Introduction

Angiogenesis is a complex morphogenetic process involving the coordinate migration of several cell types, including endothelial cells (EC), pericytes, and stromal fibroblasts ***(1–4)***. Angiogenesis is regulated by interactions between cells, soluble factors, and extracellular matrix components. The extracellular matrix (EM) in contact with vascular cells changes during angiogenesis in terms of its composition and structural organization. For example, ECs lining the lumen in a "resting" vessel are attached to a 2D substratum of specialized structure and composition (i.e., the basement membrane). Following exposure to angiogenic factors, endothelial cells migrate from their 2D environment into the surrounding 3D tissue stroma. Within this 3D macromolecular environment, the endothelial cells adopt an elongated "sprouting" phenotype and synthesise new EM components. Pericytes and fibroblasts are normally resident within a 3D macromolecular matrix, as provided by the vessel basement membrane and tissue stroma, respectively. Nevertheless, pericytes also form part of the newly formed vascular sprouts and fibroblasts surround and accompany these. In addition, vascular sprouts are commonly accompanied by inflammatory cells that produce proteases and cytokines, thereby contributing to further alterations in the composition of the microenvironment. The migration of ECs, pericytes, and adjacent fibroblasts during angiogenesis is directional. As new vessels move towards the source of angiogenic stimulus, they

From: *Methods in Molecular Medicine, Vol. 46: Angiogenesis Protocols*
Edited by: J. C. Murray © Humana Press Inc., Totowa, NJ

migrate into matrices of different and variable composition (e.g., during wound healing new vessels and fibroblasts invade a fibrin clot) ***(1–7)***.

Apart from providing a substratum for cell migration, the macromolecular matrix also contributes to the control of cell migration by a number of mechanisms, including (a) direct effects upon the activity of the cytoskeletal motile machinery, and (b) modulating cellular response to motogenic cytokines. Both of these mechanisms are dependent upon the ligation of matrix molecules by their respective cell surface receptors and the elicited signal transduction cascades. This regulatory function of the macromolecular matrix is in turn dependent upon a number of characteristics of its constituent molecules, including their (a) isoformic composition, (b) conformational state, (c) association into higher order molecular aggregates, (d) crosslinking and other forms of post-translational modification, and (e) concentration relative to other constituent molecules that may elicit agonistic or antagonistic signals ***(8–10)***.

Cell migration has commonly been studied on 2D substrata, such as plastic tissue culture dishes ("wounded" monolayer assay) and through the pores of a polycarbonate membrane separating upper and lower medium compartments (transmembrane or Boyden chamber assay). The transmembrane assay allows a distinction to be made between *chemokinesis* (defined as random cell movement in response to an isotropic distribution of soluble effector molecule) and *chemotaxis* (defined as directional cell movement in response to an imposed concentration gradient of soluble effector) ***(11)***. Matrix macromolecules may also affect cell migration by *haptotaxis* (defined as directional cell movement in response to an imposed concentration gradient of adsorbed effector molecule).

In view of the role played by the EM in the regulation of cell migration, it is important to develop model systems which will make it possible to study cell movement within a physiologically relevant 3D macromolecular environment in vitro. With this objective in mind, we devised a migration assay involving the plating of cells on the surface of a collagen gel substratum and scoring their subsequent migration down into the 3D collagen fiber matrix ***(12–14)***. We have previously shown that the motogenic response of fibroblasts to both cytokines and matrix macromolecules may be different in this collagen gel migration assay compared to that obtained in the more commonly employed transmembrane assay. ***(8,15,16)***. In a similar way, the response of ECs to angiogenic factors is critically dependent upon the biochemical composition of the substratum and the conformation of matrix molecules therein ***(17–20)***.

One potential difficulty with the initial version of the collagen gel assay is that it does not easily lend itself to distinguishing between chemokinetic and chemotactic modes of motogenic response. In order to address this difficulty, we now describe a "sandwich" variant of the collagen gel assay, which allows such a distinction to be made. In the simplest version of this assay, the *Two-Layer Sandwich Assay*, cells are plated on the surface of a native type I collagen gel. Thirty minutes after plating (i.e., after sufficient time for the cells to attach to the collagen fibers, but not spread or induce fiber reorganization) this substratum is overlaid with a second gel. As a result of this procedure, the plated cells become embedded within an isotropic 3D collagenous matrix. The subsequent movement of individual cells (either up or down from their original location) may then be assessed microscopically. Effector molecules (e.g., cytokines or matrix macromolecules) may be added equally to both upper and lower collagen gels (to identify a chemokinetic response) or to only one of these compartments (to identify a chemotactic/ haptotactic response). A second version of this assay, the *Three-Layer Sandwich Assay*, involves plating the cells within a 3D gel which is underlaid and overlaid by cell-free gels; this protocol ensures that cells are surrounded by an isotropic matrix from the beginning.

These assays (**Fig. 1**) will be described in detail for fibroblasts; with minor modifications (*see* **Notes**) these same assays can be used to study ECs, pericytes, and other cell types whose migration into 3D matrices is of biological relevance ***(12,14,21–23)***.

2. Materials

2.1. Preparation of Native Type I Collagen

1. Rat tails from healthy young rats: Cut off the tail approximately 1 cm from the base using a single edge razor blade (SLS/BDH); wipe thoroughly with 70% ethanol and freeze at –20°C in batches of 10 or as required (NB: 10 tails yield approximately 2 g of collagen). The tails may be stored at –20°C for 1 yr (*see* **Note 1**).
2. Forceps (straight artery clamp, such as Spencer Wells, 125–150 mm long), scissors, a magnetic stirrer bar, one 500 mL beaker, two 1000 mL centrifuge buckets, and filter funnels. All sterile.
3. 70% ethanol.
4. 3% Acetic acid Analar (or Aristar). Make fresh and pass through 0.2-μm filter to sterilize. Five hundred mL will be required for 10 rat tails (*see* **Note 2**).
5. 50 m*M* ethylenediaminotetraacetic acid (EDTA) with 50 m*M* sodium bicarbonate.
6. Sterile and treated dialysis tubing. Boil approx 5 m of dialysis tubing (size 8, internal diameter 32/32 from Medicell International) for 1–2 h in a solution of 50 m*M* EDTA and 50 m*M* sodium bicarbonate. The tubing is then washed outside and inside by filling it up with water through one end while keeping the other end closed and then releasing the closed end to pour out the water. This operation is repeated at least 3 times with tap water and then 3 times with distilled water. Cut into 25-cm pieces, tie a knot at one end of each piece, and autoclave in distilled water. It may then be stored at 4°C for 2–3 mo.

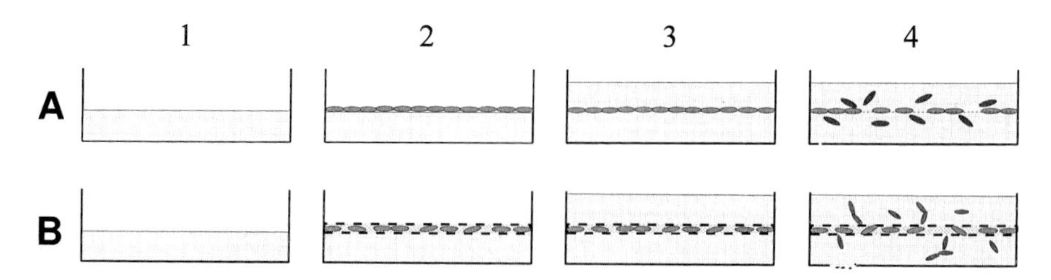

Fig. 1. Variations of the collagen gel sandwich assay. **(A)** Two-layer sandwich assay. **(B)** Three-layer sandwich assay. *1*, A 3D native type I collagen gel (*lower gel*) is cast in a tissue culture dish. *2*, Cells are plated on the surface of this gel (two-layer assay) or within a second thinner gel (three-layer assay). *3*, Cells are overlaid with another (*upper*) gel. *4*, Cells migrate from their plating location into the upper and lower gels. Effector molecules (e.g., cytokines or matrix macromolecules) may be added to upper and lower collagen gels (as well as to cell-containing gel in **(B)** to identify chemokinetic and chemotactic responses.

2.2. Preparation of 3D Collagen Gels

1. Collagen working solution at 2.2 mg/mL in distilled water (pH 4.1). May be stored at 4°C for 5–6 mo (*see* **Note 3**).
2. Medium: 10X concentrated Eagle's Minimum Essential Medium, with Eagle's salts and sodium bicarbonate (10X MEM) (Sigma). Stored at 4°C.
3. 7.5% Sodium bicarbonate solution (Sigma). Stored at 4°C.

2.3. Maintenance of Fibroblasts and Preparation of Cell Suspensions

1. Medium: Eagle's Minimum Essential Medium, with Eagle's salts and sodium bicarbonate (MEM) (Sigma). Stored at 4°C.
2. Medium additives: Nonessential amino acids, sodium pyruvate, penicillin/streptomycin (optional), glutamine (added fresh) (Sigma). Aliquot and store at –20°C. All added at 1% (v/v) final dilution.
3. Donor calf serum (DCS): Aliquot and store at –20°C (*see* **Note 4**). Thaw and heat at 56°C for 20 min to inactivate complement before adding to medium.
4. Ascorbic acid (Analar, BDH): Make up a stock solution at 50 mg/mL in phosphate buffered saline. (PBS). Sterilize by filtering. Aliquot and store at –20° C for up to 1 month and thaw only once (*see* **Note 5**).
5. Hanks' balanced salts solution without Ca^{2+}/Mg^{2+} (Sigma). Store at room temperature.
6. Trypsin 10X solution: 2.5 g porcine trypsin powder/100 mL (Sigma). Aliquot and store at –20° C.
7. Soya bean trypsin inhibitor (Sigma). Aliquot and store at –20° C.

8. ethylene glycol-bis (B-aminoethyl ether)-N,N,N′,N′-tetra acetic acid (EGTA): 0.76 g/L in PBS (Sigma). Aliquot and store at –20° C.
9. Trypsin-EGTA: 0.05% (w/v) trypsin, 2 m*M* EGTA in PBS. Stored at –20°C for 6 mo. Thaw only once. Once thawed, may be stored at 4°C for 2–3 wk.
10. Tissue culture dishes, 90-mm and 30-mm diameter (Life Technologies).
11. Collagenase (type I, CLS1, Worthington Enzymes): 0.5 mg/mL in MEM. Filter sterilized; may be aliquoted and stored at –20°C for 6 mo. Thaw only once.

2.4. Migration Assay

1. Cell suspension.
2. Collagen working solution (2.2 mg/ml) (*see* **Note 3**).
3. Donor calf serum.
4. Minimal essential medium (MEM).
5. Suspension of inert marker in MEM. For example: dry ink particles (International Market Supply) (*see* **Note 6**). In this case, a stock suspension of dry ink powder (10 mg/mL) is made in PBS containing 2 mg/ml bovine serum albumin (BSA). This is mixed well, autoclaved, and stored at 4°C. Working under sterile conditions, a 2-mL aliquot is transferred into a 20-mL universal centrifuge tube and mixed with approximately 5 mL of sterile PBS containing 2 mg/mL BSA. Mix gently for at least 2 hr at 37°C, allow to sediment by gravity, and remove supernatant when most of the particles have sedimented but the finer ones are still in suspension. Add approximately 15 mL 60% DCS-MEM, mix and allow to sediment as before. Repeat the operation three more times adding 15 mL of serum-free (SF)-MEM and mixing for 10–15 min. Finally resuspend in 15 ml SF-MEM, mix well, and separate the particles by spinning down at 100*g* for 1 min. Resuspend in 5 mL SF-MEM (approx concentration 3–4 mg/mL) and store at 4°C. To obtain the final working suspension, several 30-mm dishes containing 2 mL collagen gels and 1 mL SF-MEM are prepared as described below. These are used to check the concentration of the marker as follows: Remove the medium from the gels, agitate the particle suspension, and add 100 μL onto one gel taking care to distribute this volume homogeneously on the surface. Tilt the gel to remove the medium leaving the particles on the surface of the gel. Check the concentration of particles under the microscope, using the same microscope magnification and grid required to quantify migration. Between three and five particles should be present per field. If necessary, make various dilutions from the suspension tested and repeat the testing protocol, using duplicate gels when appropriate. Always agitate before taking aliquots.

2.5. Quantification of Cell Migration

1. PBS. May be made with tablets (Sigma) dissolved in distilled water.
2. PBS with 0.01% sodium azide.
3. 10% Formaldehyde in PBS.
4. Graticule for microscope eyepiece (Graticules Limited, Tonbridge, Kent).

3. Methods

3.1. Preparation of Native Type I Collagen Gels

3.1.1. Extraction of Collagen from Rat Tails

1. Thaw one batch of 10 tails directly into a beaker containing enough 70% ethanol to cover the tails (about 200 mL). Place the beaker in a laminar flow cabinet and do all subsequent work under sterile conditions. Have ready in the cabinet a sterile 1000-mL centrifuge bucket (with screw cap) containing approximately 100 mL 70% ethanol. When the tails have thawed (20–30 min), take one tail out of the beaker. Using two pairs of forceps, one in each hand, hold the tail at two points: one 2–3 cm from the tip end of the tail, and the second about 2.5–3.0 cm from the first, closer to the base of the tail. The tail should be held firmly but without exerting pressure. Break the tail with a quick flick of the wrists (as if to bring the two forceps together under the tail) and pull to separate the two parts. This operation will leave a short length of tail (on the forceps closest to the tip) with tendons protruding from it. Place this piece on top of the centrifuge bucket containing ethanol and cut the tendons cleanly with sharp scissors so that they fall into the bucket (*see* **Note 7**). Discard that piece and repeat the operation moving along the tail.
2. When all the tails have been processed, decant the ethanol taking care not to lose the tendons (this is quite straightforward because they tend to adhere to the sides of the bucket). Add 500 mL of sterile 3% acetic acid (approx 100 mL per gram of tendons), making sure that all the tendons are submerged. Insert the sterile magnetic stirrer bar, put the cap on the bucket and stir overnight at 4°C.
3. Centrifuge at 2500*g* at 4°C for 2.5 h (*see* **Note 8**).
4. Decant supernatant, using a sterile funnel, into sterilized dialysis tubing pre-tied at the bottom. Tie the top securely.
5. Dialyze for 2 d against distilled water, changing the water 2 times per day (*see* **Note 9**).
6. Wipe dialysis tubing with 70% ethanol and drain into a sterile centrifuge bucket.
7. Centrifuge at 2500*g* at 4°C for 2.5 h.
8. Decant fluid into a sterile 500-mL bottle.
9. Add penicillin/streptomycin solution, to a final concentration of 1% (*see* **Note 10**).
10. Store at 4°C. Check the pH of the collagen solution; this should be between pH 3.9 and 4.3 (*see* **Note 11**). Determine collagen concentration and dilute to 2.2 mg/mL (*see* **Notes 3** and **12**). Check pH again and adjust if necessary. Incubate approximately 5 mL in a tissue culture incubator for 48 h to check sterility.

3.1.2. Preparation of Serum-Free Gels

The gels are made on the day the cells are to be plated:

1. Working under sterile conditions in a laminar flow cabinet, pipet 1 mL of 10X MEM, 0.5 mL of 7.5% sodium bicarbonate solution, and 200 μL of (1X) MEM into a 20-mL universal tube (*see* **Note 13**).

2. Add 8.5 mL of collagen working solution (2.2 mg/mL) to the above mixture at 4°C, thus making a total of 10.2 mL gelling solution.
3. Mix by pipeting up and down with a 10 mL pipet, taking care to avoid air bubbles. The mixing has to be done thoroughly as well as rapidly.
4. Transfer all the gelling solution into the pipet and plate 2 mL onto each of five 35-mm dishes (*see* **Note 14**).
5. Gently tilt and swirl the dish to spread the collagen evenly.
6. Without moving the dishes further, allow the collagen to set in the laminar flow cabinet. This will take approx 5 min, but it is better not to move them for another 5–10 min. Make sure that the surface of the cabinet is level!
7. Repeat **steps 1–6** to make the total number of gels required.
8. Once the gels are set, transfer to a humidified tissue culture incubator at 37°C with 5% CO_2 and incubate for 30–60 min before plating the cells.

3.1.3. Preparation of Gels Containing 1% Serum

Follow the same **Steps 1–8** of **Subheading 3.1.2.** for making serum-free gels(above), except that in **step 1** pipet 1 mL of 10X MEM , 0.5 mL of 7.5% sodium bicarbonate solution, 100 µL of serum and 100 µL of (1X) MEM (*see* **Note 15**).

3.1.4. Preparation of Gels Containing Additional Factors

Follow the same **steps 1–8** of **Subheading 3.1.2.** for making serum-free or serum-containing gels except that the factors to be added are dissolved in the volume of MEM (100–200 µL) added in **step 1** of **Subheading 3.1.3.**.

3.1.5. Preparation of Gels Containing Cells

In this case the cells are spun down in a universal centrifuge tube and resuspended in 100 µl of MEM. In a separate universal pipet 1 mL of 10X MEM, 0.5 mL of 7.5% sodium bicarbonate solution and 100 µL of MEM (or serum) as for **step 1**. The gelling solution is mixed in this second universal (**step 2**) and immediately mixed with the cells before casting the gels. Further details are given in **Subheading 3.4.**

3.2. Maintenance and Plating of Fibroblasts

Fibroblasts may be maintained routinely in a variety of tissue culture media (*see* **Note 16**). However, the same medium should be used to plate the cells and make the gels. Therefore, if the cells are not normally maintained in MEM, either the medium used to make the gels is modified or the cells are adapted to and maintained in MEM for at least 1 passage before the assay.

Stock cultures of human skin fibroblast are cultured in MEM supplemented with 15% DCS (*see* **Note 4**), 1 m*M* sodium pyruvate, nonessential amino acids (81.4 mg/L) and 2 m*M* glutamine. The addition of penicillin (100 U/mL) and streptomycin (100 µg/mL) is optional (*see* **Note 5**). Glutamine is stored at –20°C in aliquots of convenient volume; these are thawed and added to the volume of medium required for 7–10 d. The complete growth medium is referred to by the % and type of serum (e.g., 15% DCS-MEM). When serum is omitted, this complete medium is referred to as SF-MEM.

3.2.1. Routine Passage of Cells

Stock cultures are grown in plastic (uncoated) tissue culture dishes and split at a 1/3 ratio when confluent (*see* **Note 17**). This is accomplished as follows:

1. Remove the medium from the stock culture to be passaged and wash the attached cells twice with 4–5 mL Hanks'.
2. Add a small volume of trypsin EGTA to cover the dish (e.g., 1.5 mL/90 mm dish). Incubate at 37°C until the cells begin to round and detach from the dish (about 5 min) (*see* **Note 18**).
3. Pipet up and down to detach all cells. Transfer trypsin and floating cells to a universal containing 7.5 mL of 30% DCS-MEM (soya bean trypsin inhibitor may be used if the experiment requires serum-free conditions). Wash the dish twice with 3 mL SF-MEM, transferring the washes to the same universal.
4. Distribute the total volume (e.g., 15 mL of 15% DCS-MEM containing the cells harvested from one 90-mm dish) into three dishes.
5. Cell attachment occurs rapidly, with 90 % of the cells adherent to the dish within 15–30 min. Check microscopically that this is the case. After 1–2 h remove medium with nonattached cells and replace with fresh medium. The medium is normally changed on Mondays, Wednesdays, and Fridays. The volume added per 90-mm dish is increased from 5 mL when the cells are sparse to 10 mL when confluent or over a weekend.

3.2.2. Plating Cells for Experiments

1. Bring the cells to be plated into suspension (**Subheading 3.2.1., steps 1–3**, above).
2. Take a small aliquot (0.1–0.5 mL) to determine cell numbers (e.g., with a Coulter counter).
3. Spin down cell suspension at 100*g* for 5 min.
4. Remove supernatant and resuspend the cells as required for the experiment.

3.3. The Two-Layer Sandwich Assay

For this assay the cells are plated on a 2-mL gel, and as soon as they attach they are overlaid by another 2-mL cell-free gel. An example is given using 1% serum gels (*see* **Note 19**).

1. Cast out 2mL gel containing 1% serum per 30-mm dish, allow to set in cabinet (approx 5–10 min) and then incubate for 30–60 min in tissue culture incubator (*see* **Subheading 3.1.3.** and **Note 20**).

2. Bring the cells to be plated into suspension, spin down at 100*g* for 5 min and resuspend in 1% DCS-MEM at 2.5×10^5 cells/mL (*see* **Note 21**).
3. Plate out 1 mL/gel, shaking the gel horizontally in various orientations to ensure homogeneous distribution of the cells on the gel surface.
4. Transfer gels to tissue culture incubator and monitor the extent of cell adhesion. Normally over 80% of the cells attach within 30 min.
5. After 30 min incubation (or when 80–90% of the cells have attached, but there is minimal or no spreading) the medium with nonattached cells is removed and the cultures are gently washed once with 1% DCS-MEM.
6. The position of the cell layer may be marked at this stage by adding slowly and evenly 100 μL containing an appropriate marker on top of the cells when the medium had been removed (*see* **Note 6**).
7. Tilt the gel and aspirate out the excess medium, leaving the marker on the surface of the gel.
8. Leave for another 2–5 min in the cabinet and repeat **step 7** to ensure the surface of the gel is dry.
9. Cast a 2-mL gel (containing 1% serum) on top of the monolayer and allow to set in the cabinet (5–10 min) (*see* **Note 22**).
10. Return to the tissue culture incubator.
11. Assess migration 4 d later or at different time points, as required (*see* **Note 23**).

3.4. The Three Layer Sandwich Assay

For this assay the cells (5×10^5 cells/culture) are plated within a 1-mL gel, which is underlaid and overlaid by two 2 mL cell-free gels. The cells, therefore, are from the beginning embedded within a 3D matrix. The volumes of the gels may vary. For example, the cells may be plated within a 0.5 mL gel, underlaid and overlaid by two 1.5-mL cell-free gels (*see* **Note 14** about gel volumes). As for the two-layer assay, the conditions may be serum-free or include 1–5% serum (*see* **Notes 16** and **19**). An example is described using 1% serum conditions:

1. Cast 2 mL gels containing 1% serum per 30-mm dish, allow to set in cabinet (approx 5–10 min) and then incubate for 30–60 min in tissue culture incubator (*see* **Subheading 3.1.3.** and **Note 20**).
2. To mark the position of the first boundary between cells and cell-free gel add 0.5 mL of 1% DCS-MEM per gel, remove the medium, and add 100 μL containing an appropriate marker slowly and evenly on top of the gel.
3. Tilt the gel and aspirate out the excess medium, leaving the marker on the surface of the gel.
4. Leave for another 2–5 min in the cabinet and repeat **step 3** to ensure the surface of the gel is dry.
5. Bring the cells to be plated into suspension (*see* **Notes 17, 18**, and **19**), determine the concentration of cells (by Coulter counter or hemocytometer) and calculate the volume of cell suspension that contains the number of cells required to plate

a selected number of cultures (between 5 and 10). (For example, 10 cultures at 5×10^5 cells/culture require 5×10^6 cells).

6. Aliquot into 20-mL universals the calculated volume, using as many universals as required to plate out the whole experiment. (For example, if a total of 40 cultures at 5×10^5 cells/culture are required, aliquot into four universals the volume containing 5×10^6 cells. This way, 10 cultures will be plated out of each universal. (*See* **Note 24**).
7. Spin down the cells at 100*g* for 5 min, remove supernatant, and resuspend the cell pellet in 100 µL of 1% DCS-MEM. Pipet up and down to make a single cell suspension, taking care not to form bubbles.
8. While the cells are spinning, get ready a number of universals (the same number as used for the cells) to prepare the required volume of gelling solution. The volume should match the number of cultures to be plated out of each universal from **step 6**. For example, for 10 cultures (to be plated out of each universal) we need 10 mL of gelling solution (*see* **Note 24**). In this case, aliquot 1 mL 10X MEM, 0.5 mL of sodium bicarbonate, and 100 µL serum per universal.
9. When **step 4** is completed, make gelling solution in one universal at a time. For the example given, add 8.5 mL of collagen working solution (2.2 mg/mL) to each universal from **step 8**. Mix thoroughly and transfer to universal from **step 7**, containing 5×10^6 cells in 100 µL.
10. Mix thoroughly to make a homogeneous cell suspension within the gelling solution and plate out onto 10 gels (ready from **step 4**) at 1 mL/gel. Plate evenly and tilt the gel in various orientations to ensure even distribution.
11. Allow the cell-containing gel to set in cabinet (approx 5–10 min) and then incubate for 15–20 min in tissue culture incubator.
12. Mark the position of the second boundary between cells and cell-free gel as described in **steps 2–4**.
13. Cast a 2 mL gel (cell-free) on top and allow to set in the cabinet (5–10 min).
14. Return to the tissue culture incubator.
15. Assess migration 4 d later or at different time points, as required (*see* **Note 23).**

3.5. Addition of Growth Factors and/or Matrix Molecules

In the assays described above, the cells are provided with an isotropic environment. Different concentrations of soluble factors or matrix molecules to be tested may be added to both top and bottom gels (and cell-containing gel in the three-layer assay; i.e. to test chemokinesis) or only to one of the gels (i.e., to test chemotaxis). In either case, the test substance is added to the gelling solution as described in **Subheading 3.1.4.** (*see* **Note 25**).

3.6. Quantification of Cell Migration

The cultures are fixed by adding 1 mL of 10% formaldehyde. They may then be left at room temperature for 1–2 h or stored at 4°C for 2–3 d. Before quantification, the cultures are washed twice with PBS and then 1 mL of PBS containing sodium azide is added. These cultures may be stored for months at 4°C provided that the PBS is not allowed to dry out (*see* **Note 26**). The morphology of the cells is illustrated in **Fig. 2** (fibroblasts) and **Fig, 3** (endothelial).

Cell migration into the upper and lower gels may be quantified by various means, including image analyses. Here we describe two methods using phase contrast microscopy. We routinely use the first method, which involves counting individual cells. The second method, involving point counting, is used when it is difficult to distinguish individual cells; as sometimes is the case with ECs.

3.6.1. Method 1: Percentage of Migrating Cells

An arbitrary field is defined by a graticule placed in the eyepiece of the microscope. The cultures are viewed using phase contrast optics set at the lowest magnification that allows a clear identification of the cells (normally either 10× or 20× objective and 10× eyepiece). Between 10 and 15 fields are randomly selected per culture, using the same procedure for each culture. On each of the randomly selected fields, we follow the same protocol:

3.6.2. For a Two-Layer Sandwich Assay

1. The majority of the cells plated remain as a layer at the interface between upper and lower gels and these are easily identified, even without a marker. Focus on this cell layer and count every cell included within the graticule.
2. Move the focus slowly up to the top surface of the upper gel and count every cell found within the field.
3. Focus back on the interface cell layer and move the focus slowly down until reaching the bottom of the dish. Count every cell present within the field as before.
4. When all the fields have been counted calculate the mean value of cells per field in: (1) interface, (2) upper gel and (3) lower gel, and (4) the addition of the three values (i.e., total cells per field).
5. Calculate the number of cells in upper and lower gels as percentage of the total (*see* **Note 27**).
6. Repeat the scoring in the replicate cultures and calculate the mean value (±SD) for the conditions and time point tested.

3.6.3. For a Three-Layer Sandwich Assay

1. In this assay, the majority of the cells plated remain within the 3D gel whose upper and lower boundaries had been marked with inert particles. Focus through from upper to lower cell boundaries and count the cells present within this volume and included within the graticule field.
2. Focus back on the particles that mark the upper cell boundary and move the focus slowly up till the top surface of the upper gel, counting every cell found within the field.

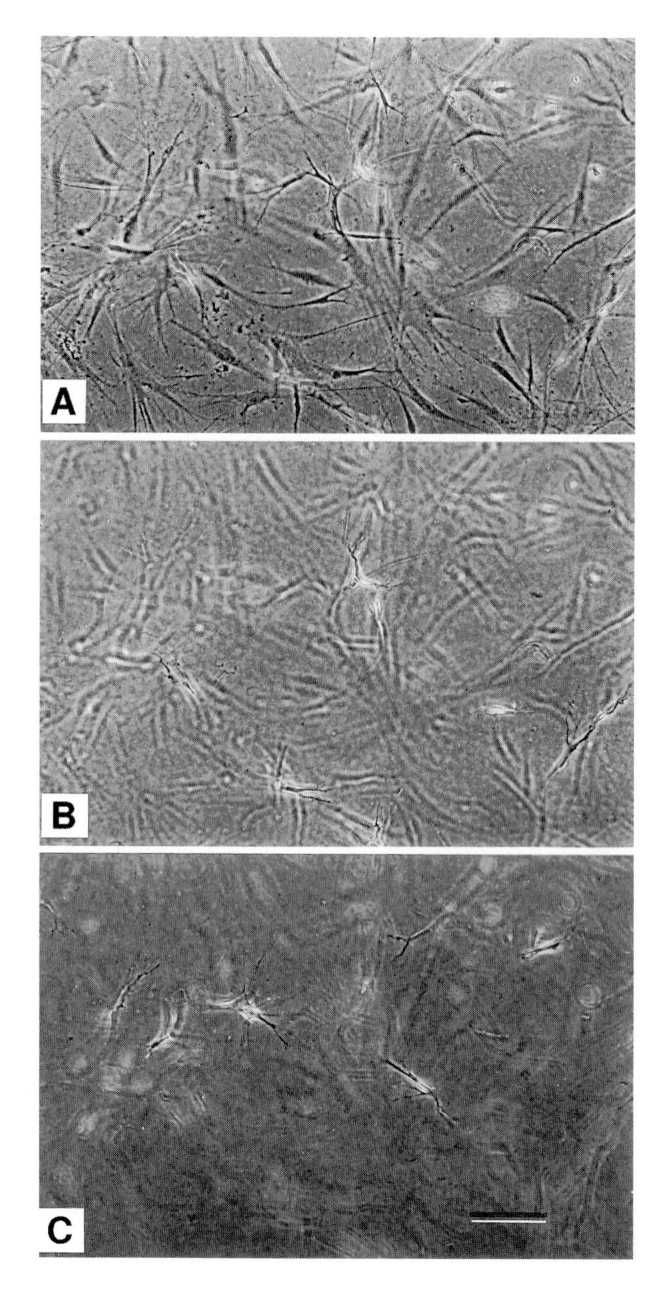

Fig. 2. Morphology of human skin fibroblasts in two-layer sandwich assay. **(A)** Cell layer at the interface between upper and lower gels (original plating location). **(B,C)** Cells that migrated into the upper gel **(B)** and lower gel **(C)** appear in focus, close to the out-of-focus interface layer. Bar, 200 μm.

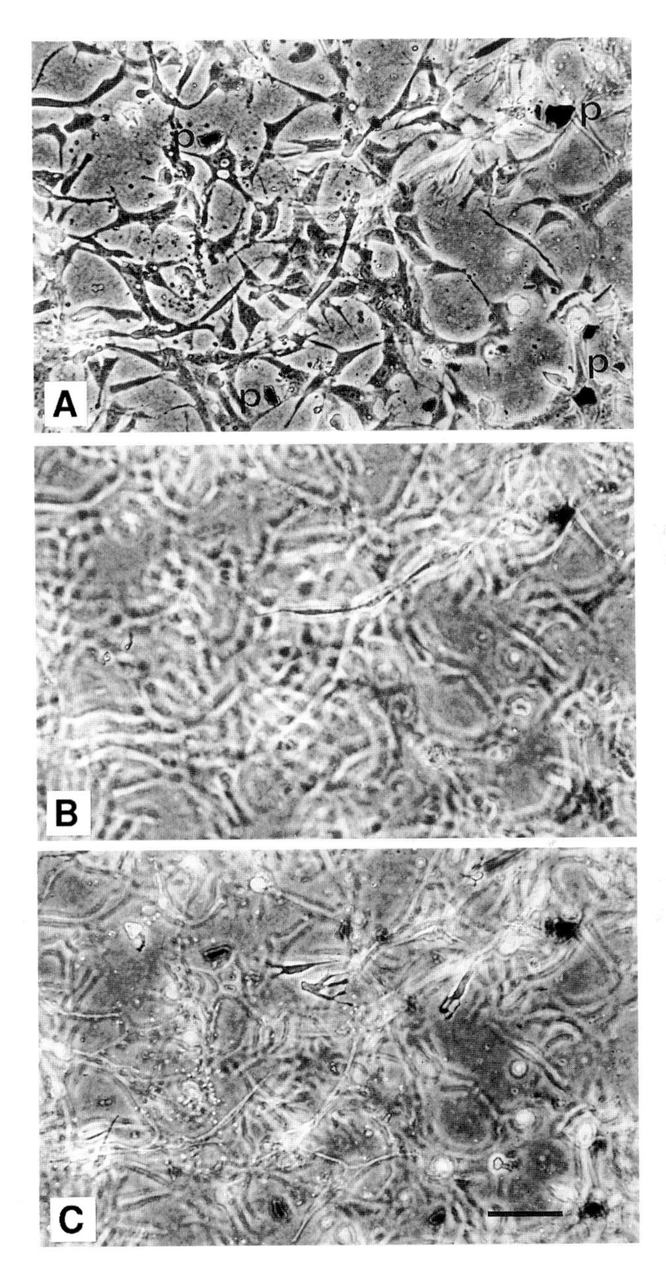

Fig. 3. Morphology of bovine aortic ECs in two-layer sandwich assay. **(A)** Cell layer at the interface between upper and lower gels (original plating location). The position of the interface layer has been marked with ink particles (*p*). **(B,C)** Cells that migrated into the upper gel **(B)** and lower gel **(C)** appear in focus, close to the out-of-focus interface layer. Bar, 100 μm.

3. Count in the same way the cells found between the lower cell boundary and the bottom of the dish.
4. When all the fields have been counted calculate the mean value of cells per field: (1) between cell boundaries, (2) in the upper gel, (3) in the lower gel and (4) the addition of the three values (i.e., total cells per field)
5. Calculate the number of cells in upper and lower gels as percentage of the total (*see* **Note 27).**
6. Repeat the scoring in the replicate cultures and calculate the mean value (±SD) for the conditions and time point tested.

3.6.4. Method 2: Point Counting of Migrating Cells

A graticule that contains a fixed number of points (e.g., a Chalkley grid with 100 points) is placed in the eyepiece of the microscope. Using the same procedure as for Method 1 above, 10–15 fields are randomly selected per culture. In each field, those cells that coincide with the points of the grid and that are in focus, are recorded. This is a classic stereological method that reflects the volume occupied by the cells. The protocol is similar to that followed for Method 1 for either two-layer or three-layer assays:

1. Score the points that coincide with cells in the original plating area.
2. Score the points that coincide with cells in the upper and lower gels. To that effect, the distance between the cell layer (on two-layer assay) or cell boundaries (on three-layer assay) and the limit surface of the gels (top of the upper gel and bottom of the dish) is divided into a fixed number (e.g., four) of equidistant focusing planes. At each focusing plane, those cells that coincide with the points of the grid and are in focus are recorded (note that a single cell may be recorded more than once, as different parts of the cell come in focus and coincide with the grid points). Therefore, for each field selected on a horizontal plane, four fields are selected in a vertical dimension.
3. Taking the mean of all fields examined, results for each culture are expressed as the percentage of points that coincide with a cell in three locations: (1) original plating area (cell layer or between upper and lower cell boundaries), (2) upper gel, and (3) lower gel. The points in upper and lower gels can then be expressed as percentage of points in the original plating area. The mean ±SD of replicate cultures is then calculated.

As for Method 1 (*see* **Note 27**) this method can be simplified by point counting only the cells in the upper and lower gels.

4. Notes

1. Young rats (e.g., 6 mo old) are preferable to old ones, possibly due to the increase in collagen crosslinking that occurs with aging. Variations also occur amongst strains, therefore it is important to use the same type of rats to avoid problems due to variations amongst collagen batches. The tails are frozen for convenience

and to facilitate excision of the tendons. In spite of carefully controlling all stages of the protocol, there are differences amongst batches of collagen. Such differences are more apparent with some cell types than with others ***(14)***. Certain batches appear to have toxic effects on ECs but not on fibroblasts or pericytes. We have observed a similar, cell-specific, toxic effect for some batches of serum and shown it to be caused by the presence of active TGFβ-1 (*see* **Note 4**). The reason for differences amongst batches prepared in the same way is not clear. It may be due to variations in collagen crosslinking or even to the presence of active TGFβ-1. It is therefore important to test in a pilot experiment the collagen batch to be used. For ECs it is better to test several batches before carrying out the assay and retest the chosen batch if it has been stored for more than 3 mo. Type I collagen preparations (bovine dermal and rat tail) suitable to make gels for tissue culture are commercially available (Collagen Bio-Materials, Becton & Dickinson, Sigma). We cannot comment on these, as we make our own collagen preparations.

2. Analar or Aristar grade acetic acid is essential for the preparation of collagen; if inferior grades are used, the collagen may set erratically or precipitate during dialysis.
3. The concentration of the collagen working solution may vary from 2.0–2.6 mg/mL, as long as the final concentration in the gels used for the assay is constant (1.7–2.2 mg/mL). This is the concentration that allows maximal migration of various cell types tested ***(14)***.
4. Either donor or fetal calf sera may be used, but batches should be tested for growth-promoting properties on the ECs, even if they are to be used for the culture of fibroblasts or pericytes. Great variations may be found among batches of sera, irrespective of their origin. We have found that some batches contain active TGFβ-1; these should not be used. However, TGFβ-1 becomes activated by prolonged storage (e.g., 6 mo at –20°C), therefore, batches should be retested when stored for more than 3–4 mo. Serum-free medium (containing growth factors) is commercially available (e.g., TCS Biologicals, Buckingham, UK). If required, it should be possible to use this, instead of serum-supplemented medium.
5. The use of ascorbic acid as an additive to the cell growth medium is also optional, but it should be noted that the composition of the matrix synthesized by the cells will be affected. We routinely add fresh ascorbic acid to endothelial and pericyte medium, but not to that for fibroblasts. When used on the latter, it becomes necessary to use collagenase, as well as trypsin, to bring the cells into suspension.
6. Several types of inert markers are commercially available (e.g., New England Nuclear Life Science Products; non-radioactive microspheres of 10-, 15-, 25-, or 50-μm diameter; Sigma, gel filtration cellulose beads from 30–50- to 250–500-μm diameter; International Market Supplies dry ink, various colours). The preparation of the working suspension may vary accordingly, as long as the aim is achieved. Namely, the particles should be sterile, immobile, easily visualized and nonreactive (with the gel, the cells, or the medium). The use of a marker is

not usually required for the two-layer assay, whereas it is so for the three-layer assay. It is therefore recommended that a marker is first tested using the two-layer assay with and without marker in order to check that it does not alter the results.

7. Cutting the tendons should be done by a second operator. Although this is not essential, it makes an enormous difference in efficiency. With two operators working together, the whole procedure takes about 1 min per tail.
8. Following this step, the supernatant may be mixed with an equal volume of 20% NaCl to precipitate the collagen. The precipitate is recovered by another centrifugation (as for **step 3**), redissolved in 0.3% acetic acid, and dialyzed (**step 4**). We normally omit this NaCl precipitation, since tests show that it did not apparently affect the final end product or results obtained.
9. The collagen tends to thicken during dialysis and occasionally it may aggregate and precipitate in the tubing. This is most likely owing to an elevated pH and can be reversed by bringing the water used for dialysis to pH 4.1 by the addition of acetic acid.
10. The addition of antibiotics is optional. Products to control fungal contamination may also be added, but Fungizone should not be used when working with ECs, because it appears to be toxic. Fungizone is not toxic for fibroblasts.
11. The collagen solution should only be stored in the refridgerator. Freezing and thawing alters the physical properties of the resultant gels. The pH of the collagen solution needs to be kept as close to 4.1 as possible (NB: pH 3.9–4.4 is acceptable). If it is lower than pH 3.9 the collagen will not set, if higher than pH 4.4 it will set too quickly for proper gel formation. *See* **Note 2** regarding time of storage.
12. A collagen standard curve is made using a known concentration of collagen diluted in distilled water acidified to pH4.1 with acetic acid. In the first instance, the concentration of collagen is assessed by freeze drying and weighing known volumes of the stock collagen solution, or using commercially available type I collagen (such as Vitrogen, 2 mg/mL). At least six dilutions are made so that the standard curve should include from 0.1–0.6 mg/mL. The absorbance is read at optical density 230 nm. The concentration of collagen in new batches is determined by diluting aliquots with distilled water (1/5, 1/10 and 1/15 is usually sufficient), measuring the optical density at 230 nm and comparing the results with the standard curve. The values obtained for each dilution that falls within the curve are used to calculate the concentration of collagen (mean ± SD). The stock collagen solution is then diluted to a working dilution of 2.2 mg/mL (*see* **Note 3**).
13. The 200 μL of serum-free MEM may contain soluble factors or matrix macromolecules as part of an experiment.
14. The temperature is very important, as the collagen solution is induced to set by raising the pH and temperature. It is usually convenient to keep the collagen solution on ice. The total volume of gelling solution (10.2 mL) may be adjusted to smaller or larger volumes if required, using the same ratios of the components listed on **steps 1** and **2**. However, we do not recommend adjustment of volumes larger than 12 mL, because it is crucial to cast the gelling solution as rapidly as

possible. In practice, we mix 10.2 mL of gelling solution in order to cast 5 × 2.0 mL gels. Alternatively one could make 10 mL and cast 1.9 mL per dish. The volume per dish may be reduced, for example 1 mL per 30-mm dish. However thin gels have a more pronounced meniscus and this produces problems regarding heterogeneous distribution of cells and their visualization. The volume can be adapted for use on smaller dishes (e.g., 24-well trays). It is always advisable to make more gels than required for the experiment, as some gels may have to be discarded (e.g., those containing air bubbles).

15. According to this protocol the final serum concentration will be 0.98%. The volume of serum may be increased if wishing to make gels with higher concentrations (e.g., add 500 µL for gels containing 4.71% serum). Similarly, the volume of (1X) MEM may be increased or decreased if required. To compare experiments done with different serum concentrations (or volume of MEM) it may become necessary to adjust the concentration of collagen working solution used in **step 2** so that the concentration of collagen in the gelling solution is the same. For example, following the protocols given, the collagen concentration in the working solution is 2.2 mg/mL and in the gelling solution 1.83 mg/mL for serum-free gels. To maintain the same concentration for 4.7% serum gels, the concentration of collagen working solution had to be 2.29 mg/mL. Such adjustments are not usually necessary.
16. Although the assay will be described using fibroblasts as the target cell population, it can also be used (with minor modifications) for ECs, pericytes, and other cell types. Details about the maintenance of endothelial cells and pericytes may be found in **refs.** ***24*** and ***25***, respectively.
17. For optimal experimental reproducibility, the stock cultures used to set up experiments should be of similar density. ECs and fibroblasts are routinely farmed when just confluent, whilst pericytes are farmed before reaching confluence.
18. If ascorbic acid is used, it may be necessary to incubate the cultures with a solution of collagenase (0.5 mg/mL in MEM) for 15–20 min prior to trypsinization.
19. The same protocol is used for serum-free or higher serum concentration conditions. In all cases the cells are plated in the same serum concentration as present in the gels. If necessary (e.g., to prevent cell aggregation) MEM may be supplemented with BSA (1–5 mg/mL) to resuspend the cells under serum-free conditions.
20. It is advisable to make a 10–20% excess cultures to allow for possible loses and to have extra controls.
21. In order to obtain a single cell suspension, the cell pellet is first resuspended in 0.5–1.0 mL of medium and pipetted up and down with a fine bore pipette. More medium is next added to bring the volume to approximately half of the final volume required. This intermediate cell suspension is mixed well and a 0.5 mL aliquot taken to count cell number and estimate viability. Based on these results the intermediate cell suspension is diluted to the final concentration to be plated. When the volume of the final cell suspension is more than 25–30 mL, we aliquot it into plastic universals (15–20 mL per universal). Each is kept closed and mixed well before plating the cells. The plating density recommended (2.5×10^5 cells/

gel) results in a high-density cell layer, particularly for fibroblasts and pericytes. Other densities may be used and should, indeed, be tested when assessing the effects of an effector molecule, as such effects may be density dependent. ECs are smaller than fibroblasts and pericytes and may be plated at higher densities if required (e.g., $5–10 \times 10^5$ cells/gel).

22. It is recommended that the second gel be cast in small batches (e.g., three gels at a time), because it must to be done with great care, with the collagen gelling solution dropped evenly on top of the cells from as close a distance as possible and relatively fast, without the cells being touched. It is convenient to practice on gels containing only the marker and no cells, checking that the marker particles are not disturbed by the addition of the gel. It is also important that the cells are allowed to attach to the first gel for the same length of time before casting the second gel; therefore this operation should be done by more than one operator if the number of cultures is large.
23. The cells are embedded in a 4 mL gel (in the two-layer assay) or in a 5 mL gel (in the three-layer assay) and no additional medium is added to the cultures. Pilot experiments should be carried out to determine the viability of the cells under the chosen conditions. We have found that fibroblasts remain healthy for at least 4 days under serum free conditions, whereas ECs require serum or growth factors present if kept for that length of time.
24. Since it is important to mix the cells with the gelling solution rapidly and to avoid making bubbles, it is easier to prepare an excess both of cells (**step 5**) and of gelling solution (**step 8**). For example: aliquot into each universal enough cells to plate onto 11 gels, prepare 12 mL of gelling solution, take 11 mL of this to mix with the cells, and aliquot 10 mL onto 10 gels (1 mL/gel).
25. It is also possible to add soluble factors in a slow-release pellet placed at a chosen distance from the cells, rather than homogeneously mixed with the cell-free gels. An example, using methyl-cellulose pellets is given below. For this variation of the assay we need:
    (a) Methyl cellulose 4000 centipoise (Sigma). 0.5% solution in distilled water (a touch of phenol red may be added). Autoclave in a bottle containing a magnetic stirrer, then stir in the cold overnight, or until dissolved.
    (b) Short Teflon rods (2.2 mm diameter) and rack

    Protocol:
    (a) Mix an equal volume of methyl cellulose with a solution containing the test material and aliquot 10–15 µL of the mixed solution onto the tops of Teflon rods. Allow to dry in a desiccator overnight.
    (b) Using very fine forceps (e.g., watchmaker's straight superfine long points) and with the help of a dentist needle if required, remove dry pellets, placing each one on a tissue culture dish.
    (c) Add 30 µL of collagen gelling solution onto each pellet to anchor it.
    (d) As soon as it sets (5–10 min) mark the area with a circle on the base of the dish and proceed to cast 2 mL of collagen gelling solution carefully on top (as for the first step in the two-layer and three-layer assay)

(e) Proceed with the rest or the assay.
(f) When the top cell-free gel is cast, place another dry pellet on top, opposite to the first one, and anchor it as described above in **steps b** and **c**.

26. It is feasible to quantify live cultures. In this case the cultures have to be taken out of the incubator one at a time, and the time recorded to ensure that: (a) every culture is out of the incubator for the same amount of time, and (b) this time does not exceed a certain limit (e.g., 15 min in every 2 h).
27. The method may be simplified by counting only the number of cells in upper and lower gels. Results are therefore expressed as the mean (± SD) number of cells per field on these gels, rather that percentages. It is also possible to determine the total number of cells per culture as follows:
(a) Wash the gels twice with Hanks'.
(b) Add 1 mL collagenase (0.5 mg/mL in MEM) per gel. The collagenase may be added to the culture dish or the gel may be transferred first to a centrifuge tube by detaching the gel from the sides of the dish and "pouring" it into the tube with the help of a spatula.
(c) When the gel has been totally lyzed (2–3 h) spin the cells down (300*g* for 10 min).
(d) Resuspend the cells in 1 mL EGTA-trypsin, pipet up and down, and incubate for 10–15 min at 37°C. If the gel was lyzed in the original culture dish, add also 1mL EGTA-trypsin to the dish (as some cells may have attached to it), incubate, and pool with the cells in the centrifuge tube.
(e) Pipet up and down to obtain a single cell suspension; neutralize the trypsin (e.g., add 2 mL 15% DCS-MEM), dilute as required, measure the total volume accurately, and take an aliquot to count the number of cells present. If inert particles had been used, these can be allowed to sediment; it may be necessary, however, to use a hemocytometer rather than an electronic particle counter.

Acknowledgement

This work was funded by the Biotechnology and Biological Sciences Research Council

References

1. Hunt, T. K. (1980) *Wound Healing and Infection: Theory and Surgical Practice.* Appleton-Century-Crofts, New York.
2. Schor, A. M. and Schor, S. L. (1983) Tumour angiogenesis: a review. *J. Exp. Path.* **141,** 385–413.
3. Schor, A. M., Schor, S. L., and Arciniegas, E. (1997) Phenotypic diversity and lineage relationships in vascular endothelial cells. In: *Stem Cells* (Potten, C. S., ed.). Academic Press Limited, London, pp. 119–146.
4. Diaz-Flores, L., Gutierrez, R., and Varela, H. (1994) Angiogenesis: an update. *Histol. Histopathol.* **9,** 807–834.

5. Clark, R. A. F. (1996) *The Molecular and Cellular Biology of Wound Repair*, 2nd ed. Plenum Press, New York.
6. Bouissou, H., Pieraggi, M., Julian, M., Uhart, D., and Kokolo, J. (1988) Fibroblasts in dermal tissue repair: electron microscope and immunohistochemical study. *Int. J Dermatol.* **27,** 564–570.
7. Ishibashi, T., Miller, H., Orr, G., Sorgente, N., and Ryan, S. J. (1987) Morphologic observation on experimental subretinal neovascularization in the monkey. *Invest. Ophthalmol. Vis. Sci.* **28,** 1116–1130.
8. Schor, S. L. (1994) Cytokine control of cell motility: modulation and mediation by the extracellular matrix. *Prog. Growth Factor Res.* **5,** 223–248.
9. Boudreau, N. and Bissel, M. J. (1996) Regulation of gene expression by the extracellular matrix. In: *Extracellular Matrix, vol. 2, Molecular Components and Interactions* (Comper, W. D., ed.) Overseas Publishers Association, Amsterdam.
10. Greiling, D. and Clark, R. A. (1997) Fibronectin provides a conduit for fibroblast transmigration from collagenous stroma into fibrin clot provisional matrix. *J. Cell. Sci.* **110,** 861–870.
11. Zigmond, S. and Hirsch, J. G. (1973) Leukocyte locomotion and chemotaxis. *J. Exp. Med.* **137,** 387–410.
12. Schor, S. L. (1980) Cell proliferation and migration on collagen substrata in vitro. *J. Cell. Sci.* **41,** 159–175.
13. Schor, S. L., Allen, T. D., and Harrison, C. J. (1980) Cell migration through three-dimensional gels of native collagen fibers: collagenolytic activity is not required for the migration of two permanent cell lines. *J. Cell. Sci.* **46,** 171–186.
14. Schor, S. L., Schor, A. M., Winn, B., and Rushton, G. (1982) The use of 3D collagen gels for the study of tumour cell invasion in vitro: experimental parameters influencing cell migration into the gel matrix. *Int. J. Cancer* **29,** 57–62.
15. Schor, S. L., Ellis, I., Dolman, C., Banyard, J., Humphries, M. J., Mosher, D. F., et al. (1996) Substratum dependent stimulation of fibroblast migration by the gelatin-binding domain of fibronectin. *J. Cell. Sci.* **109,** 2581–2590.
16. Ellis, I. and Schor, S. L. (1995) The interdependent modulation of hyaluronan synthesis by TGFβ-1 and extracellular matrix: Consequences for the control of cell migration *Growth Factors* **12,** 211–222.
17. Schor, A.M., Schor, S.L., Weiss, J.B., Brown, R.A., Kumar, S., and Phillips, P. (1980) A requirement for native collagen in the mitogenic effect of a low molecular weight angiogenic factor on endothelial cells in culture. *Brit. J. Cancer* **41,** 790–799.
18. Sutton, A. M., Canfield, A. E., Schor, S. L., Grant, M. E., and Schor, A. M. (1991) The response of endothelial cells to TGF-β1 is dependent upon cell shape, proliferative state and the nature of the substratum. *J. Cell. Sci.* **99,** 777–787.
19. Canfield, A. E. and Schor, A. M. (1995) Evidence that tenascin and thrombospondin-1 modulate sprouting of endothelial cells. *J. Cell. Sci.* **108,** 797–809.
20. Madri, J. A., Kocher, O., Merwin, J. R., Bell, L., and Yannariello-Brown, J. (1989) The interactions of vascular cells with solid phase (matrix) and soluble factors. *J. Cardiovasc. Pharmacol.* **14 (suppl 6),** S70–S75.

21. Schor, A. M. and Schor, S. L. (1986) The isolation and culture of endothelial cells and pericytes from the bovine retinal microvasculature: a comparative study with large vessel vascular cells. *Microvasc. Res.* **32,** 21–38.
22. Schor, A. M., Schor, S. L., and Allen, T. D. (1983) The effects of culture conditions on the proliferation and morphology of bovine aortic endothelial cells in vitro: reversible expression of the sprouting cell phenotype. *J. Cell. Sci.* **62,** 267–285.
23. Canfield, A. E., Allen, T. D., Grant, M. E., Schor, S. L., and Schor, A. M. (1990) Modulation of extracellular matrix biosynthesis by bovine retinal pericytes in vitro: effects of the substratum and cell density. *J. Cell. Sci.* **96,** 159–169.
24. Schor, A. M., Ellis, I., and Schor, S. L. (2001) Collagen gel assay for angiogenesis. Induction of endothelial cell sprouting. In: *Methods in Molecular Medicine: Angiogenesis Protocols* (Murray, J. C., ed.), Humana Press Inc., Totowa, NJ, pp. 145–162.
25. Schor, A. M. and Canfield, A. E (1998) Osteogenic potential of vascular pericytes in Marrow stromal cell culture. In: *Practical Animal Cell Biology Series* (Beresford, J. N. and Owen, M. E., eds.). Cambridge University Press, Cambridge, UK, pp. 128–148.

13

Rat Aortic Ring

3D Model of Angiogenesis In Vitro

Mike F. Burbridge and David C. West

1. Introduction

Angiogenesis is a necessary component of normal tissue repair, tumor growth and dissemination, and a wide variety of other inflammatory and pathological processes as well, including diabetic retinopathy, rheumatoid arthritis, and psoriasis. Consequently, the last two decades have seen extensive research into the regulation of neovascularization, particularly in tumors. Partly due to the emphasis on tumor angiogenesis, much of this effort has been devoted to the detection and characterization of angiogenic growth factors and inhibitory molecules. Thus assays have commonly been developed to measure the amount of vascular growth in vivo, or the modulation of endothelial cell (EC) proliferation, or migration in simple 2D culture systems *(1)*.

In the study of the angiogenic process and its modification, it is important to be able to develop models of in vivo angiogenesis representative of the clinical situation. Animal models are extremely complex and difficult to interpret for routine screening, and relatively expensive. Thus, in recent years a number of in vitro models have been developed. Recently, several methods for studying angiogenesis have been introduced in an attempt to quantify angiogenesis in 3D matrices in vitro *(2–5)*. Madri et al *(6)* examined the effect of a 3D collagen matrix on the phenotype of cultured capillary ECs. Their studies indicate that, compared with 2D culture systems, the cells retain endothelial properties to a greater extent and respond differently to cytokines. Other phenotypic differences have also been reported *(7)*. Both Montesano et al. *(8)* and Schor et al. *(5)* have employed collagen and fibrin gels to examine the invasive properties

From: *Methods in Molecular Medicine, Vol. 46: Angiogenesis Protocols*
Edited by: J. C. Murray

of ECs. In this system the cells invade as 'patent' vessels when stimulated by angiogenic factors, the extent of invasion being apparently related to a shift in the balance between proteases and their inhibitors.

Angiogenesis is a complex process, involving a sequence of steps in response to various stimuli, including the mobilization of the ECs from existing vessels, their migration through, and breakdown of, the surrounding extracellular matrix, and finally anastomosis with other neovessels to form functional capillaries. It is thus important that an in vitro model reflects these different stages, in order to more fully understand the process and its modification. Most in vitro models are carried out in 2D, and while useful in the study of any one of these stages of microvessel growth, they are unable to fully describe the process as a whole.

Perhaps the most interesting in vitro model is that developed by Nicosia and Ottinetti ***(4)***. In this model, rat aortic rings are embedded in a plasma clot, collagen, or fibrin gel and cultured in an optimized, defined serum-free growth medium for microvascular ECs ***(9)***. A complex network of branching and anastomosing microvessels develops from the ECs of the aortic intima ***(10)***, interspersed with fibroblasts as individual cells, thus providing a closer approximation to the angiogenic process as observed in vivo (**Fig. 1**). This process occurs in response to endogenous growth factors released upon resection of the aorta, and thus no further growth factor supplements are required in the culture medium, i.e., we have a controlled microenvironment. Potential angiogenic stimulators and inhibitors may be applied throughout the entire length of culture, or over defined periods in order to observe their impact at various stages of the angiogenic response.

No other in vitro system exhibits all the processes (except, of course, blood flow) observed in vivo. Recently this model has been employed to examine the effect of matrix constituents, cytokines, RGD peptides, antiangiogenic compounds, oxidized low-density-lipoproteins (ox-LDL), and neutralizing antibodies on vessel growth, and the model has been further developed into a 'quantitative' serum-free matrix system ***(4,11–17)***. Using the quantitative serum-free in vitro angiogenesis model, we have examined the effect of pH on vessel growth ***(18)***, elucidated the importance of vascular endothelial growth factor in the metastatic propensity of prostatic cancer cell lines ***(19)***, and employed it as a general screening method for antiangiogenic compounds. We are currently investigating the role of the matrix metalloproteinases in this model. Recent variations on the model have employed rabbit aortae ***(17)***, human placental vessels ***(20)***, and mouse embryonic kidney ***(21)***.

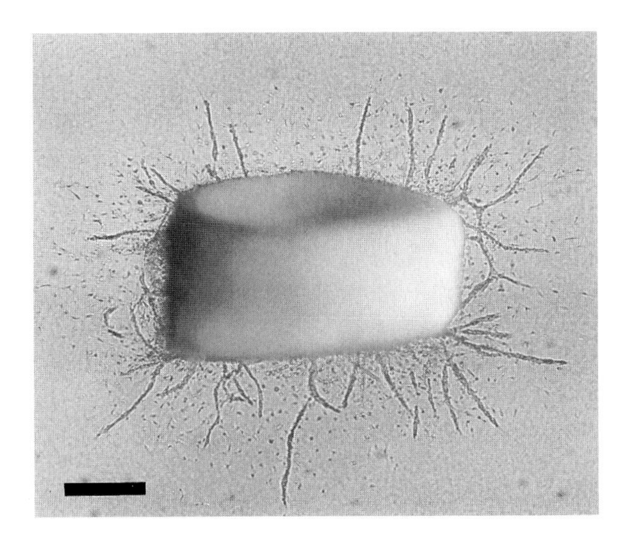

Fig. 1. A 6-d-old culture of a rat aortic ring embedded in collagen and cultured in an optimized serum-free growth medium for microvascular endothelial cells (MCDB131). ECs are seen arranged as microvessels and fibroblasts as individual cells. Bar, 500 μm.

2. Materials

2.1. Preparation of Agarose Culture Wells

1. Agarose (type VII, cell culture tested, Sigma; *see* **Note 1**).
2. Petri dishes, 100-mm diameter cell culture treated (Costar).
3. Cutters, or cork-borers, 10- and 17-mm diameter.
4. Bacteriological Petri dishes, 100-mm diameter (Falcon).

2.2. Type 1 Collagen Gel Preparation

1. Rat-tail collagen: 2 mg/mL in 0.5 *M* acetic acid (Collagen R, Serva; *see* **Note 2**).
2. 10***x*** Eagle's Minimum Essential Medium (10X MEM; Gibco), containing phenol red pH indicator.
3. 1.4% w/v Sodium bicarbonate solution (diluted from 7.5% solution, Sigma).
4. Sterile aqueous 1 *M* NaOH.

2.3. Fibrin Gel Preparation

1. Sterile bovine fibrinogen (culture-tested, Sigma) (*see* **Note 3**).
2. Sterile bovine thrombin (Sigma) (*see* **Note 3**).
3. Eagle's Minimum Essential Medium (MEM, Gibco).

2.4. Preparation of rat aortic segments

1. Male Fischer-344 rats, 8–12 wk of age, sacrificed by intraperitoneal phenobarbital or by CO_2 inhalation (*see* **Note 4**).
2. Microdissection (iridectomy) scissors and forceps.
3. Syringe (1 mL) fitted with a 23-gauge needle.
4. Ice-cold serum-free culture medium such as MEM (Gibco) or a balanced salt solution such as Dulbecco's Phosphate Buffered Saline (PBS).
5. Dissecting microscope (preferably in a laminar flow cabinet).
6. Scalpel blade.
7. Petri dish (cell culture-treated).

2.5. Culture

1. Serum-free culture medium, MCDB131 (Gibco), supplemented with glutamine (2 m*M*), penicillin (100 U/mL), streptomycin (100 µg/mL), and 25 m*M* $NaHCO_3$ (see **Notes 5** and **6**).
2. Humidified incubator with a 5% CO_2 / air atmosphere, at 37°C.

2.6. Quantification

1. Inverted microscope, such as Olympus IMT-2, with non-phase-contrast optics (*see* **Note 7**).
2. Camera and monitor and/or image analyzer (*see* **Notes 8** and **9**).

2.7. Histology

1. Lab-Tek Permanox 8-well Chamber Slides (Nalge Nunc International) (*see* **Note 10**).
2. Cyanoacrylate gel (Loctite Super Glue).
3. Fluorescein-labelled acetylated-low-density-lipoprotein (DiI-Ac-LDL, Biomedical Technologies Inc.).
4. Polyclonal sheep antibody to rat von Willebrand factor (Cedarlane) (*see* **Note 11**).
5. Fluorescence microscope, such as Olympus IMT-2.

3. Methods

3.1. Preparation of Agarose Culture Wells

1. Prepare a 1.5 % w/v sterile aqueous solution of agarose, then autoclave to dissolve and sterilize.
2. Pour 30 mL each into 100-mm cell-culture—treated Petri dishes and allow to gel at room temperature in sterile conditions 30 min (*see* **Notes 12** and **13**).
3. Punch concentric circles in the agarose (up to 15 per dish) using cork borers of 10- and 17-mm diameter.
4. Remove central portions and transfer the agarose rings to 100-mm diameter bacteriological Petri dishes with a bent spatula (*see* **Note 14**).

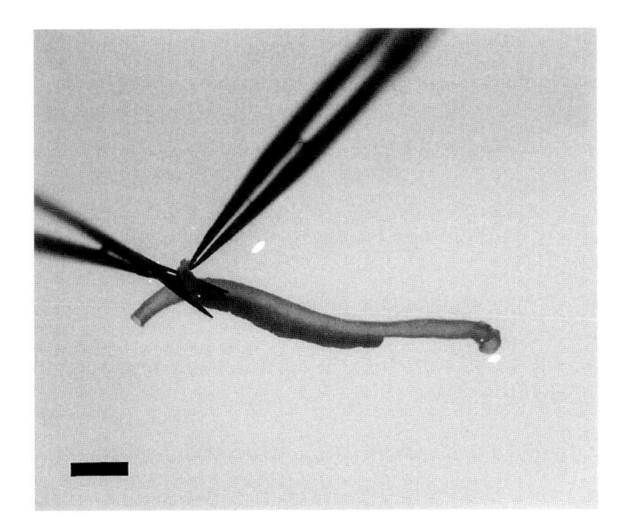

Fig. 2. The rat thoracic aorta. The fibro-adipose tissue and colateral vessels are removed with fine microdissection scissors in ice-cold culture medium. Bar, 5 mm.

3.2. Preparation of rat aortic segments

1. Remove thoracic aortas rapidly and, using fine microdissection forceps, place immediately in ice-cold culture medium or balanced salt solution (*see* **Note 15**).
2. Flush aorta gently with ice-cold culture medium (from abdominal end) using a 1 mL-syringe fitted with a 23-gauge needle until the medium runs clear and aorta is free of clotted blood (1–2 mL should suffice) (*see* **Note 16**).
3. Under a dissecting microscope, carefully remove fibroadipose tissue and co-lateral vessels with fine microdissection scissors (iridectomy scissors), **Fig. 2**. (*see* **Note 17**).
4. Using a scalpel blade cut rings 1 mm in length. Approx 30 may be obtained from each aorta (*see* **Note 18**).
5. Transfer rings to 10 mL of fresh ice-cold medium in a sterile culture tube (*see* **Note 19**).
6. Allow rings to settle to bottom of tube, pour off medium, and replace with fresh. Repeat five times to ensure adequate rinsing of rings. Swirl tube to resuspend rings, and pour medium and rings into a petri dish on ice.

3.3. Preparation of Type 1 Collagen Gels and Embedding of Aortic Rings

1. Working under sterile conditions, mix, on ice, 7.5 vol of 2 mg/mL collagen solution with 1 vol of 10X MEM (containing phenol red pH indicator), 1.5 vol of 1.4% w/v $NaHCO_3$, and a predetermined vol of 1 *M* NaOH (approx 0.1 vol) to adjust the pH to 7.4.

2. Mix gently using a magnetic stirrer, avoiding the creation of air bubbles and allowing time for the pH to equilibrate (*see* **Notes 20** and **21**).
3. Add 200 µL collagen solution to coat the bottom of each agarose well and allow to gel in an humidified incubator to avoid dehydration.
4. With sterile microdissection forceps, touch each ring gently onto a clean Petri dish (culture treated) to remove excess medium, and place in culture wells (4 per 100-mm dish) such that lumen is orientated horizontally. Do not allow rings to dry out during this step.
5. Prepare fresh collagen as described above, and gently fill each well with 200 µl to cover each aortic ring.
6. Add 30 mL of MCDB131 medium to each Petri dish and incubate at 37°C in a humidified incubator in 5% CO_2 / air atmosphere.

3.4. Preparation of Fibrin Gels and Embedding of Aortic Rings

1. Prepare fibrinogen solutions at 1.5 mg/mL by dissolving lyophilized cell culture-tested bovine fibrinogen in MEM (*see* **Note 22**).
2. Filter the solution through a 0.4 µm sterile filter to remove remaining clumps of fibrinogen molecules that would interfere with the uniform polymerization of the gel.
3. Rapidly vortex 1 vol of this fibrinogen solution with 0.02 vol of a 50 U/mL sterile solution of thrombin, and use within 10 sec (*see* **Note 23**).
4. Add 200 µL collagen solution to coat the bottom of each agarose well and allow to gel in a humidified incubator to avoid dehydration.
5. With sterile microdissection forceps, touch each ring gently onto a clean Petri dish (culture treated) to remove excess medium, and place in culture wells (4 per 100-mm dish) such that lumen is orientated horizontally. Do not allow rings to dry out during this step.
6. Prepare fresh fibrin as described above, and gently fill each well with 200 µL to cover each aortic ring.
7. Add 30 mL of MCDB131 medium, containing 300 µg/mL ε-amino-n-caproic acid (**Note 5a**), to each Petri dish and incubate at 37°C in a humidified incubator in 5% CO_2 / air.

3.5. Culture and Response to Test Materials

In the standard procedure, culture wells of 400 µL are formed by agarose rings placed in Petri dishes (4 per dish). These are filled with gels of extracellular matrix components such as fibrin or collagen, in which are embedded short segments of rat aorta (**Fig. 3**). Each 100-mm Petri dish receives 30 mL medium and is incubated at 37°C, in a humidified 5% CO_2/air atmosphere. Cultures may be maintained for a period of up to 2 wk before excessive breakdown of the collagen or fibrin matrix occurs. It is not necessary to replace the medium during culture as the large volume of medium, relative to the small tissue fragments, is sufficient to avoid depletion of medium components throughout the culture period. Indeed, if medium is replaced, regression of

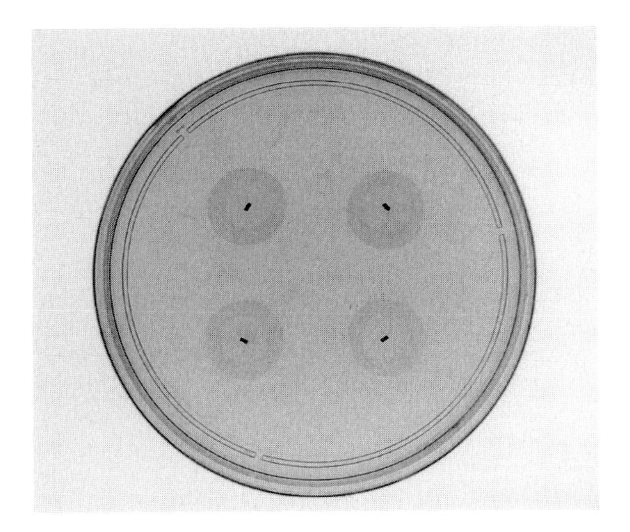

Fig. 3. Culture wells are formed by agarose rings placed in a Petri dish. Each well contains a 1 mm segment of rat aorta embedded in a collagen or fibrin gel.

microvessels (*see* **Subheading 3.5.1.**) is accelerated, probably owing to removal of endogenous growth factors. In experiments where medium replenishment is necessary to remove substances applied only for short periods of culture, the medium of control dishes should also be replaced at the same time. Cultures may be observed over a period of time, or at a defined time point depending on the information required.

3.5.1. The angiogenic response

A similar spontaneous angiogenic response is observed with collagen and fibrin cultures. Individual fibroblasts are seen to grow out of the aortic explants within 2 d of culture, forming an expanding carpet of individual cells (**Fig. 4**). By d 3, ECs begin to migrate into the matrix as microvessels, clearly visible by d 4. Growth continues throughout the first week of culture, initially in terms of an increase in number and length of microvessels, then as a lengthening and thickening of existing vessels. In the second week of culture, regression of the microvessels is observed, probably owing to a depletion of growth factors and other soluble components, and the reduction in integrity of the surrounding matrix.

3.5.2. Intra- and Inter-Aorta Variation

The spontaneous angiogenic response of each aortic ring may vary twofold. Mean values of all rings from two rats should be found to be very similar, even

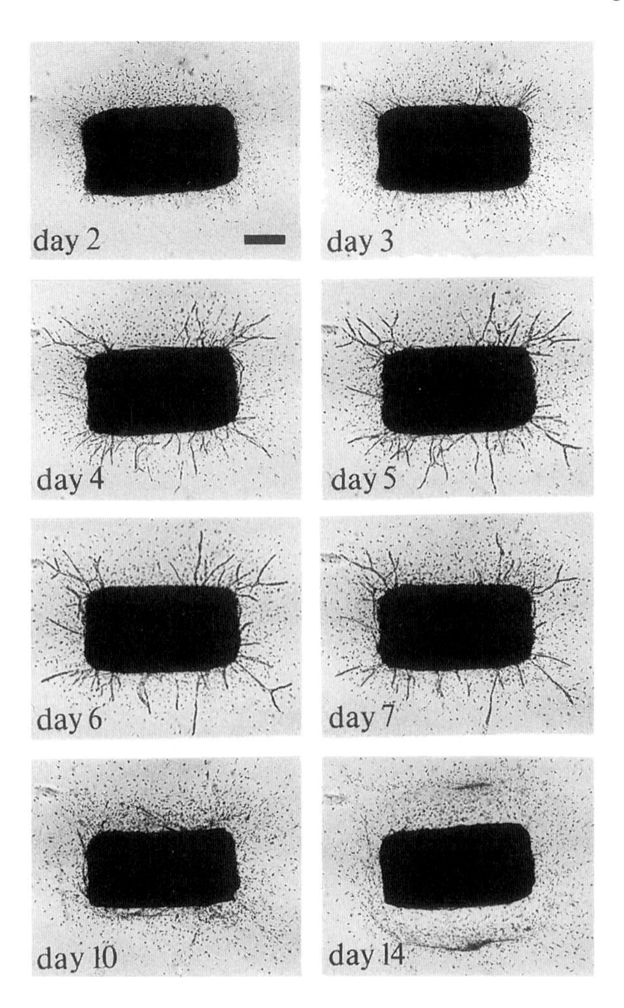

Fig. 4. The spontaneous angiogenic response of a rat aortic ring embedded in collagen. Culture was maintained for 2 wk without medium change. See text for details. Bar, 500 μm.

if rats of different ages (6 wk–6 mo) are compared. It should be borne in mind, however, that response to stimulators and inhibitors of aortic rings from rats of different ages will not necessarily be the same.

3.5.3. Interpretation of the Angiogenic Response

In view of the complexity of the vascular network observed in such a system, it is important to be able to fully quantitate the many parameters of microvessel outgrowth if a comprehensive description of the process and its modification are required. Criteria should be defined to enable valid interpre-

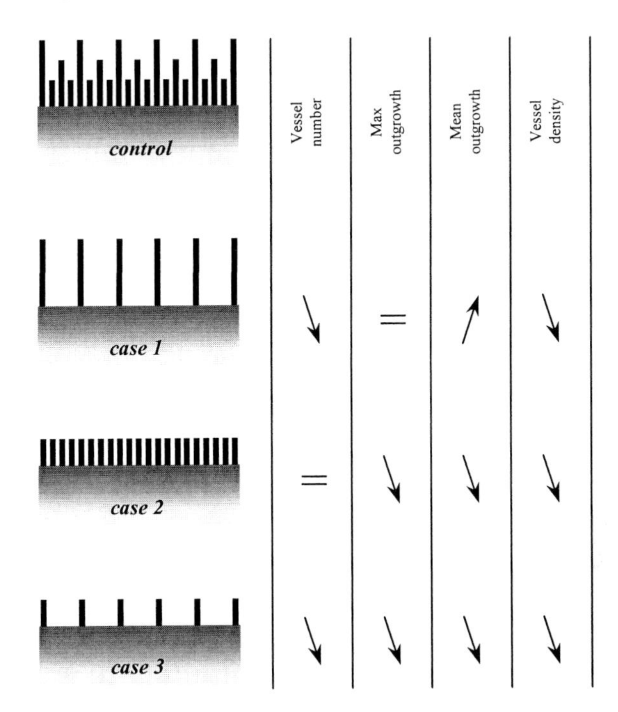

Fig. 5. Three possible cases of angiogenesis inhibition. *Top*: graphical representation of an aortic ring and its endothelial outgrowth, and below: three possible cases of inhibition (*see* **Subheading 3.5.3.**). The variations in four parameters of microvessel outgrowth for each case are shown on the right.

tation of the angiogenic response and its modification by exogenous stimulators and inhibitors.

The most obvious quantification would simply be a count of the number of microvessels. However, this parameter may not necessarily be sufficient to detect a variation in angiogenic response. In **Fig. 5**, the aortic ring is represented along with its vascular outgrowth and three possible cases of inhibition. In case 1, there is an obvious reduction in number of vessels. However, in case 2, whilst inhibition is evident, a vessel count would lead to similar results as in the control. In case 3, an accumulation of cases 1 and 2, vessel number is again reduced. Another parameter would be a measure of distance of outgrowth of the microvascular network - either that of maximum outgrowth, i.e. the distance migrated by the most distal vessel tip, or of mean outgrowth, taking into account all vessel tips. Whilst in cases 2 and 3 both maximal and mean outgrowth obviously decrease, in case 1, maximal outgrowth is unchanged, and mean outgrowth, due to the absence of intermediary length vessels, actually increases. However, only a measure of vascular density (*see* **Subheading 3.6.1.**) would give an indication of inhibition in all three cases. In other words, whilst in case 3, all four parameters would indicate inhibition, in case 2 only outgrowth (maximal or mean) and density are reduced, vessel number being unchanged, and in case 1 only two parameters would indicate a reduction in the angiogenic response. The measurement of the degree of angiogenic stimulation is simply an inverse situation and a measure of vascular density would again be most reliable. Other parameters, such as branching, individual vessel length, or total vessel length could also be measured, and it is important to bear in mind all these parameters and their possible modification when evaluating the angiogenic response. Microvascular growth curves showing the evolution of the spontaneous angiogenic response in terms of different parameters are presented in **Fig. 6**.

Angiogenesis in this model closely approximates that observed in wound healing, where a rapid growth rate of vessels and migration of fibroblasts is followed by a slowing down in growth rate as microvessels mature and the wound is healed. However, if we consider vessel growth in the case of pathological situations such as tumor growth and diabetic retinopathy, a notion of growth rate rather than of maximal density would be more applicable. It is thus important to consider whether a measure of any of these parameters at any one time point, or rather a comparison of kinetic profiles would be more relevant to the subject of investigation. Microvascular growth curves for stimulation by vascular endothelial growth factor are shown in **Fig. 7**. It is apparent that percentage stimulation calculated for this growth factor would depend both on the parameter and on the day chosen for measurement.

3.6. Quantification—Image Analysis

Examine cultures under an inverted microscope with non-phase-contrast optics. *See* **Note 7**. Transfer images to a monitor screen for image analysis.

3.6.1. Semiautomated Image Analysis

If a detailed analysis of microvascular outgrowth is required, an accurate reproduction of the microvessel network is acquired manually before analysis is performed. *See* **Note 24**.

1. Using suitable software, trace the outline of each vessel on the monitor screen.
2. Transfer the coordinates of these traced vessels to graphic software (such as Excel, Microsoft) to reproduce the pattern of microvessel outgrowth in the form of a silhouette (**Fig. 8A,B**).
3. A number of parameters may be automatically measured, including vessel number, mean vessel length, maximum and mean distance of outgrowth, and number of branchings.

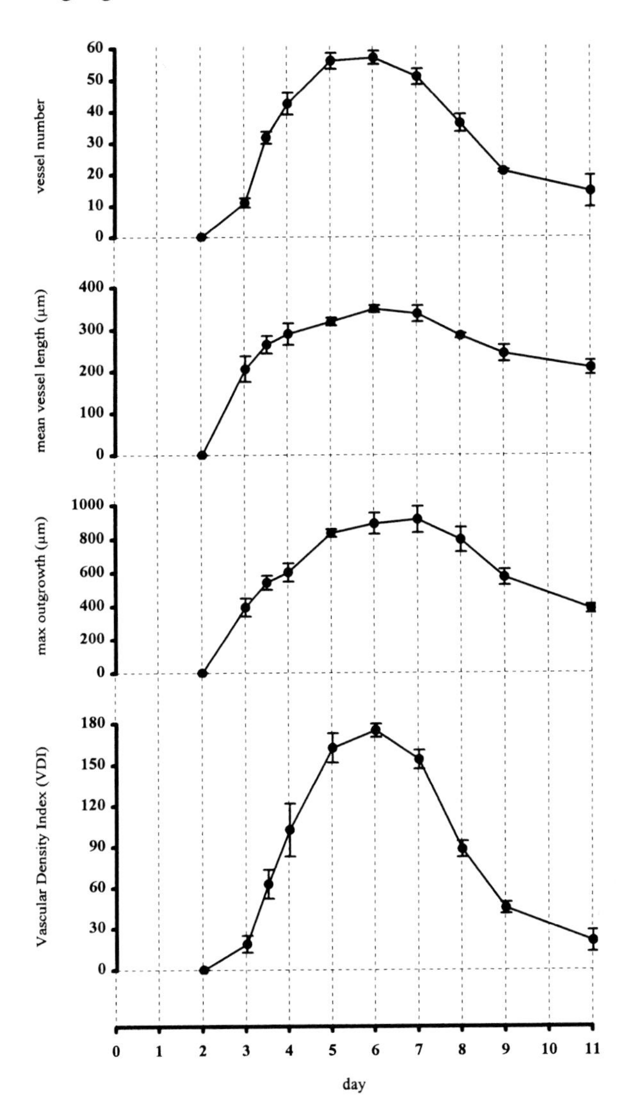

Fig. 6. Microvascular growth curves for serum-free culture of rat aorta in collagen gel in terms of different parameters. Vessel growth is expressed in terms of vessel number, mean vessel length, maximum distance of outgrowth, and the vascular density index (VDI; *see* **Subheading 3.6.1.**) as described in the text. Each point represents the mean value of 4 explants for a representative experiment. Error bars indicate SEM.

4. An imaginary grid placed over the microvessel network provides a histogram of intersections at increasing distances from the ring, and can be considered to give an 'outgrowth profile' for each culture (**Fig. 8C,D**).

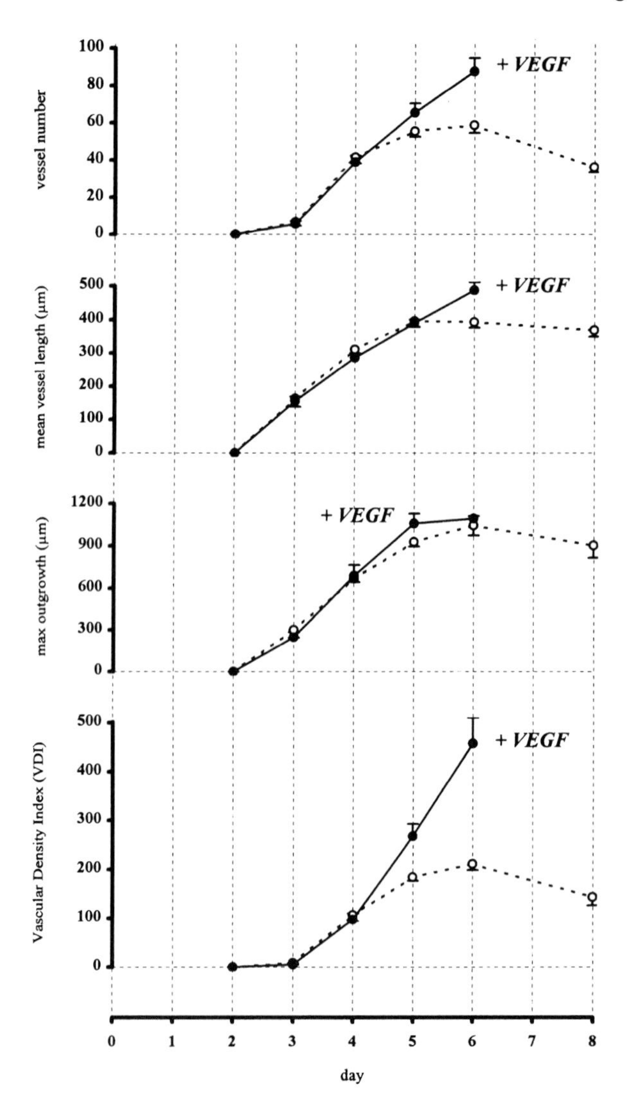

Fig. 7. Microvascular growth curves for serum-free culture of rat aorta in collagen gel in the absence (- - - -) and presence (———) of 10 ng/mL vascular endothelial growth factor. Vessel growth is expressed in terms of vessel number, mean vessel length, maximum distance of outgrowth, and the vascular density index (VDI; *see* **Subheading 3.6.1.**) as described in the text. Each point represents the mean value of four explants for a representative experiment. Error bars indicate SEM.

5. The vascular density index, or VDI, is defined as the total number of vessel intersections with a grid with lines at intervals of 100 µm (for greater precision, lines at intervals of 50 µm are used and corresponding VDI is then calculated as half the number of intersections).

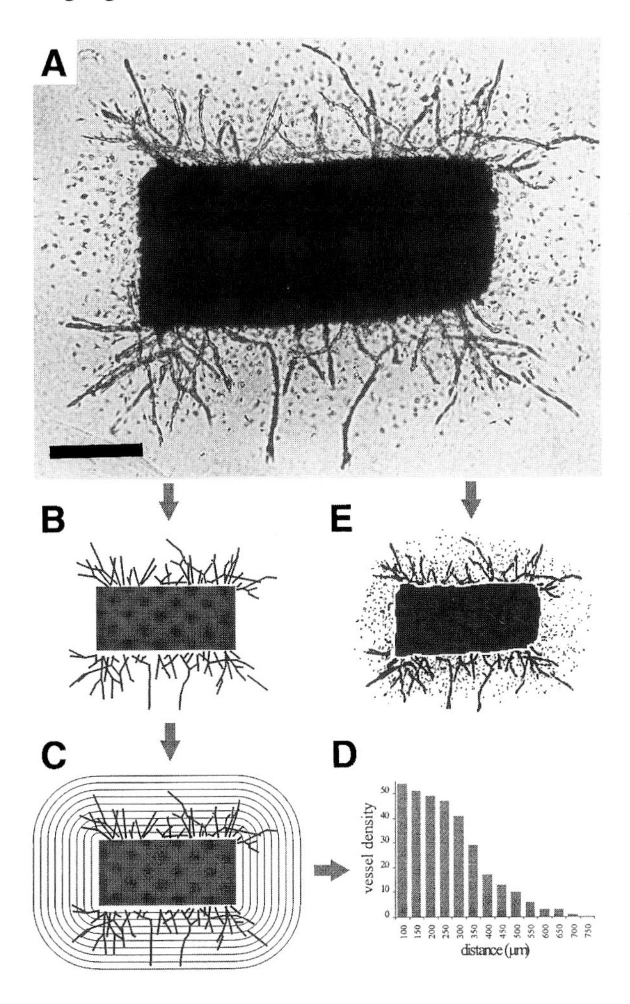

Fig. 8. Angiogenesis quantification. **(A)** Photomicrograph of aortic ring d 6 (bar, 500 µm). **(B,C,D)** Semiautomated image analysis. **(E)** Automated image analysis. *See* **Subheading 3.6.** for details.

3.6.2. Automated Image Analysis

When using the model for high-throughput screening of angiogenic or antiangiogenic molecules, a less precise analysis of cultures may be considered sufficient. Automated image analysis techniques may readily be used to isolate the vessel patterns from the surrounding fibroblasts and to quantify angiogenesis in terms of area occupied by these vessels and extent of outgrowth (**Fig. 8E**). An approximation to the number and distance of migration of fibroblasts may also be obtained by such techniques (*see* **Note 25**).

3.7 Variations on the Standard Technique

3.7.1. Repressed Cultures

The spontaneous outgrowth observed in the aortic cultures is stimulated by the endogenous growth factors released by the aortic wall upon resection. When cultured in basal media such as MEM, unsuited to serum-free growth of ECs, this outgrowth is absent. Replacement of this medium on d 5 by MCDB131 leads to minimal outgrowth. However, the addition of certain growth factors to these repressed cultures will cause marked stimulation, suggesting that cells remain viable under such conditions and that the reduced outgrowth is due rather to the removal of released endogenous growth factors along with the MEM. This technique thus provides a means by which to investigate the intrinsic effect of growth factors, bearing in mind that growth factors may work in synergy, and alone may lead to minimal response. **Figure 9** shows stimulation by vascular endothelial growth factor of such a repressed culture.

Multiwell Cultures

As described, this model is useful for the long-term maintenance and observation of aortic ring cultures. However, it may be necessary to adapt the model to situations in which smaller volumes of media are required, as is the case when using angiogenic stimulators or inhibitors the cost of which would prohibit dilutions into vol of 30 mL.

Cultures may be carried out in 12- or 24- well culture plates with minor modifications.

1. Embed the aortic segments in the collagen or fibrin gels, as outlined for standard method.
2. Carefully remove the agarose rings from the gels with the bent spatula.
3. Transfer the collagen or fibrin gels, again with the bent spatula, to wells of culture plates filled with medium (e.g., 3 mL per well for a 12-well plate).

Outgrowth in these cultures is reduced as compared to cultures in 100-mm Petri dishes, owing partially to an alteration in mechanical forces on the collagen matrix (removal of the supporting agarose enables retraction of the collagen gel), and possibly to a modification in the medium environment resulting from the reduced culture volume (agarose-free cultures in Petri dishes lead to outgrowth superior to that in 12-well plates). Angiogenesis in terms of VDI is half-maximal by d 5 as opposed to d 4 for the standard petri dish culture. This difference in outgrowth should be borne in mind when comparing results. The action of stimulators and inhibitors of angiogenesis may conceivably be altered to a certain degree by differences in growth kinetics, especially when the stability of the added compound is limited. However, the results are of course no less valid than in Petri dish culture, and such multiwell culture is useful for high throughput screening using automated image analysis. With four rats, around 100 potential antiangiogenic compounds may be screened in one such experiment (which should be repeated for confirmation).

If volumes of medium to be used are exceedingly small, it is possible to culture directly in collagen or fibrin gels in 96-well plates (2 layers of 50 μL per well). However, the small volume of medium is rapidly spent (the ratio medium/gel is also very low), and outgrowth may be less satisfactory over extended periods. For analysis, plates should be immersed in a medium bath to enable observation of cultures without image interference due to meniscus within each well.

3.8. Histological Examination

Although it is possible to fix and dehydrate cultures for paraffin embedding, massive shrinkage of the gels during the various dehydration and embedding steps leads to distorted arrangement of fibroblasts and ECs, rendering interpretation and extrapolation to the 3D cultures difficult. Moreover, due to the high content of water, frozen sections of gel are not feasible.

Specific whole cell or antigen marking using immunohistochemical or other techniques may be performed on intact cultures. To enable rapid penetration and washing out of antibodies and other markers, small fragments rather than rings of rat aorta are embedded in thin films of gel.

3.8.1. Culture Wells

1. Culture wells are formed by the silicone gasket of 8-well Lab-Tek Permanox Chamber Slides after removal of the upper chambers.
2. Attach each slide to the base of a Petri dish with a drop of cyanoacrylate gel.

3.8.2. Aortic Fragments

1. Prepare small fragments of 0.25–0.5 mm^2, by first cutting each aorta in half lengthways with a scalpel blade, then cutting each half into strips. Chop each strip into small lengths to produce square fragments.
2. Wash as for rings, remove excess medium, and place in chamber slide wells (5–10 fragments per well).
3. Carefully add 50 μL collagen or fibrin to each well without disturbing fragments and place in a humidified incubator to gel before covering with medium (**Fig. 10A**).

3.8.3. Fluorescent Staining of Viable ECs

ECs within these thin-film cultures may be stained based upon the selective uptake of fluorescence-labeled low-density-lipoprotein (DiI-Ac-LDL).

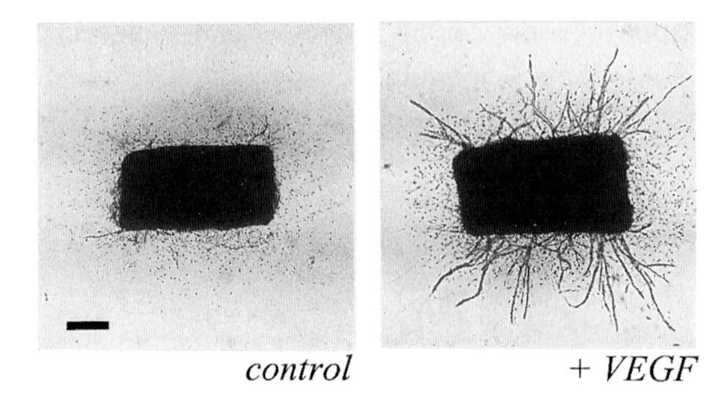

Fig. 9. Repressed cultures. Stimulation with 10 ng/mL vascular endothelial growth factor (VEGF) (photo day 9, bar; 500 μm). *See* text for details.

1. Carefully remove aortic fragments.
2. Incubate living cultures 4 h with 10 μg/mL DiI-Ac-LDL, followed by 3 successive washings of 1 h each in fresh medium.
3. Examine under a fluorescence microscope, set up to view fluorescein labelling. (**Fig. 10B**).

3.8.4. Antibody Labelling of Fixed Cultures

1. Carefully remove aortic fragments.
2. Fix in either 4% paraformaldehyde or ethanol.
3. Air dry the gels to leave preparations of a thickness of a few tens of micrometers, to which may be applied standard immunohistological staining techniques. A typical culture labelled for von Willebrand factor is shown in **Fig. 10C**.

4. Notes

1. Type VII is ideal for its melting and solidifying temperatures and rigidity.
2. Sterile type I collagen solutions may be prepared from rat tails, although this is a tedious and time-consuming procedure, and may lead to large variations between experiments when using different batches. Commercially prepared solutions of rat-tail collagen in acetic acid are now readily available, and will generally lead to more satisfactory and consistent results (e.g., 'Collagen R,' 2 mg/mL in 0.5 *M* acetic acid, Serva). Other commercially available sources of collagen (e.g., bovine skin) have been found to give gels of a less rigid consistency. For rat-tail collagen, a concentration of 1.5 mg/mL is optimal; lower concentrations lead to excessively labile gels, whereas higher concentrations may lead to a reduction in outgrowth from the aortic explants.
3. Fibrinogen and thrombin may be from a variety of origins; human or bovine preparations give similar results and are relatively inexpensive.
4. The rat is the chosen species for these experiments because it provides rings of a practical size, and because rings from mouse aortas have been found not to give

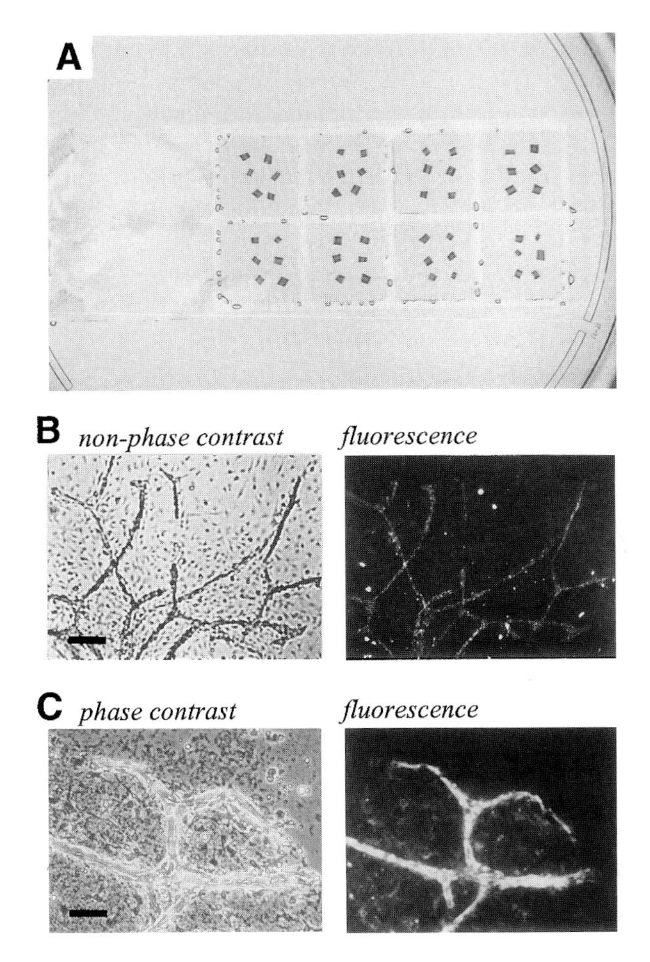

Fig. 10. Fluorescent staining of ECs. **(A)** Aortic fragments are cultured in wells formed by the silicone gasket of Lab-Tek Chamber Slides. **(B)** Fluorescence-labeled DiI-Ac-LDL uptake (bar, 200 μm). **(C)** Fluorescent marking for von Willebrand factor (bar, 100 μm). Cultures are fixed in ethanol 15 min at room temperature and allowed to air dry. Nonspecific binding sites are blocked 30 min in 10% fetal calf serum. Preparations are then incubated 2 h with 10 μg/mL polyclonal sheep antibody to rat von Willebrand factor (Cedarlane), 1 h with biotin conjugated rabbit antisheep IgG (Rockland) and 1 h with ExtrAvidin-R-phycoerythrin conjugate (Sigma).

rise to satisfactory vessel outgrowth, possibly owing to a reduced local concentration of growth factors.

5. MCDB131 is a culture medium optimized for the culture of microvascular ECs and enables growth in serum-free conditions (*see* Knedler and Ham ***(9)***). This commercially available medium contains a concentration of 14 m*M* $NaHCO_3$,

which will equilibrate at pH 7.4 in atmospheric CO_2, concentration of 2%. If conditions of 5% CO_2 are used, pH may be corrected by adding $NaHCO_3$ to a final concentration of 25 m*M*, or by adding NaOH until equilibrium is reached at pH 7.4, the net result in terms of $NaHCO_3$ concentration and buffering capacity being the same [Freshney *(22)*]. Maintenance of pH at 7.4 is of utmost importance for optimal growth.

6. Where fibrin gels are used, 300 μg/ml ε-amino-n-caproic acid should be added to culture medium to prevent spontaneous fibrinolysis.
7. For optimal contrast and depth of field, reduce the aperture as far as possible.
8. Use a black and white camera for optimal image resolution, especially when automated image analysis is required.
9. The image analyzer used in our laboratory is the BIOCOM Visiolab™ 2000, using specialized software developed specifically for analysis of the 3D aortic ring cultures.
10. The use of Permanox plastic ensures minimal autofluorescence when using fluorescent markers.
11. In view of the large species heterogeneity of von Willebrand factor, antibodies raised to the human antigen may be unsuccessful, even though species cross-reactivity is claimed.
12. Cell culture-treated dishes are preferred to ensure reduced adherence of agarose for easier manipulation.
13. At this stage, agarose may be stored at 4°C for up to three days; any longer, and agarose tends to dehydrate.
14. Using bacteriological plastic here improves adherence of agarose to the base of the dish so wells do not become detached during culture.
15. The rat aortic endothelium is exceptionally well retained to the intima, and unlike that of larger species, the formation of clotted blood will not lead to a stripping off of ECs if aortas are rinsed within 15 min of removal from the animal.
16. If more than one aorta is used in an experiment, this stage should be completed for each aorta before moving on to the next.
17. It is not necessary to keep Petri dishes on ice during dissection procedures under the dissecting microscope. Frequent replenishment of ice-cold medium is sufficient to maintain integrity of the aorta and its cellular components. Total procedure may last up to 4 h with no detrimental effects on the subsequent spontaneous angiogenic response.
18. The optimal length of aortic rings is found to be 1 mm. Shorter lengths are more difficult to handle and position correctly, and give rise to a lower level of outgrowth, probably because there is a lower local concentration of growth factors. Longer lengths, although giving satisfactory outgrowth, would obviously lead to a lower yield per aorta.
19. While all preceding steps may be performed on the workbench in a culture laboratory, from here on work under a laminar-flow hood.
20. The polymerization of the collagen gel will vary with pH.

21. Prepared collagen solutions may be kept on ice up to 30 min; at room temperature they will gel rapidly (3 min).
22. As for collagen, a concentration of 1.5 mg/mL is found to be optimal.
23. Take care to work at pH 7.4 for satisfactory gel polymerization.
24. The time required for the analysis of each aortic ring is between 2 and 5 min, depending on the extent of outgrowth.
25. Care should be taken when using such automated image analysis techniques because artifacts in the captured image (such as incomplete removal of fibroadipose tissue, or bubbles, or other defects in the collagen or fibrin gels) may significantly alter results if not carefully monitored. Quantification is also highly sensitive to illumination level during image capture.

Acknowledgments

D. C. W. acknowledges the support of the North West Cancer Research Fund and Servier. Dr. Gordon Tucker and Prof. Ghanem Atassi have contributed greatly to the development of these techniques.

References

1. Auerbach, R., Auerbach, W., and Polakowski, I. (1991) Assays for angiogenesis: a review. *Pharmacol. Ther.* **51,** 1–11.
2. Montesano, R. and Orci, L. (1985) Tumor-promoting phorbol esters induce angiogenesis in vitro. *Lab. Invest.* **42,** 469–477.
3. Madri, J. A., Pratt, B. M., and Tucker, A. M. (1988) Phenotypic modulation of endothelial cells by transforming growth factor beta depends upon the composition and organization of the extracellular matrix. *J. Cell Biol.* **106,** 1375–1384.
4. Nicosia, R. F. and Ottinetti, A. (1990) Growth of microvessels in serum-free matrix culture of rat aorta—A quantitative assay of angiogenesis in vitro. *Lab. Invest.* **63,** 115–122.
5. Schor, A. M., Ellis, I., and Schor, S. L. (1999) Collagen gel assay for angiogenesis: induction of endothelial cell sprouting. In: *Methods in Molecular Medicine: Angiogenesis Protocols* (Murray, J. C., ed.), Humana Press Inc., Totowa, NJ, pp XX.
6. Madri, J. A., Merwin, J. R., Bell, L., Basson, C. T., Kocher, O., Perlmutter, R., and Prinz, C. (1992) Interactions of matrix components and soluble factors in vascular responses to injury. In: *Endothelial Cell Dysfunctions*, (Simionescu, N. and Simionescu, M., eds.), Plenum Press, New York, pp11–30.
7. Pröls, F., Loser, B., and Marx, M. (1998) Differential expression of osteopontin, PC4, and CEC5, a novel mRNA species, during in vitro angiogenesis. *Exp. Cell Res.* **239,** 1–10.
8. Montesano, R., Pepper, M. S., and Orci, L. (1990) Angiogenesis in vitro: morphogenetic and invasive properties of endothelial cells. *News. Physiol. Sci.* **5,** 75–79.

9. Knedler, A. and Ham, R. G. (1987) Optimized medium for clonal growth of human microvascular endothelial cells with minimal serum. In Vitro *Cell Dev. Biol.* **23,** 481–491.
10. Nicosia, R. F., Bonanno, F., and Villaschi, S. (1992) Large-vessel endothelium switches to a microvascular phenotype during angiogenesis in collagen gel culture of rat aorta. *Artherosclerosis* **95,** 191–199.
11. Nicosia, R. F. and Bonanno, F. (1991) Inhibition of angiogenesis in vitro by arg-gly-asp-containing synthetic peptide. *Am. J. Pathol.* **138,** 829–833.
12. Nicosia, R. F., Nicosia, S. V., and Smith, M. (1994) Vascular endothelial growth factor, platelet-derived growth factor and insulin-like growth factor-1 promote rat aortic angiogenesis in vitro. *Am. J. Pathol.* **145,** 1023–1029.
13. Nicosia, R. F., Lin, Y. J., Hazelton, D., and Qian, X. H. (1997) Endogenous regulation of angiogenesis in the rat aorta model—role of vascular endothelial growth factor. *Am. J. Pathol.* **151,** 1379–1386.
14. Derringer, K. A. and Linden, R. W. A. (1998) Enhanced angiogenesis induced by diffusible angiogenic growth factors released from human dental pulp explants of orthodontically moved teeth. *Eur. J. Orthod.* **20,** 357–367.
15. Wakabayashi, T., Kageyama, R., Naruse, N., Tsukahara, N., Funahashi, Y., Kitoh, K., and Watanabe, Y. (1997) Borrelidin is an angiogenesis inhibitor; Disruption of angiogenic capillary vessels in a rat aorta matrix culture model. *J. Antibiot.* **50,** 671–676.
16. Bocci, G., Danesi, R., Benelli, U., Innocenti, F., Di Paolo, A., Fogli, S., and Del Tacca, M. (1998) Inhibitory effect of suramin in rat models of angiogenesis in vitro and in vivo. *Cancer Chemother. Pharmacol.* **43,** 205–212.
17. Chen, C. H., Cartwright, J., Li, Z., Lou, S., Nguyen, H. H., Gotto, A. M., and Henry, P. D. (1997) Inhibitory effects of hypercholesterolemia and Ox-LDL on angiogenesis-like endothelial growth in rabbit aortic explants—Essential role of basic fibroblast growth factor. *Arterioscler. Thromb. Vasc. Biol.* **17,** 1303–1312.
18. Burbridge, M. F., West, D. C., Atassi, G., and Tucker, G. C. (1999) The effect of extracellular pH on angiogenesis in vitro. *Angiogenesis* **3,** 281–288.
19. Burbridge, M. F. (2000) The rat aortic ring model of angiogenesis in vitro as an assay for angiogenic modulators. Ph.D. Thesis, University of Liverpool, Liverpool, UK.
20. Brown, K. J., Maynes, S. F., Bezos, A., Maguire, D. J., Ford, M. D., and Parish, C. R. (1996) A novel in vitro assay for human angiogenesis. *Lab Invest.* **75,** 539–555.
21. Antes, L. M., Villar, M. M., Decker, S., Nicosia, R. F., and Kujubu, D. A. (1998) A serum-free in vitro model of renal microvessel development. *Am. J. Physiol.* **274,** F1150–F1160.
22. Freshney, R. I. (1987) In: *Culture of Animal Cells. A Manual of Basic Technique.* Alan R. Liss, Inc., New York.

14

In Vitro Matrigel Angiogenesis Assays

M. Lourdes Ponce

1. Introduction

A variety of in vivo and in vitro methods have been used to study angiogenesis, the process of blood vessel formation. Two widely accepted but technically difficult assays include the cornea implant assay and the chick chorioallantoic membrane assay. The cornea assay requires special equipment and a skilled person to implant beads containing the test compound in the eyes of animals; only a small number of samples can be tested due to cost and time. The chorioallantoic membrane assay requires a large number of samples on account of the variability of the system and its difficulty in quantitation. In our laboratory, we have developed a quick and highly reliable method for testing numerous compounds for angiogenic and/or antiangiogenic activity. The method is based on the differentiation of ECs on a basement membrane matrix, Matrigel, derived from the Engelbreth-Holm-Swarm tumor *(1)*. ECs from human umbilical cords as well as from other sources differentiate and form capillary-like structures on Matrigel in the presence of 10% bovine calf serum (BCS) and 0.1 mg/mL of endothelial cell growth supplement (ECGS) *(2)*, which is a mixture of both acidic and basic fibroblast growth factor (**Fig. 1**, **Panel C**). The formation of tube-like vessels under these conditions can be used to assess compounds that either inhibit or stimulate angiogenesis. In the assay, substances that affect angiogenesis, such as the laminin-1 peptide containing the IKVAV sequence, disturb the formation of capillary-like structures and the distinctive morphological cell characteristics resulting are indicative of the potential activity of the compound (**Fig. 1**, **Panel D**). However, it can not be determined from the morphological characteristics whether a compound is angiogenic or antiangiogenic.

From: *Methods in Molecular Medicine, Vol. 46: Angiogenesis Protocols*
Edited by: J. C. Murray © Humana Press Inc., Totowa, NJ

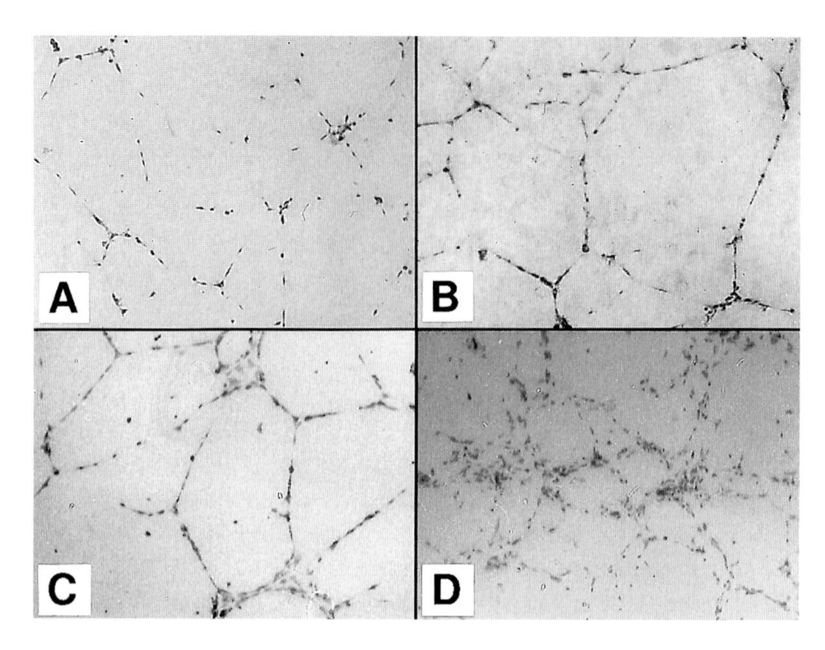

Fig. 1. Endothelial cells on Matrigel. HUVEC were plated on Matrigel in the presence of HUVEC media containing either 5% **(A,B)** or 10% BCS **(C,D)**. Cells incubated with 25 ng/mL of hepatocyte growth factor (HGF) formed tubes in serum reduced media **(B)** as compared to their control **(A)**. Cells that were plated with 0.1 mg/mL of a laminin-1 peptide that contains the IKVAV sequence did not form complete tube-like structures **(D)** as compared to their negative control devoid of peptide but containing 10% BCS **(C)**. Original magnification ×5.

Tubes do not form when both serum and ECGS concentrations are reduced (**Fig. 1**, **Panel A**). This allows one to determine whether a substance is angiogenic. Under these circumstances human umbilical vein endothelial cells (HUVEC) do not completely differentiate; instead, they form incomplete, short tube-like structures. In the presence of angiogenic compounds, such as fibroblast growth factor or hepatocyte growth factor, the ECs differentiate into well-defined tube-like structures (**Fig. 1 Panel B**). An additional advantage of these assays is that they can be scaled down and 48- or even 96-well plates can be used if many compounds need to be tested or their quantities are limited. The use of either one or both of these assays can help to quickly and efficiently identify compounds with angiogenic or antiangiogenic activity. Using both of these assays, a number of new compounds that affect angiogenesis have been identified including hepatocyte growth factor (HGF), haptoglobin, estrogen, IP-10, and numerous laminin-1 peptides among others ***(3–6)***. It should be noted that the activity in all cases should be confirmed by an additional assay.

2. Materials

1. Matrigel (Sigma Chemical Co., Saint Louis, MO or Becton Dickinson, Inc., Rutherford, NJ).
2. 24-Well tissue-culture plates.
3. Pipetmen and sterile tips.
4. Human umbilical vein endothelial cells (HUVEC) prepared from umbilical veins or commercially obtained from Clonetics Corporation, San Diego, CA (*see* **Note 1**).
5. RPMI 1640 media (1 L, Life Technologies, Gaithersburg, MD).
6. Table centrifuge.
7. 50-mL Conical tubes.
8. Hematocytometer.
9. Trypsin-EDTA solution containing 0.05% trypsin, 0.53 m*M* ethylenediaminotetraacetic acid in Hanks' Balanced Solution (Life Technologies, Gaithersburg, MD).
10. Incubator with 95% air/5% CO_2 mixture.
11. HUVEC medium:
    - 500 mL RPMI 1640
    - 100 mL of bovine calf serum (BCS), defined and supplemented (HyClone Laboratories Inc., Logan, Utah)
    - 0.1 g ECGS (Collaborative Biomedical Products, Bedford, MA)
    - 2,500 U of heparin sodium (Fisher Scientific, Fair Lawn, NJ)
    - 5 mL of fungizone solution (Life Technologies, Gaithersburg, MD)
    - 1.25 mg gentamicin (Life Technologies, Gaithersburg, MD)
    - 5 mL of penicillin/streptomycin solution (10,000 U/mL of penicillin G sodium and 10 mg/mL streptomycin sulfate in 0.85% saline; Life Technologies, Gaithersburg, MD).
12. Diff-Quik fixative and solution II (Baxter Scientific Products, McGaw, IL)

3. Methods

3.1. Testing Angiogenic Compounds

1. Allow Matrigel to thaw on ice at 4°C; once thawed keep on ice at all times or it will solidify at room temperature.
2. Under sterile conditions, coat 24-well plates with 300 μL per well of Matrigel without introducing air bubbles. Let the plates sit at room temperature for at least 15 min to allow gelling of Matrigel.
3. Rinse a confluent 150-mm dish or equivalent containing 3–4 × 10^6 ECs in 5 mL of RPMI 1640 medium.
4. Add 5 mL of trypsin-EDTA solution and incubate until the cells detach from the plate (approx 5 min).
5. Place the cells in a 50 mL conical sterile tube and add 5 mL of HUVEC media to neutralize the trypsin.
6. Centrifuge the tube on a table top centrifuge for 5 min at 50*g*; remove the supernatant, tap the pellet gently and resuspend the pellet in 2 mL HUVEC medium.
7. Set aside a 0.2 mL cell aliquot as control.

8. Count a 10 μL aliquot in a hemocytometer or use a Coulter Counter to determine the total number of cells in the tube.
9. Adjust the volume with HUVEC media (containing 20% BCS) in order to have 2×10^6 cells per mL (*see* **Note 2**).
10. To reduce simultaneously the concentration of serum to 5% and the number of cells to 0.5×10^6 per mL, add an amount of RPMI 1640 media equal to three times the volume present in the tube.
11. Plate 192 μL of the cell mixture, containing 48,000 cells, into each Matrigel-coated well.
12. Add the test compound in a total volume of 8 μL (*see* **Note 3**).
13. Controls that will contain 10% serum are similarly prepared from the aliquot that was set aside, except that 1×10^6 cells per mL (instead of 2×10^6 cells per mL) are suspended in HUVEC media (20% serum) and are subsequently diluted to double their volume with RPMI 1640 medium.
14. Gently swirl the dishes to evenly mix the test compound(s) with the cells.
15. Incubate dishes overnight at 37°C in a 5% CO_2/95% air incubator.
16. The next day gently aspirate the medium from each well, add approx 200 μL of Diff-Quik fixative per well and incubate for 30 sec.
17. Aspirate the fixative and stain the cells for 2 min with Diff-Quik solution II that has been diluted 1/1 with water (see **Notes 4** and **5**).
18. Observe tube structures under a microscope and photograph.

3.2. Screening Compounds That Influence Angiogenesis

This method is a variation of the assay described in **Subheading 3.1.** and can be used for screening compounds that either induce or inhibit angiogenesis.

1. Follow protocol as in **Subheading 3.1.** up to **step 8**.
2. Suspend cells in HUVEC media containing 20% BCS to yield 1×10^6 cells per mL.
3. Dilute cell suspension with an equal volume of RPMI 1640 medium lacking serum to reduce by half the amount of serum and the number of cells per mL.
4. Continue assay as described in **Subheading 3.1.** Controls should include cells in medium containing 5 and 10% serum (*see* **Notes 6–8**).

4. Notes

1. HUVEC should not be older than 6 passages because they are primary cell cultures and tend to lose their differentiated characteristics. They should always be handled with gloves since they are of human origin.
2. The amount of HUVEC media in which the cells are suspended in **step 9** will vary from experiment to experiment according to the total number of cells harvested from each plate.
3. Several concentrations of each compound should be tested at least in triplicate. Include 6 wells for controls (3 positive and 3 negative).

4. Staining of the Matrigel with Diff-Quik solution II for longer than 2 min or without previous dilution will excessively stain the Matrigel making it difficult to observe the tubular structures.
5. Dishes containing stained cells can be stored wrapped with plastic for a week or longer at 4°C. Pictures should be taken within a few days since the Matrigel tends to dry up.
6. To perform assays on 48-well dishes, wells are coated with 200 μL of Matrigel and 24,000 cells are plated in 150 μL of medium. Ninety-six well plates are coated with 100 μL of Matrigel and plated with 14,000 cells in 100 μL of medium.
7. Unknown samples should be tested using both serum concentrations.
8. HUVEC can form tubes even in the absence of serum in some commercially available batches of Matrigel. It is recommended that a control experiment be carried out using low serum concentrations each time a new Matrigel lot is used.

References

1. Kubota, Y., Kleinman, H. K., Martin, G. R., and Lawley, T. J. (1988) Role of laminin and basement membrane in the morphological differentiation of human endothelial cells into capillary-like structures. *J. Cell Biol.* **107,** 1589–1598.
2. Grant, D. S., Kinsella, J. L., Fridman, R., Auerbach, R., Piasecki, B. A., Yamada, Y., et al. (1992) Interaction of endothelial cells with a laminin A chain peptide (SIKVAV) in vitro and induction of angiogenic behavior in vivo. *J. Cell Physiol.* **153,** 614–625.
3. Grant, D. S., Kleinman, H. K., Goldberg, I. D., Bhargava, M. M., Nickoloff, B. J., Kinsella, J. L., et al. (1993) Scatter factor induces blood vessel formation in vivo. *Proc. Natl. Acad. Sci. USA* **90,** 1937–1941.
4. Cid, M. C., Grant, D. S., Hoffman, G. S., Auerbach, R., Fauci, A. S., and Kleinman, H. K. (1993) Identification of haptoglobin as an angiogenic factor in sera from patients with systemic vasculitis. *J. Clin. Invest.* **91,** 977–985.
5. Ponce, M. L., Nomizu, M., Delgado, M. C., Kuratomi, Y., Hoffman, M. P., Powell, S., et al. (1999) Identification of endothelial cell binding sites on the laminin γ1 chain. *Circ. Res.* **84,** 688–694.
6. Malinda, K. M., Nomizu, M., Chung, M., Delgado, M., Kuratomi, Y., Yamada, Y., et al. (1999) Identification of laminin α1 and β1 chain peptides active for endothelial cell adhesion, tube formation, and aortic sprouting. *FASEB J.* **13,** 53–62.

IV

Associated Techniques

15

Microvessel Endothelial Cells from Human Adipose Tissues

Isolation, Identification, and Culture

Peter W. Hewett

1. Introduction

Human endothelial cells (EC) can be readily prepared from large vessels such as the aorta, and umbilical *(1)* and saphenous veins. ECs derived from these vessels have proven an abundant, convenient, and useful tool for the investigation of many aspects of endothelial biology. However, the majority of the body's endothelium (>95%) forms the microvasculature, and it is at the level of the blood/tissue interface that many of the phenomena occur that we wish to investigate. The high degree of functional diversity that the endothelium demonstrates between organs, and within the different vascular beds of a given organ, is manifest in considerable morphological, biochemical, and molecular differences *(2–6)*. This apparent heterogeneity has highlighted the need for reliable methods for microvessel EC isolation and culture from a variety of tissues of different species in order to establish more realistic in vitro models.

Many techniques have been developed to enrich ECs from tissue homogenates either directly or after a period in culture, and/or to selectively promote their proliferation in vitro *(6)*. Most methods use as their starting point tissue homogenization and digestion, and most are often hampered by low EC yield and the contamination of cell populations that readily adapt to culture. Some tissues are inherently better suited to microvessel EC isolation such as brain and adipose. Wagner and Matthews *(7)* first utilized adipose from the rat epididymal fat pad for the isolation of microvessel endothelium. The major

From: *Methods in Molecular Medicine, Vol. 46: Angiogenesis Protocols*
Edited by: J. C. Murray © Humana Press Inc., Totowa, NJ

advantage of this tissue is the difference in buoyant densities between adipocytes and the stromal component that permits their separation by centrifugation. Similarly, high microvessel density, ease of separation, and the resistance of the parenchyma to growth in culture have made brain a relatively accessible source of microvessel endothelium ***(2,8)***. Nonetheless the isolation of ECs from more complex tissues that contain a wide variety of cell types, many of which readily adapt to cell culture, has until recently remained a significant technical problem.

The use of super-paramagnetic beads (Dynabeads™) coupled to endothelial-specific ligands to select the ECs from mixed cell populations represents a major advance in the purification of ECs. The original technique described by Jackson and colleagues utilized the lectin *Ulex europaeus* agglutinin-1 (*UEA*-1), which binds specifically to α-fucosyl residues of glycoproteins on the surface of endothelial cells ***(9,10)***. However, *UEA*-1 also binds to some epithelia and mesothelial cells ***(1,11)***. We refined this technique by coupling antibodies raised against platelet EC adhesion molecule (PECAM-1/CD31) ***(13)***, a pan-endothelial marker ***(4,5)***, to Dynabeads (PECA-beads), and have used these to prepare microvessel ECs from a variety of human tissues ***(12)***. Dynabeads conjugated to antibodies against other endothelial markers, including CD34 and E-selectin (endothelial cell leukocyte adhesion molecule-1/ELAM-1) ***(14)***, have also been used to isolate microvessel ECs from various tissues.

The identification of new markers such as VE cadherin, ICAM-2, the endothelial restricted tyrosine kinases (Flt-1, Flt-4, KDR/Flk-1, Tie-1 and Tie-2/Tek), and the increasing number of commercially available antibodies raised against them have greatly improved the ease of EC characterization ***(5)***. Many EC markers crossreact with subpopulations of hematopoietic cells underlining the common developmental origins of these cell types. However, due to their limited viability in vitro these cells do not represent a significant problem for EC culture. Mesothelial cells are a potentially difficult contaminant of EC cultures isolated from tissues such as omentum, because they exhibit similar morphology and marker expression. However, the absence of PECAM-1 and E-selectin expression in human mesothelial cells readily distinguishes them from ECs ***(11)***. The use of PECA-beads in principle should eliminate mesothelial cell contamination of endothelial isolates from sites such as the lung and omentum that are surrounded by mesothelium ***(12)***.

In this chapter we describe the use of super-paramagnetic beads (Dynabeads™) coupled to anti-PECAM-1 antibodies (PECA-beads) to isolate microvessel endothelial cells from human adipose tissue obtained and methods for the routine culture of these cells. In addition, we briefly discuss some of the characteristics of these cells and outline the adaptation of this basic technique to isolate microvessel endothelial cells from human lung and stomach ***(12,18)***.

2. Materials

2.1. Equipment for EC Isolation, Characterization, and Culture

2.1.1. Hardware

1. A class II laminar flow cabinet is essential for all procedures involving the processing of tissue and cultured cells in order to maintain sterility and protect the operator.
2. Phase contrast microscope for observing cell cultures.
3. Light microscope equipped with epifluorescence for immunocytofluorescent characterization of the cells.
4. Scalpels, scissors, and forceps are required for the isolation procedure and should be sterilized prior to use by autoclaving at 121°C for 30 min.

2.1.2. Sterile Plasticware

1. Tissue culture flasks (25 and 75 cm^2).
2. Large plastic dishes (e.g., Bioassay dishes, Nunc, Naperville, IL, USA).
3. 30 mL universal tubes.
4. 50 mL centrifuge tubes.
5. Lab-Tek multiwell glass chamber slides (Nunc, Naperville, IL, USA).

2.1.3. Antibodies

There are many commercially available antibodies against endothelial markers.

1. Monoclonal antibodies against human PECAM-1, E-selectin (e.g., clones 9G11 and 13D5 respectively, R&D Systems, Abingdon, Oxon, UK), and vWF (e.g., clone F8/86 Dako, High Wycombe, Bucks, UK).
2. Fluorescein isothiocyanate (FITC)-conjugated goat antimouse secondary antibodies (Sigma, Poole, Dorset, UK).

2.2. Preparation of Anti-PECAM-1 Coated Magnetic Beads (see Note 1)

Mix 0.1–0.2 mg of mouse anti-PECAM-1 monoclonal antibody in sterile calcium-magnesium free Dulbecco's phosphate buffered saline (PBS/A) containing 0.1% bovine serum albumin (BSA) (PBS/A+0.1% BSA) per 10 mg of Dynabeads-M450 (Dynal UK, Wirral, UK) precoated with antimouse IgG_2 (*see* **Note 2**). Incubate on a rotary stirrer for 16 h at 4°C. Remove free antibody by washing four times for 10 min, and then overnight in PBS/A+0.1% BSA. PECA-beads maintain their activity for more than 6 mo if sterile and stored at 4°C. However, it is necessary to wash the beads with PBS/A plus 0.1% BSA to remove free antibody prior to use.

Magnet: A suitable magnet is required for the cell selection system employed. We use the Dynal Magnetic Particle Concentrator-1 (MCP-1; Dynal), which will accept 30 mL universal tubes.

100-μm nylon filters: Cover the top of a polypropylene funnel ~10 cm with 100 μm nylon mesh filter (Lockertex, Warrington, Cheshire, UK) and sterilize by autoclaving.

2.3. Solutions for Cell Microvessel EC Isolation and Culture

1. 10% BSA solution: Dissolve 10 g of bovine serum albumin (BSA) in 100 mL of PBS/A, filter sterilize, and store at 4°C.
2. 2% Antibiotic/antimycotic solution: Dilute antibiotic/antimycotic solution (Sigma, Poole, Dorset, UK) containing 10,000 U/mL penicillin, 10 mg/mL streptomycin and 25 μg/mL amphotericin B in PBS/A.
3. Collagenase (type-II) solution: Dissolve type-II collagenase (Sigma) at 2000 U/mL in Hanks' basal salt solution (HBSS) containing 0.5% BSA, sterile filter, aliquot, and store at –20°C.
4. Trypsin/EDTA solution: Dilute 2.5 % stock solution of trypsin derived from porcine pancreas (Sigma) in sterile PBS/A to give a 0.25% solution and add 1 m*M* (0.372 g/L); ethylenediaminetetraacetic acid (EDTA), ensure that it has fully dissolved, sterile filter, aliquot and store at –20°C. Trypsin/EDTA solution should be stored at 4°C while in use.
5. Gelatin solution: Dilute stock (2%) porcine gelatin solution (Sigma) in PBS/A to give a 0.2% solution and store at 4°C. To coat tissue culture dishes add 0.2% gelatin solution (~10-mL/75-cm^2 flask) and incubate for 1 h at 37°C or overnight at 4°C. Remove the gelatin solution immediately prior to plating the cells.
6. Growth medium (*see* **Note 3**): Supplement M199 (with Earle's salts) with 14 mL/L of 1 *M* N-[2-hydroxyethyl] piperazine-N′-[2-hydroxy-propane] sulphonic acid (HEPES) solution, 20 ml/l of 7.5 % sodium hydrogen carbonate solution, 20 mL/L 200 m*M* L-glutamine solution and mix 1/1 with Ham F12 nutrient mix (Sigma). To 680 mL of M199/Ham F12 solution add 20 mL of penicillin (100 U/mL), streptomycin (100 mg/mL) solution (Sigma), 1500 U/L of heparin (Leo Laboratories, Princess Risborough, Bucks, UK), 300 mL of iron-supplemented calf serum (*see* **Note 4**) (CS) (Hyclone, Logan, Utah, USA), 5 ng/mL basic fibroblast growth factor (bFGF) and 20 ng/mL epithelial growth factor (EGF; *see* **Note 5**; PeproTech EC Ltd, London, UK). Store at 4°C for no more than 1 month.

3. Methods

3.1. Isolation of Human Adipose Microvessel Endothelial Cells

3.1.1. Collection of Tissue

A suitable large sterile container is required for the collection of adipose tissue obtained during breast or abdominal (*see* **Note 6**) reductive surgery. The fat can be processed immediately or stored for up to 48 h at 4°C.

3.1.2. Isolation of Adipose Microvessel Endothelial Cells

1. Working under sterile conditions in a class II cabinet, place the tissue on a large sterile dish (e.g., Bioassay dish, Nunc) and wash with 2% antibiotic/antimycotic solution. Avoiding areas of dense (white) connective tissue (which are often prevalent in breast tissue) and large (visible) blood vessels, scrape the fat free from the connective tissue fibers with two scalpel blades.
2. Chop the fat up finely and aliquot 10–20 g into sterile 50 mL centrifuge tubes. Add 10 mL of PBS/A and 5–10 mL of the type-II collagenase solution. Shake the tubes vigorously to further break up the fat and incubate with end-over-end mixing on a rotary stirrer at 37°C for approximately 1 h. Following digestion the fat should have broken down and no spicules should be evident.
3. Centrifuge the digests at 500 g for 5 min, discard the fatty (top) layer, and retain the cell pellet with some of the lower (aqueous) layer. Add PBS/A and recentrifuge 500 g for 5 min.
4. Re-suspend the cell pellet in 10% BSA solution and centrifuge (200 g, 10 min). Discard the supernatant and repeat the centrifugation in BSA solution. Wash the pellet with 50 mL of PBS/A. Viewed under the light microscope the tissue digest should contain obvious microvessel fragments in addition to single cells and debris.
5. Resuspend the pellet obtained in 5 mL of trypsin/EDTA solution and incubate for 10–15 min with occasional agitation at 37°C. Add 20 mL Hanks' balanced salt solution (HBSS) containing 5% CS (HBSS + 5%CS) and mix thoroughly to neutralize the trypsin. We have found it advantageous to break up the microvessel fragments and cell clumps further with trypsin/EDTA as this reduces the number of contaminating cells coisolated with ECs during PECA-bead purification.
6. Filter the suspension through 100 μm sterile nylon mesh to remove large fragments of sticky connective tissue. Centrifuge the filtrate 700 g, 5 min and resuspend the resulting pellet in ~1–2 mL of ice-cold HBSS + 5%CS.
7. Add approximately 50 μL of PECA-beads and incubate for ~20 min at 4°C with occasional agitation (*see* **Note 7**). Add HBSS + 5%CS to a final volume of 10–12 mL, mix thoroughly, and select the microvessel fragments using the MCP-1 (Dynal) for 3 min. Repeat the cell selection process a further 3–5 ×, i.e., resuspend in 12 mL of HBSS + 5%CS, wash, and reselect the microvessel fragments (*see* **Note 8**).
8. Suspend the selected cells in growth medium and seed at high density onto 0.2% gelatin-coated 25 cm^2 tissue culture flasks and incubate at 37°C in a humidified atmosphere of 5% CO_2 with the lids loose.

3.1.3. Adipose Microvessel Cells in Culture

Following the PECA-bead selection procedure there should be obvious small microvessel fragments and single cells coated with Dynabeads under light

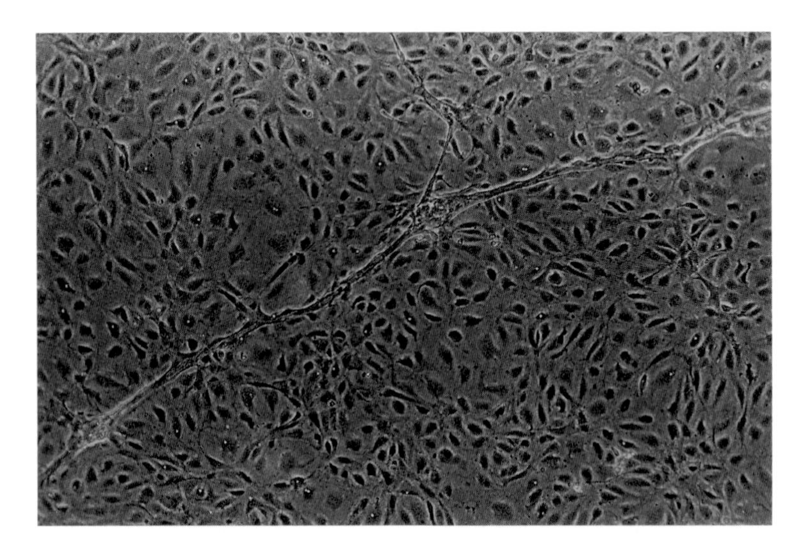

Fig. 1. Photomicrographs of postconfluent monolayer of HuMMEC (passage 2) demonstrating typical cobblestone morphology and tube formation on the surface of the monolayer.

microscopy (*see* **Note 9**; **Fig. 1**). After 24 h the cells adhere to the flasks and also start to grow out from any microvessel fragments present to form distinct colonies. Human mammary microvessel endothelial cells (HuMMEC) isolated using this technique grow to confluence forming cobblestone contact-inhibited monolayers within 10–14 days depending on the initial seeding density. The life span of cultures appears to be dependent on the individual tissue. We have successfully cultured these cells to passage 8 without observable change in morphology but routinely use them for experiments between passage 3 to 6.

3.1.4. Endothelial Cell Morphology in Culture

Cobblestone morphology is very typical of ECs derived from many tissues and they are usually readily distinguished from the fibroblastoid contaminating cell populations (*see* **Note 10**). However, a more elongated morphology with cells forming 'swirling' monolayers is often observed following stimulation of ECs with growth factors. When cultured on gel matrices such as Matrigel ECs will form "capillary-like" networks within a few hours of plating. This phenomenon also occurs in many types of microvessel ECs after several days at confluence. However, it should be noted that the formation of "capillary-like" structures is not an exclusive property of ECs in culture.

3.1.5. Cell Culture

Subculture: Maintain the microvessel ECs at 37°C, 5% CO_2 changing the medium every 3–4 d. When confluent HuMMEC are routinely subcultured with trypsin/EDTA solution, onto 0.2% gelatin-coated dishes at a split ratio of 1/4.

1. Discard the old medium and wash the EC monolayer with 5–10 mL of PBS/A to remove any remaining serum. Add a few mL of trypsin/EDTA solution, wash it over the cell monolayer, remove the surplus leaving the cells just covered and incubate at 37°C. Monitor the cells under the microscope until they have fully detached (*see* **Note 11**). This can be readily achieved by striking the flask sharply to dislodge and break up cell aggregates.
2. Add sufficient growth medium to achieve a split ratio of 1/4 and plate the cells onto gelatinized flasks.

Cryopreservation of cells: Suspend ECs at approx 2×10^6/mL in growth medium containing 10% (v/v) dimethyl sulphoxide (Sigma) and dispense into suitable cryovials. Cool the vials to –80°C at 1°C/min and store under liquid nitrogen (*see* **Note 12**).

3.1.6. Maintaining the Purity of Endothelial Cultures

To maintain the purity of cultures it may be necessary to reselect the ECs with PECA-beads and/or minor manual 'weeding' performed under sterile conditions in a flow hood using a light microscope. Reselection with PECA-beads can be performed as described above (**Subheading 3.1.2., item 7**) following removal of the cells from flasks using trypsin/EDTA solution (**Subheading 3.1.5.**). Provided that there are clear morphological differences between contaminating cells and the endothelial cells (*see* **Note 10**), it is relatively straightforward although time consuming to remove them. Manual weeding should be performed with the stage of a phase contrast microscope within a class II cabinet to ensure sterile conditions. A needle or Pasteur pipet is used to carefully remove contaminating cells from around EC colonies. The medium is discarded and the adherent cells washed with several changes of sterile PBS/A to remove all remaining dislodged contaminating cells.

3.2. Characterization of Endothelial Cells

There are many different criteria on which EC identification may be based and these have been extensively reviewed ***(2,3,5,6,19)***. Many of these markers/ properties are not unique to ECs and several may be required to confirm endothelial identity. ECs may demonstrate heterogeneity in expression of cell markers and therefore a lack of a given endothelial marker may not preclude the endothelial origin of isolates *(2)*. It is often useful to demonstrate the absence of markers, such as stress fibers staining with smooth muscle α-actin and the intermediate filament protein, desmin, that are characteristic of smooth muscle cells and pericytes *(20)*, both potential endothelial culture contaminants.

3.2.1. Key Endothelial Cell Markers

Over recent years, more specific endothelial markers have emerged, such as PECAM-1 *(13)*, E-selectin *(14)*, ICAM-2, and VE cadherin *(5)*, and they have increased the ease of endothelial characterization. Here we focus on von Willebrand Factor (vWF), PECAM-1, and E-selectin, which we believe to be useful for the rapid identification of ECs. For more detailed literature on these endothelial markers there are several reviews that cover in depth the characterization of ECs *(2,3,5,6,19)*.

vWF is only expressed at significant levels in ECs, platelets, megakaryocytes, and syncytiotrophoblast. In ECs it is stored in the rod-shaped Weibel-Palade bodies that produce the characteristic punctate perinuclear staining. These organelles are present in human ECs isolated from large vessels *(1,6)* but have been reported to be scarce or absent in capillary endothelium from various species *(2,4)*. However, typical granular perinuclear staining for vWF has been reported in human kidney, dermis *(9)*, synovium, lung, stomach, decidua, heart, adipose tissue, and brain *(8)* microvessel ECs.

PECAM-1 is constitutively expressed on the surface of ECs (>10^6 molecules/cell), and to a lesser extent in platelets, granulocytes, and a sub-population of CD8+ lymphocytes *(13)*. PECAM-1 staining of ECs in vitro is characterized by typical intense membrane fluorescence at points of cell-cell contact (**Fig. 2**).

E-selectin expression appears to be unique to ECs *(14)*. Although it is not constitutively expressed by the majority of ECs, E-selectin expression is induced following stimulation with inflammatory cytokines such as tumor necrosis factor-α (TNFα) or interleukin-1β (IL-1β) *(14)*. Absent on unstimulated controls, intense E-selectin expression is induced by pretreatment for 4 h with 10 ng/mL of TNFα reaching a maximum at 4–8 h before returning to background levels.

3.2.2. Immunocytofluoresent Characterization of Endothelial Cells

Immunocytoflourescence represents a simple rapid technique to characterize ECs. Outlined below is a staining protocol that we have routinely used for this purpose.

1. *Preparation of ECs on glass slides*. Multiwell glass chamber slides are extremely useful for this purpose as multiple tests can be performed on the same slide con-

Fig. 2. Immunofluorescent staining for platelet EC adhesion molecule-1 (PECAM-1/CD31) in HuMMEC.

serving both reagents and cells. The cells are cultured on chamber slides that have been pretreated for 1 h with 5 μg/cm^2 bovine fibronectin (Sigma) in PBS/A or 0.2 % gelatin (**Subheading 2.3. item 5**). When sufficient cells are present discard the medium and wash the cells twice with PBS/A prior to fixation. A range of fixatives can be employed depending on the activity of the antibody required. We routinely use acetone fixation that is suitable for most antibodies or formaldehyde. For acetone fixation, place the slides (*see* **Note 13**) in cold acetone (–20°C, 10 min), air dry and store frozen. Alternatively, fix cells in 3% formaldehyde solution for 30 min at room temperature. As formaldehyde does not permeabilize the plasma membrane 0.1% Nonidet P-40 or Triton X-100 can be added to detect an internal antigen.

2. Immunocytoflourescent staining
   a. Warm up slides to room temperature and wash with PBS/A (2X 5 min). To prevent nonspecific binding of the secondary antibody block slides for 20 min with 10% normal serum from the species in which the secondary antibody was raised.
   b. Incubate slides with predetermined or the manufacturers' recommended dilution of primary antibody in PBS/A for 60 min at room temperature.
   c. Wash slides with PBS/A (3X 5 min) and incubate with the appropriate FITC-labelled secondary antibody at 1/50 dilution in PBS/A for 30 min, 1 h at room temperature.
   d. Wash slides in PBS/A (3X 5 min), mount in 50% (v/v) glycerol in PBS/A. Stained slides can be stored for several months in the dark at 4°C.

Controls: To avoid false positives generated by nonspecific binding of secondary antibodies it is essential to include a negative control of cells treated as described above but with PBS/A substituted for the primary antibody. It is also useful to include control cell types such as fibroblasts or smooth muscle cells and previously characterized ECs to act as negative and positive controls respectively.

E-selectin: Most ECs do not constitutively express E-selectin and so must be incubated with a suitable inflammatory cytokine such as TNFα or IL-1β in growth medium for 4–6 h to induce its expression. We routinely use 10 ng/mL recombinant human TNF-α. Unstimulated cells should also be included as controls.

3.2.2. Other Properties of Adipose Microvessel Endothelial Cells

We have extensively examined the expression of many endothelial markers in our isolated microvessel ECs. These cells possess typical endothelial cell characteristics including scavenger receptors for acetylated low density lipoprotein, and functional angiotensin-converting enzyme at high levels. There has been considerable interest in the endothelial-restricted receptor tyrosine kinases (RTK) involved in EC differentiation and proliferation. Vascular EC growth factor/vascular permeability factor (VEGF) ***(16)*** is a key angiogenic factor and is unique in that it demonstrates pleiotropic activities specifically on ECs, including stimulation of proliferation and induction of procoagulant activity. All the microvessel EC types that we have cultured express the VEGF receptor family Flt-1, Flt-4, and KDR/Flk-1 ***(16,17)*** and proliferate and express increased tissue factor in response to VEGF. Similarly the Tie family of receptors (Tie-1 and Tie-2/Tek) are also expressed on these cells ***(17)***. As more reliable antibodies become available against these endothelial receptors they should be useful for EC characterization and selection.

3.3. Isolation of Endothelial Cells from Other Vascular Beds

We have adapted this basic method for the selection of adipose microvessel ECs for the isolation of ECs from other tissues. Here we briefly outline modifications that have been made for the isolation of human lung ***(12)*** and stomach ECs ***(18)***.

3.3.1. Microvessel Endothelial Cells from Human Lung

Although rich in ECs lung is composed of many cell types and is often more difficult to obtain. We have successfully used normal lung from transplant donors and diseased tissue from transplant recipients. To ensure that only microvascular ECs are harvested, a thin strip of tissue at the periphery of the lung is used. As the amount of tissue available is usually limited, and the subsequent yield of cells very low after Dynabead selection, it is better to select ECs after a few days in culture before they became overgrown with contaminating cells.

1. Cut small peripheral sections of lung (3–5 cm long, ~1 cm from the periphery) and wash in antibiotic/antimycotic solution.
2. Dissect the underlying tissue from the pleura and chop it up very finely using a tissue chopper.
3. Wash the 'mince' above a sterile 20-µm nylon mesh to filter out red blood cells and debris.
4. Incubate the retained material overnight in growth medium containing 2 U/mL dispase (Boerhinger Mannheim, Mannheim, Germany) on a rotary stirrer at 37°C.
5. Pellet the digest, resuspended in ~5 ml of trypsin/EDTA solution and incubate at 37°C for 15 min.
6. Add growth medium and remove fragments of undigested tissue by filtration through 100 µm nylon mesh.
7. Pellet and resuspended the cells in growth medium and plate onto gelatin-coated dishes.
8. Monitor the cultures daily, trypsinize, and select the ECs using PECA-beads before they became overgrown by contaminating cells.

3.3.2. Human Stomach Microvessel Endothelial Cells

For isolating microvessel ECs cultured from stomach biopsies and whole stomach from organ donors ***(18)***:

1. Expose the stomach mucosa and wash with antibiotic/antimycotic solution.
2. Dissect the mucosa from the underlying muscle, chop into 2–3 mm pieces and incubate in 1 m*M* EDTA in HBSS at 37°C in a shaking water bath for 30 min.
3. Transfer the pieces of mucosa to collagenase (type-II) solution containing 0.1% BSA for 60 min, and then trypsin/EDTA solution for 15 min, in a shaking water bath at 37°C.
4. Using a blunt dissecting tool scrape the mucosa and submucosa from white fibrous tissue.
5. Suspend the mucosal tissue in HBSS + 20% CS and wash through 100 µm nylon mesh.
6. Centrifuge the filtrate (700*g*, 5 min) and resuspend the pellet in ~12 mL of HBSS + 5% CS. Proceed with PECA-bead selection (**Subheading 3.1.2., step 5**).

4. Notes

1. We have found PECA-beads to be more reliable for purification of ECs ***(12)*** than using tosyl-activated Dynabeads directly coated with *UEA*-1 ***(9)***.
2. Precoated Dynabeads (and CELLelection™ beads; *see* **Note 8**) from Dynal are available carrying various secondary antibodies and are very convenient. However, antiimmunoglobulin-coated beads can be prepared using Tosyl-activated Dynabeads-M50 as follows: Incubate the secondary antibody (150 µg/mL) in

0.17 *M* sodium tetraborate buffer (pH 9.5; sterile filtered) with tosyl-activated Dynabeads-M450 for 24 h on a rotary stirrer at room temperature. Wash the beads 4X for 10 min and then overnight in PBS/A+0.1% BSA on a rotary stirrer at 4°C before proceeding to coat them with the primary antibody as described (**Subheading 2.2.**).

3. We have found that this M199/Ham F12 based recipe works well for a range of ECs but researchers may wish to optimize the medium further. There are a number of specialized EC growth media described in the literature such as MCDB131. Among several commercial sources of optimized EC medium, Clonetics (Clonetics Corp., San Diego, CA) supply a low-serum medium based on MCDB131.
4. Iron-supplemented calf serum (CS) supports the growth of ECs very well and represents an economical alternative to fetal calf serum.
5. We routinely use recombinant basic fibroblast growth factor (bFGF) and epidermal growth factor (EGF) instead of endothelial growth supplement because we have found them to be more consistent in supporting EC proliferation.
6. Omental adipose tissue obtained through general abdominal surgery can also be used with this method. Care should be taken to remove the fat from the omental membranes that are covered with a layer of mesothelium. Using PECA-beads we have not found mesothelial cell contamination to be a problem.
7. The cell PECA-bead suspension is incubated at 4°C during the purification steps to minimize nonspecific phagocytosis of Dynabeads.
8. We do not routinely remove Dynabeads following cell selection. However, it may be necessary to remove the Dynabeads if for example the cells are required for flow cytometry. This problem can be overcome by using CELLelection™ Dynabeads (Dynal) coated with the anti-PECAM-1 antibody to select the ECs. In this system antibodies are conjugated to the Dynabead via a DNA linker that can be cleaved with DNase-1 to release the beads from the cells following selection.
9. Dynabeads are internalized within –24 h of selection and are diluted to negligible numbers/cell by the first passage through cell mitosis ***(9)***. Consistent with the original observations of Jackson and colleagues ***(9,16)*** with *UEA*-I-coated Dynabeads we have not observed any adverse effects on the adherence, proliferation, or morphology of ECs following PECA-bead selection ***(12)***.
10. The major contaminating cell population that we have observed in unselected adipose EC cultures demonstrate a distinct fibroblastic morphology.
11. We have found that it is important for EC viability to rapidly remove the ECs from the surface of flasks because they seem to be very sensitive to trypsin exposure. It has been suggested that the use of a trypsin inhibitor to neutralize the tryptic activity immediately following detachment from the flask may prolong life span of ECs. We have used mung bean trypsin inhibitor (Sigma) for this purpose although we have not examined its effect on endothelial viability.
12. We have not observed an obvious decrease in EC viability following storage in liquid nitrogen for over 6 yr.

13. The plastic wells must be removed from the multiwell slides as acetone rapidly dissolves the plastic. The gasket should be left in place to keep reagents separate during the staining procedure.

Acknowledgments

I am extremely grateful to the plastic surgery departments at Nottingham City Hospital and Mount Vernon Hospital for providing breast adipose tissue.

References

1. Jaffe, E. A., Nachman, R. L., Becker, C. G., and Minidi, C. R. (1973) Culture of human endothelial cells derived from umbilical veins: identification by morphological and immunological criteria. *J. Clin. Invest.* **52,** 2745–2756.
2. Kumar, S., West, D. C., and Ager, M. (1987) Heterogeneity in endothelial cells from large vessels and microvessels. *Differentiation* **36,** 57–70.
3. Zetter, B. R. (1988) Endothelial heterogeneity: Influence of vessel size, organ localisation and species specificity on the properties of cultured endothelial cells In: *Endothelial Cells*, vol. 2 (Ryan, U. S., ed.) CRC Press: Boca Raton, Florida.
4. Kuzu, I., Bicknell, R., Harris, A. M., Jones, M., Gatter, K. G., and Mason, D. Y. (1992) Heterogeneity of vascular endothelial cells with relevance to diagnosis of vascular tumors. *J. Clin. Pathol.* **45,**143–148.
5. Garlanda, C. and Dejana, E. (1997) Heterogeneity of endothelial cells: Specific markers. *Arterioscler. Thromb. Vasc. Biol.* **17,** 1193–1202.
6. Hewett, P. W. and Murray, J. C. (1993) Human microvessel endothelial cells: isolation, culture and characterization. *In Vitro Cell. Dev. Biol.* **29A,** 823–830.
7. Wagner, R. C. and Matthews, M. A. (1975) The isolation and culture of capillary endothelium from epididymal fat. *Microvasc. Res.* **10,** 286–297.
8. Dorovini-Zis, K., Prameya, R., and Bowman, P. D. (1991) Culture and characterization of microvessel endothelial cells derived from human brain. *Lab. Invest.* **64,** 425–436.
9. Jackson, C. J., Garbett, P. K., Nissen, B., and Schrieber, L. (1990) Binding of human endothelium to *Ulex europaeus* -1 coated dynabeads: application to the isolation of microvascular endothelium. *J. Cell Science* **96,** 257–262.
10. Holthöfer, H., Virtanen, I., Kariniemi, A-L., Hormia, M., Linder, E., and Miettinen, A. (1982) *Ulex europaeus* 1 lectin as a marker for vascular endothelium in human tissues. *Lab. Invest.* **47,** 60–66.
11. Hewett, P. W. and Murray, J. C. (1994) Human omental mesothelial cells: a simple method for isolation and discrimination from endothelial cells. *In Vitro Cell Dev. Biol.* **30A,** 145–147.
12. Hewett, P. W. and Murray, J. C. (1993) Immunomagnetic purification of human microvessel endothelial cells using Dynabeads coated with monoclonal antibodies to PECAM-1. *Eur. J. Cell Biol.* **62,** 451–454.

13. Newman, P. J., Berndt, M. C., Gorski, J., White II, G. C., Lyman, S., Paddock, C., and Muller, W. A. (1990) PECAM-1 (CD31) cloning and relation to adhesion molecules of the immunoglobulin gene superfamily. *Science,* **247,** 1219–1222.
14. Bevilacqua, M. P., Pober, J. S., Mendrich, D. L., Cotran, R. S., and Grimbone, M. A. (1987) Identification of an inducible endothelial-leukocyte adhesion molecule. *Proc. Natl. Acad. Sci. (USA)* **84,** 9238–9242.
15. Hewett, P. W., Murray, J. C., Price, E. A., Watts, M. E., and Woodcock, M. (1993) Isolation and characterization of microvessel endothelial cells from human mammary adipose tissue. *In Vitro Cell. Dev. Biol.* **29A,** 325–331.
16. Neufeld, G., Cohen, T., Gengrinovitch, S., and Poltorak, Z. (1999) Vascular endothelial growth factor (VEGF) and its receptors. *FASEB J.* **13,** 9–22.
17. Mustonen, T. and Alitalo, K. (1995) Endothelial receptor tyrosine kinases Involved in angiogenesis. *J. Cell. Biol.* **129,** 895–898.
18. Hull, M. A., Hewett, P. W., Brough, J. L., and Hawkey, C. J. (1996) Isolation and culture of human gastric endothelial cells. *Gastroenterol.* **111,** 1230–1240.
19. Ruiter, D. J., Schlingemann, R. O., Rietveld, F. J. R., and de Waal, R. M. W. (1989) Monoclonal antibody-defined human endothelial antigens as vascular markers. *J. Invest. Dermatol.* **93,** 25S–32S.
20. Diaz-Flores, L., Gutiérrez, R., Varela, H., Rancel, N., and Valladares, F. (1991) Microvascular pericytes: A review of their morphological and functional characteristics. *Histol. Histopathol.* **6,** 269–286.

16

Transfection and Transduction of Primary Human Endothelial Cells

Stewart G. Martin

1. Introduction

The endothelium is involved in a number of normal physiological processes (regulating circulating levels of vasoactive agents, blood/gas exchange, regulating cellular traffic between intavascular and extravascular compartments of tissues, maintenance of the blood brain barrier, and so forth) and pathophysiological conditions characterized either by increased angiogenesis (arthritis, diabetic retinopathy, atherosclerosis, tumor growth, and metastasis) or inadequate angiogenesis [failure of ulcers to heal (inadequate wound healing in general), myocardial infarction, limb ischaemia secondary to arterial occlusive diseases and others]. Its location immediately adjacent to the blood stream and the fact that it has a large surface area makes it an attractive target for gene-transfer/gene-therapy strategies.

A large number of techniques exist for transferring nucleic acids into mammalian cells—both transfection (the process of introducing nucleic acids into eukaryotic cells by nonviral means) and transduction (viral mediated transfer) have been used with a large degree of success in nonhuman and immortalized-human endothelial cells (ECs). Unfortunately reports of successfully transfecting primary human endothelial cells (PHEC) are inconsistent and, when successful, efficiencies are usually low. The reasons for such variability and low transfection efficiencies are unclear but could be related to an intrinsic resistance of PHEC to take up and express foreign genes, the cytotoxicity of the transfection agent and/or procedure, differences in the tissue culture procedure (cellular passage number, growth conditions, and so on) and/or type of PHEC used [there is considerable morphologic, antigenic and biochemical heterogeneity between cells from different organs and from different vascular beds

From: *Methods in Molecular Medicine, Vol. 46: Angiogenesis Protocols*
Edited by: J. C. Murray © Humana Press Inc., Totowa, NJ

within the same organ *(1)*]. Transfection techniques, although generally yielding lower gene transfer efficiencies than viral transduction, are less labor intensive and time consuming. Such techniques are particularly well suited for screening purposes with transduction subsequently being used to achieve higher rates of transfer of genes that are of particular interest. **Table 1** shows some typical PHEC gene-transfer efficiencies obtained by various investigators using a variety of different techniques.

Numerous variables can act to influence the overall transfection frequency obtained, e.g., the proportion of cells that have taken up the nucleic acid and are expressing protein, toxicity of the procedure, concentration and purity of the nucleic acid used, initial cell number and degree of confluency, the transfection procedure itself (i.e., concentration of reagents, duration of exposure, among others), promoter and transgene sequences, vector sequences, the expression period given before selection or assay and the sensitivity of detection method. The differences in transfection efficiencies obtained by different laboratories may be related to differences in one or more of such variables. In general, transfection procedures should be optimized in each laboratory for each individual cell type by use of a reporter gene.

Although a number of techniques have been used by the author, the only ones that have yielded consistent and reproducible results are, as can be seen from **Table 1**, ExGen 500, electroporation, and adenoviral transduction. Each of these will be discussed in turn.

2. Materials

2.1. ExGen 500 Transfection

ExGen500 (TCS Biologicals) is a polycation polymer that our laboratory has had a certain amount of success with (~5–10% transfection efficiency of human umbilical vein endothelial cells (HUVEC), Martin et al., unpublished observations). It is a 22 kDa fully linear member of the polyethylenimine (PEI) family, complexes with DNA, and is rapidly endocytosed by cells. The high buffering capacity of the polycation protects the DNA whilst in the acidic endosomes and also leads to extensive endosomal swelling and rupture, thereby providing an escape mechanism for the polycation/DNA complex. In contrast to branched PEIs the linear structure makes the complexes insensitive to up to 30% serum during transfection [an important consideration since certain primary HEC rapidly undergo apoptosis in response to serum or growth factor deprivation *(14)*].

1. Early passage, pooled population of PHEC. Plated 24 hr prior to transfection procedure to ensure approximately 60–80% confluency at time of transfection (i.e., approx 1–2 × 10^5 HUVEC per well of a 6-well plate).

Table 1
Reports of Transfection Protocols Successfully Used in PHEC

Cell type[a]	Gene-transfer Method	Efficiency (if known)	Reference
HUVEC	$CaPO_4$ DEAE Dextran	0.1%–10%	Hanemaaijer et al. *(2)*, Sun et al. *(3)*, Tanner et al. *(4)*
HUVEC	Lipofection	5%	Sun et al *(3)*
	(Lipofectin)	5–16.6%	Siphehia and Martucci *(5)*
	(γAP-DRLIE/DOPE)	20.28%	Tanner et al *(4)*
HUVEC	Receptor mediated lipofection (Apo E—Lipofectin)	22.6%	Siphehia and Martucci *(5)*
HUVEC / HuMMEC	Polycation polymers (ExGen 500)	5–10%	Martin (unpublished results)
HUVEC	Electroporation	—	Nathwami et al. *(6)*
		—	Schwachtgen, et al. *(7)*
HUVEC / HuMMEC		2–4%	Martin (unpublished results)
HUVEC	Biolistic particle delivery	4%	Tanner et al. *(4)*
HUVEC	Adenovirus	98%	DeMartin et al. *(8)*
		88%	Lemarchand et al. *(9,10)*
		90%	Martin (unpublished results)
HUVEC	Retrovirus	57%	Inaba et al. *(11)*
		89%	Eton et al. *(12)*
HUVEC	Adeno-Associated Virus	4.6%	Lynch et al. *(13)*
HOMEC		12.6%	

[a]HUVEC, human umbilical vein endothelial cells; HuMMEC, human mammary microvessel endothelial cells, HOMEC, human omentum microvessel endothelial cells

2. ExGen 500 reagent (TCS Biologicals).
3. DNA: high-purity nucleic acid should be used (e.g., ion exchange column, Qiagen, 2x CsCl density centrifugation, etc. Experiments yield best results when plasmid DNA of the highest purity is used, presence of endotoxin significantly decreases transfection efficiencies. This cautionary note applies to all transfection procedures.
4. Phosphate buffered saline (PBS) and Hanks' balanced salt solution (HBSS).
5. Optimem serum free media (Gibco BRL). note: Certain PHEC will rapidly undergo apoptosis if deprived of serum or growth factors. If such a line is being used optimization of the procedure with varying concentrations of serum is required.
6. Microfuge tubes.

7. Vortex mixer.
8. Small volume pipets (e.g., Gilson's).

2.2. Electroporation

A number of laboratories have successfully employed electroporation to transfect HUVEC but even though careful optimization of the electrical pulse and field strength parameters [i.e., voltage, capacitance, ionic strength of solution (+/– serum)] is usually carried out, high levels of cytotoxicity invariably result. In comparison to other techniques, large amounts of DNA are usually required but unfortunately this is the only technique by which certain laboratories have obtained consistent and reproducible results. Certain laboratories [Schwachtgen and colleagues *(7)*] report that higher transfection efficiencies can be obtained with electroporation if cells are synchronised and held in G2/M phase of the cell cycle, i.e., whilst the nuclear membrane is melted. We have not noticed any improvement in transfection efficiencies if such synchronization is carried out.

1. PHEC (Note: due to the cytotoxicity associated with electroporation a large starting population of cells is required).
2. Electroporator.
3. Electroporation cuvets (4 mm).
4. High-purity DNA.
5. PBS and complete PHEC culture medium.
6. Pasteur and small volume pipets (e.g., Gilson's)
7. T75 tissue culture flasks.
8. Bench top centrifuge.

2.3. Adenoviral Transduction

The usual method of making an adenoviral vector is to replace the viral E1A or E1B gene with the gene of interest but substitutions and deletions of viral E3 and late genes have also been used. In E1/E3 double-deleted vectors up to 7.5 kb of nucleic acid can be inserted. E1 deletion decreases viral oncogenic and lytic properties yet still enables replication-defective infectious viral particles to be produced in a helper/producer cell line. The helper cell line usually used for viral production is the 293 line (a human embryonic kidney cell line transformed with sheared Ad5 DNA). The 293 cells contain 11% of the Ad5 genome and can be used to produce titers of up to 10^{12}–10^{13} pfu/mL. A useful alternative to 293 cells that yields higher titers in a shorter overall time is the 911 line ***(15)***, which was derived from human embryonic retinoblasts. Kits are now commercially available to construct E1- and/or E3-deficient vectors (Microbix Biosystems Inc., Toronto) based upon those developed by Graham and colleagues ***(16,17)***. A bacterial plasmid is transfected, along with a shuttle plasmid containing the DNA of interest, into a suitable helper cell line thereby allowing the production of infectious, nonreplicative viral particles. A number of different plasmids are available that allow for cloning into the E1 or E3 region for gene expression from the human CMV promoter. Promoterless vectors are available if promoter characterization studies are required. It takes approx 2 mo to produce recombinant infectious nonreplicative adenovirus using such kits. The process of viral production, propagation, and transduction is described, in detail, elsewhere ***(18)***. Recently, a number of biotechnology companies have started producing rapid adenoviral expression systems in which the timescale for viral production is cut down to less than three weeks (e.g., Adeno-X™ expression system, CLONTECH Laboratories). They also eliminate the requirement for plaque purification.

As adenoviruses infect both nondividing and dividing cells the purpose of the gene transfer strategy dictates which viral strategy should be pursued, i.e., if only actively dividing cells were to be transduced then a retroviral vector would be preferential, whereas if a high level of transduction of all cells is required then adenoviral vectors are extremely useful (selectivity of expression can subsequently be obtained by use of tissue specific or inducible promoters). Although the level of transduction is generally much higher with adenoviral vectors than with retroviruses, giving a correspondingly high initial level of gene expression, vector expression is usually negligible by 2–6 wk, either owing to the transduced cell becoming overloaded by vector, the episomal maintenance of the vector (as opposed to stable integration of the whole genome), or in vivo transduced cells disappearing because of the host immune responses to viral proteins. A number of approaches are currently being investigated in an attempt to reduce the immune response, including the development of second and third-generation vectors ***(19,20)***, gene transfer to the thymus ***(21)***, and use of nonhuman adenoviral vectors ***(22)***.

Certain laboratories insist, for safety reasons, that adenoviral work be carried out in dedicated locations within a laboratory and that stringent decontamination procedures be followed. Prior to starting such work, guidance from the requisite health and safety committee should be sought. As with the above techniques, regulations and guidelines concerning transfer of genetic material into mammalian cells should be understood prior to commencement of the work—regulations vary from laboratory to laboratory.

3. Methods

3.1. ExGen 500

1. Seed cells (from a pooled, early passage PHEC population) 24 h prior to transfection to ensure approximately 70% confluence (e.g., HUVEC, $1–2 \times 10^5$ cells per

well of a 6-well plate. Note: all subsequent procedures will assume that the protocol is conducted in such a format.). Incubate overnight at 37°C, 5% CO_2, in a humidified atmosphere.
2. Aspirate off media and replace with 1 mL Optimem 1 h prior to commencing transfection.
3. Dilute 4 µg of plasmid DNA in HBSS to give final volume of 50 µL; vortex gently. In a separate microfuge tube add 40 µL HBSS to 10 µL of ExGen 500; vortex gently. Add ExGen 500 solution to DNA (order of addition is critical); vortex gently and incubate at room temperature for 15 min with intermittent shaking.
4. Add the 100 µL of ExGen/DNA solution to the respective tissue culture well (dropwise whilst gently shaking) and incubate for 2 h at 37°C, 5% CO_2, in a humidified atmosphere. Remember to include a control well of ExGen transfection reagent with no DNA. Following an adequate expression period (usually varies between 24–72 h depending upon the plasmid and cell type being used) transfected cells are now ready for subsequent manipulations, e.g., determination of transfection efficiency.

3.2. Electroporation

1. For each time point (i.e., expression time, electroporation voltage or capacitance, or different DNA concentration and so forth): PHEC (from logarithmically growing population, ~80% confluent), trypsinized, pelleted, and washed with PBS, pelleted and resuspended at 1×10^7 cells per 0.5 mL (i.e., 2×10^7 per mL) in PBS.
2. Add 50 µg DNA to 0.5 mL of cells in electroporation cuvet. Incubate on ice for 10 min.
3. Electroshock cells (although in our hands best results were obtained using 300 V, 300 µF, it is recommended that various voltage and capacitance settings be carried to optimize the procedure).
4. Using a Pasteur pipet, plate cells into a T75-cm^2 flask (rinse cuvet with growth medium to ensure all cells are obtained). Incubate at 37°C, 5% CO_2, in a humidified atmosphere for an adequate expression period.

3.3. Estimation of Transfection Efficiency

Ideally both positive and negative controls should be incorporated into transfection optimization experiments. COS cells (African Green Monkey kidney cells) can be used as a positive control cell line with vectors driven by the SV40 promoter. As a result of being transformed with portions of the SV40 virus, COS cells contain the large T antigen which functions in SV40 replication. The combination of large T antigen and SV40 origin results in a higher copy number of such vectors in COS cells, which in turn may result in increased expression of the reporter gene compared to vectors lacking the SV40 origin. An empty vector and an extract, prepared from mock-transfected cells, should also be assayed for the presence of endogenous reporter-protein activity in cultured cells and acts as a suitable negative control.

Transfection efficiencies can vary significantly depending upon the promoter used to drive the reporter gene. Even supposed strong ubiquitous promoters vary in their capacity to drive expression of the same gene in the same vector construct, e.g., Tanner and colleagues ***(4)*** showed that the CMV promoter caused a significant increase in expression of biolistically transfected heat stable human placental alkaline phosphatase (hpAP) in HUVEC when compared to the RSV promoter. The usefulness of the SV40 promoter in maximizing reporter expression in HUVEC is questionable. Certain individuals report that it is a weak reporter in such cells (E. D. G. Tuddenham and J. McVey, personal communication), whereas others find it a suitable promoter for their purposes. If maximal unrestricted expression is required then the correct choice of ubiquitous promoter is essential, a number of reporter plasmids are commercially available that differ only in the promoter sequence used. If expression limited to the endothelium is required then tissue-specific promoters, such as the von Willebrand promoter ***(23–25)***, the KDR promoter ***(26)***, the Flt-1 promoter ***(27)*** or the ICAM-2 promoter ***(28)***, among others can be utilized. Inducible promoters such as the E-selectin ***(29,30)***, P-selectin ***(31,32)***, and the VCAM-1 promoter ***(33,34)*** can also be useful in certain endothelial gene transfer strategies.

Another major factor that determines overall efficiencies is the choice of reporter gene and detection method ***(35)***. A number of genes are used for such purposes, having certain advantages and disadvantages when compared with each other (i.e., ease of detection, sensitivity of detection method, and so forth).

References

1. Kumar, S., West, D. C., and Ager, M. (1987) Heterogeneity in endothelial cells from large vessels and microvessels. *Differentiation* **36,** 57–70.
2. Hanemaaijer, R., Arts, J., le Clerq, L., Kooistra, T., and van Hinsbergh, V. (1994) Plasminogen activator regulation studies following transfection of uman endothelial cells. *Fibrinolysis* **8,** 19–21.
3. Sun, B., Plumpton, C., Sinclair, J. H., and Brown, M. J. (1994) In vitro expression of calcitonin gene-related peptide in human endothelial cells transfected with plasmid and retroviral vectors. *Neuropeptides* **26,** 167–173.
4. Tanner, F. C., Carr, D. P., Nabel, G. J., and Nabel, E. G. (1997) Transfection of human endothelial cells. *Cardiovasc. Res.* **35,** 522–528.
5. Siphehia, R. and Martucci, G. (1995) High-efficiency transformation of human endothelial cells by Apo E-mediated transfection with plasmid DNA. *Biochem. Biophys. Res. Comm.* **214,** 206–211.
6. Nathwami, A. C., Gale, K. M., Pemberton, K. D., Crossman, D. C., Tuddenham, E. G. D., and McVey, J. H. (1994) Efficient gene transfer into human umbilical vein endothelial cells allows functional analysis of the human tissue factor gene promoter. *Brit. J. Haematol.* **88,** 122–128.

7. Schwachtgen, J-L., Ferreira, V., Meyer, D., and Keririou-Nabias, D. (1994) Optimisation of the transfection of human endothelial cells by electroporation. *BioTechniques* **17,** 882–887.
8. DeMartin, R., Raidl, M., Hofer, E., and Binder, B. R. (1997) Adenovirus-mediated expression of green fluorescent protein. *Gene Ther.* **4,** 493–495.
9. Lemarchand, P., Jaffe, H. A., Danel, C., Cid, M. A., Kleinman, H. K., Stratford-Perricaudet, L. D., et al. (1992) Adenovius-mediated transfer of a recombinant human α_1-antitrypsin cDNA to human endothelial cells. *Proc. Nat. Acad. Sci. (USA)* **89,** 6482–6486.
10. Inaba, M., Toninelli, E., Vanmeter, G., Bender, J. R., and Conte, M. S. (1998) retroviral gene transfer: Effects on endothelial cell phenotype. *J. Surg. Res.* **78,** 31–36.
11. Eton, D., Terramani, T. T., Wang, Y., Takahashi, A. M., Nigro, J. J., Tang, L., and Yu, H. (1999) *J. Vasc. Surg.* **29,** 863–873.
12. Lynch, C. M., Hara, P. S., Koudy, J., Dean, R. H., and Geary, R. L. (1997) Adeno-associated virus vectors for vascular gene delivery. *Circ. Res.* **80,** 497–505.
13. Araki, S., Shimada, Y., Kaji, K., and Hayashi, H. (1990) Apoptosis of vascular endothelial cells by fibroblast growth factor deprivation. *Biochem. Biophys. Res. Comm.* **168,** 1194–1200.
14. Fallaux, F. J., Kranenberg, O., Cramer, S. J., Houweling, A., van Ormondt, H., Hoeben, R. C., and van der Eb, E. B. (1996) Characterisation of 911: A new helper line for the titration and propagation of early region 1-deleted adenoviral vectors. *Hum. Gene Ther.* **7,** 215–222.
15. Graham, F. L. (1984) Covalently closed circles of human adenovirus DNA are infectious. *EMBO J.* **3,** 2917–2922.
16. Bett, A. J., Haddara, W., Prevec, L., and Graham, F. L. (1994) An efficient and flexible system for construction of adenovirus vectors with insertions or deletions in early region-1 and region-3. *Proc. Nat. Acad. Sci. (USA)* **91,** 8802–8806.
17. Yeh, P., Dedieu, J-F., Orsini, C., Vigne, E., denefle, P., and Perricaudet (1996) Efficient dual transcomplementation of adenovirus E1 and E4 regions from a 293-derived cell line expressing a minimal E4 functional unit. *J. Virol.* **70,** 559–565.
18. Graham, F. L. and Prevec, L. (1991) Manipulation of adenovirus vectors. In: *Methods in Molecular Biology*, vol. 7: *Gene Transfer and Expression Protocols.* (Murray, E. J., ed.) The Humana Press Inc., Totowa, NJ, pp. 109–128.
19. Wang, Q. and Finer, M. H. (1996) Second-generation adenovirus vectors. *Nat. Med.* **2,** 714–716.
20. Yeh, P., Dedieu, J-F., Orsini, C., Vigne, E., Denefle, P., and Perricaudet (1996) Efficient dual transcomplementation of adenovirus E1 and E4 regions from a 293-derived cell line expressing a minimal E4 functional unit. *J. Virol.* **70,** 559–565.
21. DeMatteo, R. P., Raper, S. E., Ahn, M., Fisher, K. J., Burke, C., Radu, A., et al. (1995) Gene transfer to the thymus: A means of abrogating the immune response to recombinant adenovirus. *Ann. Surg.* **222,** 229–242.
22. Nguyen, T. T., Nery, J. P., Rocha, C. E., Carney, G. E., Spindler, K. R., and Villarraeal, L. P. (1999) Mouse adenovirus (MAV-1) expression in primary human

endothelial cells and generation of a full-length infectious plasmid. *Gene Ther.* **6,** 1291–1297.
23. Feirrera, V., Assouline, Z., Schwachtgen, J-L., Bahnak, B. R., Meyer, D., and Kerbiriou-Nabias, D. (1993) The role of the 5′-flanking region in the cell-specific transcription of the human von Willebrand factor gene. *Biochem. J.* **293,** 641–648.
24. Jahroudi, N. and Lynch, D. C. (1994) Endothelial-cell-specific regulation of von Willebrand factor gene expression. *Mol. Cell. Biol.* **14,** 999–1008.
25. Ozaki, K., Yoshida, T., Ide, H., Saito, I., Ikeda, Y., Sugimura, T., and Terada, M. (1996) Use of von Willebrand factor promoter to transduce suicidal gene to human endothelial cells, HUVEC. *Hum. Gene Ther.* **7,** 1483–1490.
26. Patterson, C., Wu, Y. X., Lee, M. E., Devault, J. D., Runge, M. S., and Haber, E. (1997) Nuclear protein interactions with the human KDR/flk-1 promoter in vivo—Regulation of Sp1 binding is associated with cell type-specific expression. *J. Biol. Chem.* **272,** 8410–8416.
27. Morishita, K., Johnson, D. E., and Williams, L. T. (1995) A novel promoter for vascular endothelial cell growth factor receptor (flt-1) that confers endothelial-specific gene expression. *J. Biol. Chem.* **270,** 27,948–27,953.
28. Cowan, P. J., Tsang, D., Pedic, C. M., Abbott, L. R., Shinkel, T. A., d'Apice, A. J. F., and Pearse, M. J. (1998) The human ICAM-2 promoter is endothelial cell-specific in vitro and in vivo and contains critical SP1 and GATA binding sites. *J. Biol. Chem.* **273,** 11,737–11,744.
29. Whelan, J., Ghersa, P., Vanhuijsduijnen, R. H., Gray, J., Chandra, G., and Talbot, F. (1991) An NFκβ-like factor is essential but not sufficient for cytokine induction of endothelial leukocyte adhesion molecule-1 (ELAM-1) gene transcription. *Nuc. Acid Res.* **19,** 2645–2653.
30. Smith, G. M., Whelan, J., Pescini, R., Ghersa, P., Delamarter, J. F., and Vanhuijsduijnen, R. H. (1993) DNA-methylation of the E-selectin promoter represses NFκβ transactivation. *Biochem. Biophys. Res. Comm.* **194,** 215–221.
31. Pan, J. and McEver, R. P. (1993) Characterisation of the promoter fro the human P-selectin gene. *J. Biol. Chem.* **268,** 22,600–22,608.
32. Pan, J., Xia, L., and McEver, R. P. (1998) Comparison of promoters for the murine and human P-selectin genes suggestes a species-specific and conserved mechanism for transcriptional regulation in endothelial cells. *J. Biol. Chem.* **273,** 10,058–10,067.
33. Iademarco, M. F., McQuillan, J. J., Rosen, G. D., and Dean, D. C. (1992) Characterisation of the promoter for vascular cell adhesion molecule-1 (VCAM-1). *J. Biol. Chem.* **267,** 16,323–16,329.
34. Neish, A. S., Kachigian, L. M., Park, A., Baichwal, V. R., and Collins, T. (1995) Sp1 is a component of the cytokine-inducible enhancer in the promoter of vascular cell adhesion molecule-1. *J. Biol. Chem.* **270,** 28,903–28,909.
35. Martin, S. G. and Murray, J. C. (2000) Gene-transfer systems for human endothelial cells. *Adv. Drug Deliv. Rev.* **41,** 223–233.

17

Vascular Smooth Muscle Cells

Isolation, Culture, and Characterization

Richard C. M. Siow and Jeremy D. Pearson

1. Introduction

In blood vessel development (vasculogenesis) smooth muscle cells (SMCs) are derived by differentiation of mesenchymal cells under the influence of mediators secreted by the endothelial cells (ECs) composing newly formed vessels. In angiogenesis, SMCs can be formed in the same way or by proliferation of existing SMCs *(1)*. Key events in the development of atherosclerotic lesions and restenosis of arteries are now recognized to include vascular SMC migration, hyperplasia, and hypertrophy *(2)*. This has led to an increase in studies of SMC function in response to growth factors, extracellular matrix, and lipoproteins, under controlled in vitro conditions, to address the cellular mechanisms in atherogenesis. Most of the methodology used for vascular smooth muscle culture has been developed for such studies *(3)*.

Vascular SMCs in vivo are capable of expressing a range of phenotypes, which fall within a continuous spectrum between 'contractile' and 'synthetic' states *(4,5)*. The adult vasculature predominantly consists of the contractile-state SMC, whose main function is contraction and the maintenance of vascular tone. They exhibit a musclelike appearance, with up to 75% of their cytoplasm containing contractile filaments. However, in culture, these cells are able to revert to a synthetic phenotype normally found in embryonic and young developing vessels. These proliferative cells synthesize extracellular matrix components such as elastin and collagen, and consequently contain large amounts of rough endoplasmic reticulum and Golgi apparatus but few myofilaments in their cytoplasm *(6)*. The modulation of vascular SMC from a con-

From: *Methods in Molecular Medicine, Vol. 46: Angiogenesis Protocols*
Edited by: J. C. Murray © Humana Press Inc., Totowa, NJ

tractile to a synthetic phenotype is an important event in atherogenesis and restenosis, resulting in myointimal thickening and arterial occlusion. This may arise from damage to the endothelium and exposure of SMC to blood components and stimuli released from ECs and SMCs, lymphocytes, macrophages, and platelets *(2)*.

The methodology used to isolate vascular SMC can determine the initial phenotype of cells obtained *(3)*. The two main techniques commonly employed in SMC isolation from vascular tissue are enzymatic dissociation, which readily yields a small number of SMC in the contractile phenotype, and explantation, which yields larger amounts of SMC after 1–2 wk, in the synthetic proliferative phenotype. This chapter describes these two alternative techniques for vascular SMC isolation, subculture passaging, and characterization for cell culture studies.

2. Materials

2.1. Culture Medium and Solutions for Maintaining SMC Cultures

1. The most commonly used growth medium in SMC culture is Dulbecco's modified Eagle's medium (DMEM) containing 4500 mg/L glucose (Gibco); however, Medium 199 (Sigma) is also suitable (*see* **Note 1**). The following additions to the basal medium are necessary prior to use (final concentrations): 2 m*M* L-glutamine, 40 m*M* bicarbonate, 0.002 % phenol red, 100 U/mL penicillin, 100 μg/mL streptomycin and 10% (v/v) fetal calf serum (FCS, Sigma). Sterile stocks of these components are usually prepared and stored as frozen aliquots as described below. The complete medium can be stored at 4°C for up to 1 mo and is prewarmed to 37°C prior to use in routine cell culture. Culture medium without the FCS component is used during the isolation procedures.
2. Hanks' balanced salt solution (HBSS, Sigma) is used as a tissue specimen collection medium. The following additions, from sterile stock solutions, are necessary prior to use (final concentrations): 100 μg/mL Gentamicin, 0.025 *M* HEPES, 20 m*M* bicarbonate, and 0.001% phenol red. The HBSS can be stored in aliquots at 4°C for up to 2 wk.
3. L-Glutamine (200 m*M*) stock solution: Dissolve 5.84 g L-glutamine (Sigma) in 200 mL tissue culture grade deionized water and sterilize by passing through a 0.22 μm filter. Aliquots of 5 mL are stored at –20°C and 4 mL used in 400 mL of medium.
4. Bicarbonate (4.4%, 0.52 *M*)-Phenol red (0.03%) solution: Dissolve 44 g $NaHCO_3$ and 30 mg phenol red in 1000 mL tissue culture grade deionized water, and sterilize by autoclaving for 10 min at 115°C. Aliquots of 15 mL are stored at 4°C for up to 6 mo and 2 aliquots used in 400 mL of medium.
5. Penicillin and streptomycin stock solution (80X concentrate): Dissolve 480 mg penicillin (G sodium salt, Sigma) and 1.5 g streptomycin sulfate (Sigma) in 200 mL tissue culture grade deionized water and sterilize by passing through a 0.5 μm

prefilter and a 0.22 μm filter. Aliquots of 5 mL are stored at –20°C and 1 aliquot used in 400 mL of medium.

6. Gentamicin solution (80X concentrate): Dissolve 750 mg gentamicin sulfate (Sigma) in 100 mL tissue culture grade deionized water and sterilize by passing through a 0.22 μm filter. Aliquots of 5 mL are stored at –20°C and 1 aliquot used in 400 mL of HBSS.
7. HEPES solution (1 *M*): Dissolve 47.6 g of HEPES (Sigma) in 200 mL tissue culture grade deionized water and sterilize by passing through a 0.22 μm filter. Aliquots of 5 mL can be stored at –20°C and 2 aliquots used in 400 mL of HBSS.
8. Trypsin solution (2.5 %): Trypsin from porcine pancreas (Sigma) is dissolved (2.5 g/100 ml) in PBS-A and sterilized by passing through a 0.22 μm filter. Aliquots of 10 mL are stored at –0°C.
9. Ethylenediaminetetraacetic acid (EDTA) solution (1%): EDTA disodium salt is dissolved (500/mg 50 mL) in tissue culture grade deionized water and sterilized through a 0.22 μm filter. Aliquots of 5 mL are stored at 4°C.
10. Trypsin (0.1 %)-EDTA (0.02 %, 0.5 m*M*) solution is prepared by adding 10 mL Trypsin (2.5%) and 5 mL EDTA (1%) to 250 mL sterile PBS-A. This solution is prewarmed to 37°C before use to detach cells from culture flasks and stored at 4°C for up to 2 mo.

2.2. Enzymatic Dissociation of SMC

1. Collagenase, Type II (Sigma). Dissolve collagenase in serum-free medium (3 mg/mL) on ice. Particulate material is removed by filtering the solution through a 0.5-μm prefilter and then sterilize by passing through a 0.22-μm filter. This enzyme solution can be stored long term as 5–10 mL aliquots at –20°C until use.
2. Elastase, type IV from porcine pancreas (Sigma). Immediately before use, elastase is dissolved in serum-free medium (1 mg/mL) and the pH of solution adjusted to 6.8 with 1 *M* HCl. This solution is then sterilized by passing through a 0.22 μm filter and kept on ice.

2.3. Immunofluorescence Microscopy

1. Mouse monoclonal antibody to α-smooth muscle actin (Sigma).
2. Normal rabbit serum (Sigma).
3. Fluorescein isothiocyanate conjugated rabbit antimouse antibody (Dako).
4. Methanol (100 %).

2.4. Equipment

All procedures should be carried out in a Class II laminar flow safety cabinet using aseptic technique. Dissection equipment should be thoroughly washed and sterilized by immersion in 70% ethanol or by autoclaving at 121°C for 20 min.

1. 25 cm^2 and 75 cm^2 tissue culture flasks and 90-mm Petri dishes.
2. Sterile Pasteur, 5- and 10-mL pipets.

3. Sterile 30-mL universal containers and 10-mL centrifuge tubes.
4. Scalpel handles and blades, small scissors, watchmaker's forceps, and hypodermic needles.
5. Cork board for dissection covered with aluminum foil, both sterilized by thorough spraying with 70 % ethanol.
6. Sterile conical flasks of various sizes.
7. Lab-Tek chamber slides (Nunc).

3. Methods

3.1. Collection of Tissue Samples

In our laboratory, cells are routinely isolated from human umbilical arteries, a readily available source of human vascular SMC. As soon as possible after delivery, the whole umbilical cord is placed in the HBSS collection medium and stored at 4°C. Cords collected and stored in this way can be used for SMC isolation up to 48 h after delivery. SMC can also be isolated from arteries following harvesting of ECs from the corresponding umbilical vein *(7)*. Immediately prior to proceeding with SMC isolation, 5-cm lengths of the umbilical artery should be carefully dissected out from the cord, with minimal surrounding connective tissue remaining, and stored in new collection medium. Other common sources of arterial tissue are from rat or porcine aortae, which should be carefully dissected out from the animal and stored in the collection medium at 4°C as soon as possible (*see* **Note 2**).

3.2. Isolation of SMCs by Enzymatic Dispersion

1. As much surrounding connective tissue as possible should be dissected away from around the artery and the tissue washed with new HBSS collection medium. The artery is then placed on the sterile dissection board and covered with HBSS to keep it moist.
2. The artery is fixed to the dissection board at one end using a hypodermic needle and then cut open longitudinally, using small scissors, with the luminal surface upward.
3. The endothelium is removed along the whole length of the artery by scraping the cell layer off with a sterile scalpel blade and the tissue remoistened with HBSS.
4. The thickness and nature of the arterial wall will vary depending on the source of tissue; however, in general the arterial intima and media are peeled into 1–2-mm-width transverse strips using watchmaker's forceps and a scalpel. Muscle strips are transferred in to a 90-mm Petri dish containing HBSS. This procedure is repeated for the whole surface of the vessel.
5. Most of the HBSS medium is then aspirated off and the muscle strips cut into 1–2-mm cubes using scissors or a scalpel blade. The cubes are then washed in new HBSS and transferred into a sterile conical flask of known weight and the mass of tissue measured to determine volume of enzyme solution needed for digestion.

6. Collagenase solution in serum-free culture medium is next added to the tissue to give a ratio of tissue (g) to enzyme solution (ml) of 1/5 (w/v). The flask is then covered with sterile aluminum foil and shaken in a water bath at 37°C for 30 min.
7. Elastase solution is then prepared and directly added to the solution containing the tissue and collagenase. The flask is returned to the shaking water bath at 37°C and every 30 min during the following 2–5 h the suspension is mixed by pipeting in the sterile safety cabinet.
8. At each 30 min interval, a 10 µL sample of suspension is transferred to a hemocytometer to check for the appearance of single cells. This is repeated until the tissue is digested and there are no large cell aggregates visible. Digestion should not proceed for longer than 5 h to avoid loss of cell viability. (*See* **Note 3**)
9. The cell suspension is finally divided into 10-ml centrifuge tubes and centrifuged at 50–100*g* for 5 min. The supernatants are carefully aspirated and 2–5 mL of prewarmed complete culture medium added to resuspend the cells using a sterile pipette. Cells are then seeded into 25-cm^2 culture flasks at a density of about 8×10^5 cells ml^{-1} and placed into a 37°C, 5% CO_2 incubator with the flask cap loose.
10. Viable SMC should adhere to the flask wall within 24 h; the medium is then replaced with fresh pre-warmed complete culture medium. Half of the culture medium is replaced every 2–3 d until a confluent SMC monolayer is obtained.

3.3. Isolation of SMCs by Explant Culture

Explant cultures are suitable if only limited vascular tissue is available. The sample is first treated exactly as described in **steps 1–3** of **Subheading 3.2.**, and then the following procedure is adopted:

1. The artery is cut into 2-mm cubes with a scalpel blade and placed on to the surface of a 25-cm^2 culture flask with a sterile Pasteur pipet and its luminal surface in contact with the flask wall. The artery should be kept continually moist with the HBSS and a small drop of serum containing medium should be placed on each cube when placed in the flask.
2. The cubes are distributed evenly on the surface with a minimum of 12–16 cubes per 25-cm^2 flask. The flask is then placed upright and 5 mL serum containing medium added directly to the bottom of the flask before transferring into a 37°C, 5% CO_2 incubator with its cap loose. To facilitate adherence of the explanted tissue to the culture plastic substrate, the flask is kept upright for 2–4 h in the incubator before the flask is carefully placed horizontally such that the medium completely covers the attached muscle cubes.
3. The explants should be left undisturbed for 4 d, inspected daily for infections (*see* **Note 4**). Every 4 d any unattached explant cubes should be removed and half the medium replaced with fresh prewarmed serum containing medium. Cells will initially migrate out from the explants within 1–2 wk.
4. After 3–4 wk sufficient density of SMC around the explants should allow removal of the tissue. Using a sterile Pasteur pipet, the cubes are gently dislodged from

the plastic flask surface and aspirated off with the culture medium. The culture medium is replaced and cells are then left for a further 2–4 d to proliferate and form a confluent SMC monolayer (*see* **Note 5**).

3.4. Subculture of SMCs

1. Once a confluent monolayer has been attained in a 25-cm^2 flask by either isolation method, the SMC can be subcultured (passaged) into further 25-cm^2 flasks or a 75-cm^2 flask. The culture medium is removed and cells are washed twice with prewarmed sterile PBS-A to remove traces of serum.
2. Prewarmed trypsin/EDTA solution (0.5 mL for 25 cm^2 flask or 1 mL for a 75-cm^2 flask) is added to cover the cells and the flask incubated at 37°C for 2–4 min. The flask is then examined under the microscope to ensure cells have fully detached. This can also be facilitated by vigorous tapping of the side of the flask 3–6 times to break up cell aggregates (*see* **Note 6**).
3. Serum containing medium (5 mL) is added to stop the action of the trypsin, which can reduce SMC viability through prolonged exposure. The cell suspension is then drawn up and down a sterile Pasteur pipet 4–6 times to further break up any cell clumps.
4. The cells are then transferred into new culture flasks at a split ratio of 1/3 and sufficient serum containing medium added to the new flasks (5 mL in a 25-cm^2 and 10 mL in a 75-cm^2 flask). The flasks are returned to the 37°C, 5 % CO_2 incubator and the culture medium changed as described above.

Smooth muscle cells can be passaged 10–20 times, depending on species, before their proliferation rate significantly decreases. Phenotypic changes of enzyme dispersed SMC to the ‘proliferative’ state occurs following passaging, the extent of which depends on SMC species and seeding density ***(4,5)***.

3.5. Characterization of SMCs

Confluent monolayers of vascular SMCs exhibit a characteristic ‘hill and valley’ morphology in culture ***(8)***. Isolated primary cells can be identified by their positive reaction with antibodies against α-smooth muscle actin, which has been shown to be a specific marker for SMCs ***(9)***, and negative staining for von Willebrand factor or lack of uptake of acetylated low density lipoproteins, both EC-specific markers ***(7,10)***. The following procedure, described here in brief, can be used to identify SMCs by their positive staining with a fluorescein isothiocyanate (FITC)-labeled antibody to smooth muscle α-actin.

1. SMCs are subcultured into Lab-Tek slide wells and characterized after 48 h.
2. The culture medium is removed from the wells and cells gently washed three times with serum-free culture medium before being fixed with ice-cold methanol (100%) for 45 sec, and then further washed three times with ice-cold PBS-A.

3. Cells are then incubated with a mouse monoclonal antismooth muscle α-actin antibody at 1/50 dilution with PBS-A for 60 min at room temperature. As a negative control, some cells are incubated with PBS-A only at this stage.
4. The primary antibody or PBS-A is then removed and cells washed three times with PBS-A and incubated for 5 min with normal rabbit serum at 1/20 dilution.
5. After a single wash with PBS, cells are then further incubated for 30 min at room temperature with FITC-conjugated rabbit anti-mouse IgG (Dako) diluted 1/50 in PBS-A.
6. Finally, cells are washed three times with PBS-A and viewed under a microscope equipped for epifluorescence with appropriate filters for FITC.

Visualization of positive staining with this technique should reveal cells with a 3D network of long, straight and uninterrupted α-actin filaments running in parallel to the longer axis of the cells and an underlying row of parallel filaments along the smaller cell axis, with no cytoplasmic staining between filaments.

3.6. Cryopreservation

Vascular SMCs can be cryopreserved with a recovery <50%. Explant cultures of SMCs do not appear to be adversely affected by freezing; however, enzyme dispersed SMC may have a reduced proliferation rate and passaging efficiency on thawing. Although the following protocol is suggested, other techniques of cryopreservation are also available.

1. SMCs should be detached from a 75-cm^2 culture flask as described in **Subheading 3.4.**
2. Following centrifugation of the cell suspension for 5 min at 1000 rpm, the supernatant is aspirated and the cell pellet resuspended well in serum-containing culture medium with an additional 10% (v/v) dimethyl sulfoxide (DMSO) and transferred to a suitable cryovial.
3. To facilitate slow freezing, the cryovial is then placed at 4°C for 1 h, transferred to –20°C for 30 min and –70°C for 1 h before being immersed into liquid nitrogen for long term storage.
4. To thaw cells, the cryotube should be rapidly warmed to room temperature and the cell suspension transferred to a 10 mL centrifuge tube. Prewarmed serum-containing culture medium is added and the cells centrifuged at 50–100*g* for 5 min.
5. The supernatant is aspirated to remove the DMSO, the cell pellet resuspended well in prewarmed serum-containing medium, and then transferred into a 25-cm^2 flask. Cells should be passaged once prior to use in experiments.

3.7. Summary

The two techniques for SMC isolation described yield cells with very different proliferative properties in culture. If larger quantities of SMC in culture are required, the explant isolation technique is recommended, although initially slower to yield cells. Enzymatic dispersion may provide more cells in the contractile phenotype, but the initial yield may be low and subcultures do not proliferate through as many passages.

4. Notes

1. Other "smooth muscle cell-optimized" culture media are commercially available, such as Smooth Muscle Basal and Growth Media (Clonetics), which are based on MCDB131 medium. These media may help to promote more rapid outgrowth of cells from tissue explants, but it should be noted that they may contain components which could alter SMC function, such as insulin, basic fibroblast growth factor (bFGF) and hydrocortisone.
2. For maximal SMC yield and viability, cells should be isolated as soon as possible after harvesting the vessels. This also reduces the risk of infections resulting from handling and excision of tissues under nonsterile conditions.
3. If the SMC yield and viability is low following enzyme dispersion, soybean trypsin inhibitor, at a final concentration of 0.1 mg/mL, can be added to the enzyme solution to inhibit the action of nonspecific proteases that may contaminate commercial elastase.
4. Should infections frequently occur following explantation, additional antibiotics and fungicides can be added to the culture medium for the initial 24 h following isolation and then the medium replaced. Gentamicin (25 μg/mL) and Amphotericin B (2 μg/mL) are commonly used and 5 mL aliquots of these can be stored at –20°C as 2.5 mg/mL and 0.2 mg/mL stocks, respectively.
5. When the explanted tissue is removed, SMCs can also be detached by trypsinization and redistributed evenly in the same flask as described in **Subheading 3.4.** This is advisable if cells have grown in a very dense pattern around the explants and will facilitate obtaining confluent monolayers of cells.
6. The trypsin/EDTA solution should be prewarmed to 37°C only immediately prior to use and not left in a heated water bath for extended periods to prevent loss of activity. Cells should not need incubation with trypsin/EDTA at 37°C for longer than 5–7 min to detach from the flask and this may indicate that a fresh solution should be prepared.

Acknowledgements

The research in our laboratory is supported by the British Heart Foundation and the Ministry of Agriculture, Fisheries, and Food (UK). We are grateful to the midwives at St. Mary's Hospital, Imperial College of Science, Technology and Medicine, University of London for supply of human umbilical cords.

References

1. Risau, W. (1997) Mechanisms of angiogenesis. *Nature* **386,** 671–674.
2. Ross, R. (1993) The pathogenesis of atherosclerosis: A perspective for the 1990s. *Nature* **362,** 801–809.

3. Campbell, J. H. and Campbell, G. R. (1993) Culture techniques and their applications to studies of vascular smooth muscle. *Clin. Sci.* **85,** 501–513.
4. Thyberg, J., Hedin, U., Sjolund, M., Palmberg, L., and Bottger, B. A. (1990) Regulation of differentiated properties and proliferation of arterial smooth muscle cells. *Arteriosclerosis* **10,** 966–990.
5. Owens, G. K. (1995) Regulation of differentiation of vascular smooth muscle cells. *Physiol. Rev.* **75,** 487–517.
6. Campbell, J. H. and Campbell, G. R. (1987) Phenotypic modulation of smooth muscle cells in culture. In: *Vascular Smooth Muscle Cells in Culture*, Vol. 1 (Campbell, J. H. and Campbell, G. R., eds.), CRC Press Inc., Boca Raton, FL, pp. 39–55.
7. Morgan, D. L. (1996) Isolation and culture of human umbilical vein endothelial cells. In: *Methods in Molecular Medicine: Human Cell Culture Protocols* (Jones, G. E., ed.), Humana Press Inc., Totowa, NJ, pp. 101–109.
8. Thyberg, J., Nilsson, J., Palmberg, L., and Sjolund, M. (1985) Adult human arterial smooth muscle cells in primary culture. Modulation from contractile to synthetic phenotype. *Cell Tissue Res.* **239,** 69–74.
9. Skalli, O., Ropraz, P., Trzeciak, A., Benzonana, G., Gillessen, D., and Gabbiani, G. (1986) A monoclonal antibody against α-smooth muscle actin. *J. Cell Biol.* **103,** 2787–2796.
10. Voyta, J. C., Via, D. P., Butterfield, C. E., and Zetter, B. R. (1984) Identification and isolation of endothelial cells based on their increased uptake of acetylated-low density lipoprotein. *J. Cell Biol.* **99,** 2034–2040.

18

Bovine Retinal Microvascular Pericytes

Isolation, Propagation, and Identification

Ramesh C. Nayak and Ira M. Herman

1. Introduction

The growth of new capillaries from existing vessels (angiogenesis) is of fundamental importance in wound healing and in pathological situations such as proliferative diabetic retinopathy *(1)*, rheumatoid arthritis *(2)*, and tumor growth. Consequently, considerable interest in vascular cell biology has arisen in apparently disparate clinical and experimental fields. Held in common, however, is the hope that an understanding of the cellular and molecular mechanisms that regulate angiogenesis will lead to novel therapeutic agents and targets.

Central to achieving a comprehensive mechanistic understanding of angiogenesis is the ability to isolate and propagate the cell types that form the capillary wall. The capillary is composed of two cell types, the endothelial cell (EC) and the pericyte (also known as the mural cell) *(3)*. The capillary tube that will act as the conduit for blood is formed by the ECs that subsequently become associated with pericytes at the abluminal surface of the endothelium. This close apposition of pericyte and endothelial cell allows a molecular "dialog" to take place between these cell types enabling each to tightly regulate the other's function in response to local stimuli. These regulatory interactions have recently been modeled in tissue culture *(3)*.

The first descriptions of methods for the isolation of both microvascular pericytes and ECs were published in the middle to late 1970s with the methods for pericyte isolation preceding those for ECs *(4–8)*. Meezan et al. reported, in 1974, the isolation of metabolically active microvessels from bovine retina and cerebral cortex *(4,5)*. This led Buzney et al. to use these methods of microvessel

From: *Methods in Molecular Medicine, Vol. 46: Angiogenesis Protocols*
Edited by: J. C. Murray © Humana Press Inc., Totowa, NJ

preparation to isolate and culture pericytes from simian, bovine, and human retina *(6)*. In 1977, Del Vecchio et al. *(7)* isolated capillary fragments from rat adrenal glands, which was followed in 1978 by the demonstration by Folkman et al. *(8)* that viable ECs could be recovered from isolated capillary fragments, cloned, and propagated on gelatin coated plates in conditioned media. Inexplicably, the reports from Meezan et al. and Buzney et al. did not appear to stimulate interest in tissue culture studies of the pericyte, perhaps because the importance of the pericyte in regulating capillary function had yet to become appreciated *(3)*. In 1983, using the same initial strategy of isolating viable capillary fragments, Gitlin and D'Amore were able to isolate and selectively propagate bovine retinal microvascular pericytes *(9)*. That report and the recognition of the importance of the pericyte in controlling EC growth *(3)*, combined with development of biochemical markers for the identification of pericytes *(10,11)*, have stimulated interest in studying the pericyte in vitro. Since then, several methodological procedures have been published for pericyte isolation from various tissues *(9,10,12–15,25)*. The majority of reports describe the isolation of pericytes from retina and most are a variation of the work described above. Most procedures begin with enrichment of microvessel fragments by sieving collagenase-digested tissue over nylon meshes of defined mesh size and then explanting the recovered microvessels under selective tissue culture conditions. The method we detail here has been in use in our laboratories for over 15 yr and is a variant of the methods discussed above.

2. Materials

2.1. Dissection and Harvesting of Retinal Tissue

1. 20 bovine eyes, globes intact (reject any that are pierced or perforated).
2. The following materials should be autoclaved the day before the isolation is to be performed: 3 one-litre beakers (openings covered with aluminium foil), 1 rubber policeman, 1 curved scissors, 2 straight scissors, 1 toothed forceps, 1 straight forceps, 3 scalpels, and 2 sterile fields.

   Sterile fields are prepared by taking an approx 75-cm-long strip of aluminium foil and placing within it a slightly shorter strip of absorbent paper towel (such as is used in roll dispensers). The edges of the aluminium foil are folded in over the paper and the field is then folded into overlapping thirds by folding the ends in toward the center. The field is then wrapped in aluminium foil, and autoclaved to sterilize.
3. Betadine (10% povidone-iodine solution).
4. Penicillin-streptomycin-fungizone (antibiotic-antimycotic) stock solution (Gibco Life Technologies, Grand Island, NY, Catalog Number 15240-062).
5. Sterile phosphate-buffered saline (PBS), pH 7.15.
6. One pack of sterile 10-cm-diameter tissue culture Petri-dishes (Becton Dickinson Labware, Franklin Lakes, NJ, Catalog number Falcon 3003).

7. Pasteur pipets (Fisher Scientific, Springfield, NJ, Catalog number 13-678-20C) drawn to a fine caliber.
8. Binocular dissecting microscope.
9. Vacuum line and fluid trap.

2.2. Isolation and Explant of Retinal Microvessel Fragments

1. 20 mL of 0.1%Type II Collagenase (Worthington Biochemical Corp., Freehold, NJ, Catalog number CLS 2; 250-300 units/mg) in Dulbecco's Modified Eagle's Medium (DMEM), low glucose formulation supplemented with 2 m*M* glutamine (Gibco Life Technologies, Grand Island, NY, Catalog number 12320-024), sterilized by filtration through a 0.22 µm millex GS filter (Millipore, Bedford, MA, Catalog number SLGS 025 OS).
2. Sterile 20 mL syringes (Becton Dickenson and Co., Franklin Lakes, NJ; Catalog number 309661).
3. Four 30-µm and 100-µm Nitex nylon meshes (SEFAR AMERICA, Depew, NY, Catalog numbers 3-30/21 and 3-100/47) cut into circles and fitted into 25-mm-diameter Swinnex membrane filter holders (Millipore, Bedford, MA, Catalogue number SX00 025 00) and autoclaved in paper autoclave bags.
4. A 37°C oven with a rotary shaking table.
5. A 37°C tissue culture incubator with a humidified atmosphere of 5% CO_2 in air.
6. Growth Medium: 10% heat inactivated bovine serum (Hyclone Laboratories Inc., Logan, UT, Catalog number SH30072.03) in DMEM supplemented with 2 m*M* glutamine and 1% PSF.

2.3. Weeding of Primary Cultures

1. Inverted optics phase-contrast microscope.
2. Sterile Pasteur pipets, PBS, and growth medium.
3. Fine tip marker pen.

2.4. Propagation of Pericyte Cultures

1. Trypsin-EDTA stock solution (Gibco Life Technologies, Grand Island, New York, Catalog number 25300-054)

2.5. Identification of Pericytes in Cultures

1. Micro coverglasses, round, number 1 (0.13–0.17 mm thickness), 18-mm diameter (VWR Scientific, Boston, MA, Catalog number 48380-046).
2. Twelve-well tissue culture cluster dish (Costar, Cambridge, MA; Catalog number 3512).
3. Paraformaldehyde fixative: To a small side-arm flask in the fume hood add 4.0 g of paraformaldehyde to 50 mL of distilled water and six drops of 6 *M* KOH while stirring and gently heating on a combined magnetic stirrer/hot plate until dissolved. Dissolve 1 g of DMEM powder and 0.76 g of sodium bicarbonate in 50 mL of distilled water and correct pH to 7.4. Combine the paraformaldehyde and DMEM solutions in a 1/1 ratio and correct pH if necessary.

4. Lysis buffer for permeabilization of cells: See accompanying table

Lysis Buffer *(27)*

Component	Stock solution	To make 20 mL
0.1% Triton X-100	100%	20 mL
50 m*M* HEPES, pH 6.9	0.25 *M*	4.0 mL
50 m*M* PIPES, pH7.1	0.25 *M*	4.0 mL
1 m*M* $MgCl_2$	1.0 *M*	20 mL
0.1 m*M* EGTA	0.1 *M*	20 mL
75 m*M* KCl	4.0 *M*	375 mL
H_2O	100%	11.6 mL

5. Precleaned glass microscope slides (Catalog number 12-550-11, Fisher Scientific, Pittsburgh, PA).
6. Nail Polish (any colour).
7. Cell Type Markers:
   a. Monoclonal antibody 3G5 (pericyte marker) is not available commercially. The hybridoma cell line is available from the American Type Culture Collection, Rockville, MD Catalog number ATCC CRL 1814.
   b. Antismooth muscle actin antibodies are available from Sigma, St. Louis, MO, (Catalog number A2547) and from Biomedical Technologies Inc., Stoughton, MA (Catalog number BT 561).
   c. Anti-α-actin antibody is available from Sigma-Aldrich, St. Louis, MO (Catalog number A5441).
   d. Fluoresceinated phalloidin is available from Molecular Probes Inc., Eugene, OR (Catalog number F432) and from Sigma-Aldrich, St. Louis, MO (Catalog number P5282).
   e. Anti-glial fibrillary acidic protein antibodies are available from Biomedical Technologies Inc., Stoughton, MA (Catalog number BT 575).
   f. Anti-cytokeratin 18 antibodies are available from Sigma-Aldrich, St. Louis, MO (Catalog number C 1399).
   g. Antibodies to Factor VIII related antigen (von Willebrand Factor) are available from Sigma-Aldrich, St. Louis, MO (Catalog number F 3520).
   h. DiI-acetylated LDL is available from Biomedical Technologies Inc., Stoughton, MA (Catalog number BT).

2.6. Cryopreservation of Pericytes

1. Nunc cryotube vials (VWR Scientific, Boston, MA; Nunc number 368632, VWR Catalog number 66021-987).
2. Dimethyl sulphoxide (Sigma-Aldrich, St. Louis, MO; Catalog number D5879).
3. Prepare sterile freezing solution by adding 10 mL of DMSO to 30 mL of DMEM followed by addition of 10 mL of bovine serum in a sterile 50 mL tube.
4. 0.4% Trypan blue solution (Sigma-Aldrich, St. Louis, MO; Catalog number T8154).

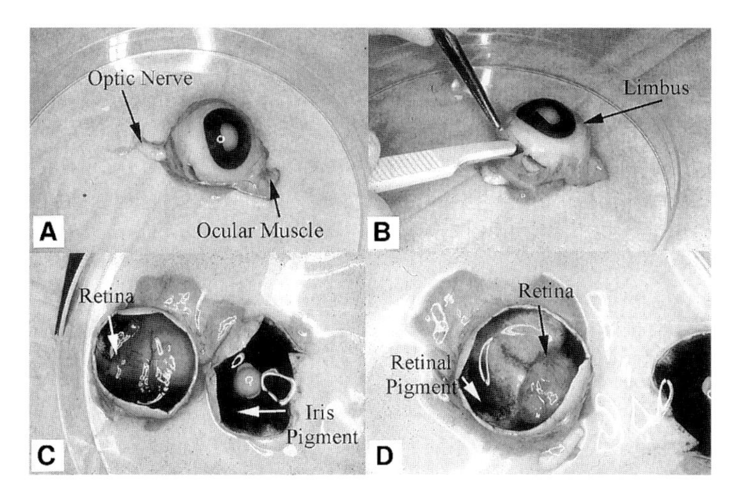

Fig. 1. Dissection of bovine eye to harvest retinae. **(A)** Bovine eye showing optic nerve and partially trimmed ocular muscles; **(B)** circumferential incision posterior to the limbus; **(C)** eye cup with vitreous removed showing the retina. The retina has been gently moved away from the underlying pigment epithelium on the left side. **(D)** The retina has been totally separated from the underlying retinal pigment epithelium but remains attached via the optic nerve head.

3. Methods

3.1. Dissection and Harvesting of Retinal Tissue

1. All procedures are performed in a sterile hood (Biocontainment safety cabinet (BL2) Biohazard Level 2). Twenty intact bovine eyes are obtained from a local slaughterhouse and processed within 4–6 h of death. The eyes are immersed in 300 mL of 10% betadine (povidone-iodine solution) and 2% penicillin-streptomycin-fungizone (PSF) in PBS at room temperature for 15 minutes in a large sterile beaker.
2. Also prepare two large sterile beakers containing 2% PSF in PBS and transfer the eyes from the betadine soak and rinse sequentially in the PSF baths.
3. Dissect the eyes on a sterile field one eye at a time (**Fig. 1**). Trim off most of the external ocular muscle tissue with sterile straight scissors, leaving enough tissue to grip with a toothed forceps. Grip the eye by the ocular muscle or optic nerve with a toothed forceps and make an incision 2–3 mm behind the limbus (the limbus is the point on the eye where the radius of curvature changes markedly and is found approximately at the iris). Extend the incision circumferentially with a curved scissor to open the globe. Pull the front of the eye forward thus pulling the vitreous humor out of the eye cup (*see* **Note 1**).
4. Gently dislodge the retina (*see* **Note 2**) with a sterile rubber policeman and detach from the optic nerve head with sterile fine scissors. Place the retina in a

sterile 10-cm Petri dish containing 2% PSF in PBS prewarmed to 37°C (2 mL/ retina).

5. Repeat the eyeball dissection and retina excision for all the eyes and place the retinae together in PSF/PBS buffer.
6. Examine retinal tissue for contamination with retinal pigment epithelium (small fragments of black tissue) with the aid of a dissecting microscope placed in the hood. Clean dissected retinae of pigmented tissue with a fine scissors and forceps. Remove small free-floating fragments of pigmented tissue with a sterile Pasteur pipet already drawn to a fine caliber in a bunsen burner flame, sterilized by immersion in 70% ethanol for 30 min, and attached to a vacuum source with an inline fluid trap. Excize any large vessels that are observed with fine scissors and forceps.

3.2. Isolation and Explant of Retinal Microvessel Fragments

1. Transfer the cleaned retinae to another 10-cm Petridish containing PSF/PBS, rinse, and then transfer to a dry sterile Petri dish (5 retinae/dish) and mince into a homogeneous slurry by cross-cutting the tissue using two sterile scalpels. The slurry from each batch of 5 retinae is transferred to a sterile 15-mL conical tube by adding 5 mL PSF/PBS and transferring the suspension with a sterile 10-mL pipet.
2. Pellet the retinal slurry by low-speed centrifugation (540g_{max}, 3,000 rpm @ r_{max} = 160 mm) for 5 min in a tabletop centrifuge. Aspirate off the buffer and estimate the pelleted tissue volume. Add 5 vol of 0.1% collagenase (Worthington Type II in DMEM) per vol tissue and incubate in a 37°C incubator for 1 hr with constant agitation on a horizontally rotating table shaker at 250 rpm.
3. Centrifuge the digested retinal slurry at 540g_{max} (3,000 rpm @ r_{max} = 160 mm) for 5 min and remove and discard the supernatant. Resuspend the digest in growth medium.
4. Take up the digested slurry with a 10-mL sterile disposable pipet and expel against the wall of the centrifuge tube ten times to break up large aggregates.
5. Take up the triturated digest from 5 retinae into a sterile disposable 20-mL syringe and pass through a sterile 100 μm Nitex mesh in a sterile Swinnex membrane holder (*see* **Note 3**).
6. Pass the 100-μm filtrate over a 30-μm Nitex mesh and collect the 30-μm filtrate in a sterile centrifuge tube.
7. Scrape the material retained on the meshes separately into growth medium and plate into two 10-cm Petri dishes per fraction per 5 retinae. Also plate the 30-μm filtrate into two 10-cm Petri dishes per 5 retinae of starting tissue. Place the petri dishes in a 37°C incubator with a humidified atmosphere of 5% CO_2 in air.
8. After 24 h remove the medium and floating debris by aspiration, wash the attached cells with PBS, and add fresh growth medium to the plates. Add 7 mL of PBS to each Petri dish and aspirate off, leaving 1.5–2.0 mL on the dish. Replace the lid of the dish and lift the dish and bang down sharply onto the counter top three times to release tightly attached debris which is then aspirated off. This step

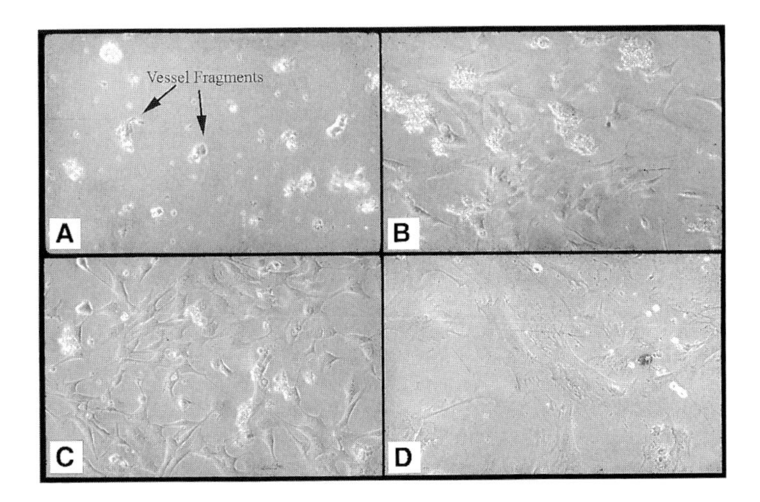

Fig. 2. Explant culture of isolated retinal microvessels. **(A)** Appearance of vessel fragments immediately after plating; **(B,C)** appearance of cells crawling out of the vessel fragments after several days in culture; and **(D)** pericytes exhibiting typical morphology, i.e., large flattened cells with multiple processes.

is repeated twice more followed by addition of fresh growth medium to the dishes, which are then returned to the incubator.

Immediately after plating, microvessel fragments can be observed by phase contrast microscopy (**Fig. 2A**) and after one to several days cells can be seen to have migrated out of the vessel fragments (**Fig. 2B,C,D**) and attached and spread on the bottom of the Petri dish.

3.3. Weeding of Primary Cultures (see Notes 4 *and* 13*)*

1. Remove nonpericyte cell types from primary cultures by "weeding." Nonpericyte cells are identified by morphology (*see* **Subheading 3.5.1.**). Perform weeding while the cells are still sparse.
2. Identify areas of the Petri dish that contain cells that do not exhibit pericyte morphology by microscopy, ring with a marker pen on the bottom of the Petri dish, and transfer to a sterile hood.
3. Remove and replace the medium with 2 mL of PBS and scrape the ringed areas with a sterile Pasteur pipet or hypodermic needle. Follow by washing with PBS. Check the scraped areas to ensure removal of nonpericyte cells by phase contrast microscopy and rescrape if necessary. When the weeding is completed add growth medium and return the plates to the incubator.
4. Continue to monitor the morphology of the cells, if cells of nonpericyte morphology are seen repeat the weeding process.

3.4. Propagation of Pericyte Cultures

When the primary pericyte cultures are near confluence, they may be further propagated by trypsinization and passage to additional plates (*see* **Note 5**).

1. Wash near-confluent primary pericyte cultures with PBS, then add 2 mL of trypsin/EDTA solution per 10-cm Petri dish.
2. The pericytes will detach from the dish in 5–10 min at room temperature and faster at 37°C. When the cells have detached (monitor by phase contrast microscopy) an equal volume of growth medium is added and the cell suspension is transferred to a sterile centrifuge tube.
3. Pellet the cells by centrifugation at $540g_{max}$ (3,000 rpm @ r_{max} = 160 mm) for 5 min.
4. Remove and discard the supernatant. Resuspend the pellet in growth medium for replating.
5. Replate cells at a 1/3 ratio (each Petri dish of primary culture is replated on three new Petri dishes or on a culture vessel with triple the surface area).
6. Remove the growth medium the following day and replace with fresh medium to remove dead cells.
7. When these cultures are near confluence, the process is repeated (usually 3–4 d later).

3.5. Identification of Pericytes (*see* Notes 6–12)

Pericytes may be identified by morphological criteria and also through analysis of antigen expression detected by immunofluorescence with specific antibodies (**Figs. 3–5**).

3.5.1. Cell Morphology

Pericytes in culture are large, well-spread cells with extremely irregular edges (**Fig. 2D**) and prominent stress fibers (intracellular actin bundles). At confluence, pericytes form multicellular nodules that may often be connected by strands of cells. These morphological characteristics distinguish them from ECs, which are polygonal and at confluence present a cobblestone-like appearance (**Fig. 3A**). The subconfluent morphology of ECs is shown in **Fig. 3E**. At sparse densities some ECs attempt to differentiate, which is seen morphologically as formation of processes that attempt to form a lumen (**Fig. 3F**). Smooth muscle cells generally have a compact spindle morphology at sparse densities and fibroblasts have an elongated spindle morphology, often with filopodial extensions. Retinal pigment epithelial cells (RPE) contain dark pigment granules, which can be used to identify them, a feature that can be rapidly lost in culture but is usually present in primary cultures from animals that are not albinos. RPE may be seen as individual cells with a circular perimeter and thick cables of intermediate filaments (cytokeratins) visible within the cell (**Fig. 3B,C,D**). In small colonies the cell margins are indistinct or invisible by

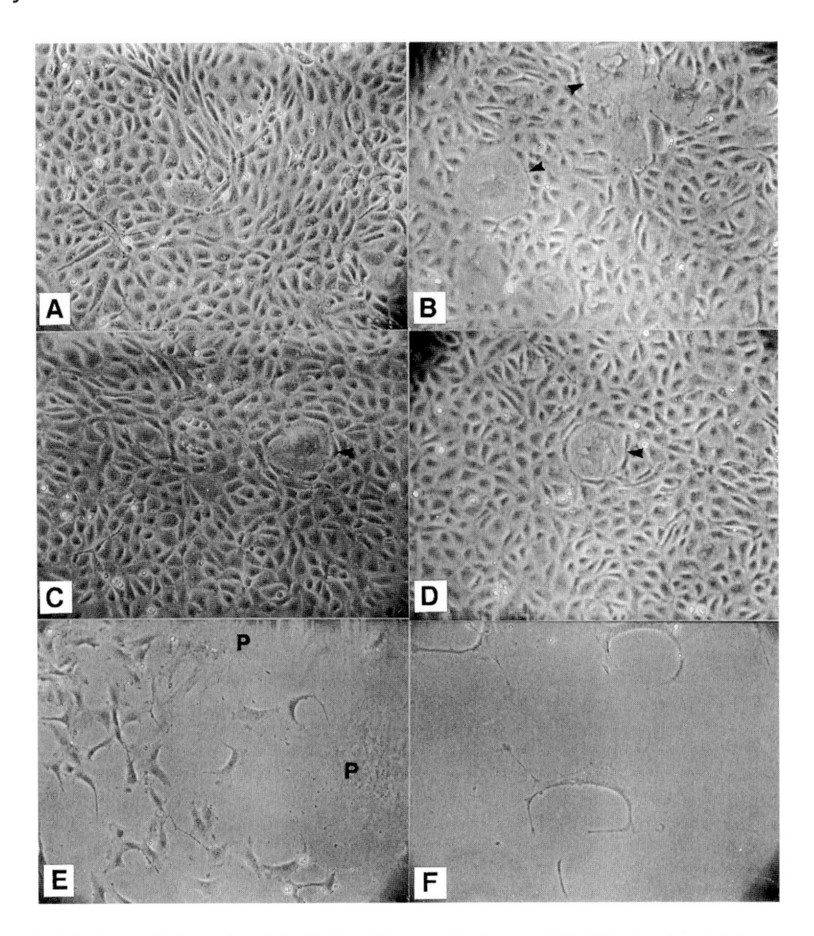

Fig. 3. Endothelial and epithelial cell morphology. **(A)** Typical cobblestone morphology of confluent bovine retinal microvascular endothelial cells; **(B,C,D)** retinal pigment epithelial cell contamination in confluent monolayers of ECs; and **(E,F)** subconfluent EC morphology. At sparse cell densities ECs may attempt to differentiate. This can be seen morphologically as an attempt to form a lumen (c-shaped cells). P, pericyte.

phase-contract microscopy but the nuclei are distinct. Large colonies may detach and be seen as floating rafts of RPE cells. Colonies that remain attached may undergo dome formation, which gives the appearance of blistering of the monolayer. Good photographic examples of the various morphologies of RPE cells in culture can also be seen in the literature ***(16)***. However, with careful avoidance of pigmented tissue at the retina harvesting stage and diligent "weeding" of the primary cell cultures we very rarely observe RPE contamination in the cell cultures.

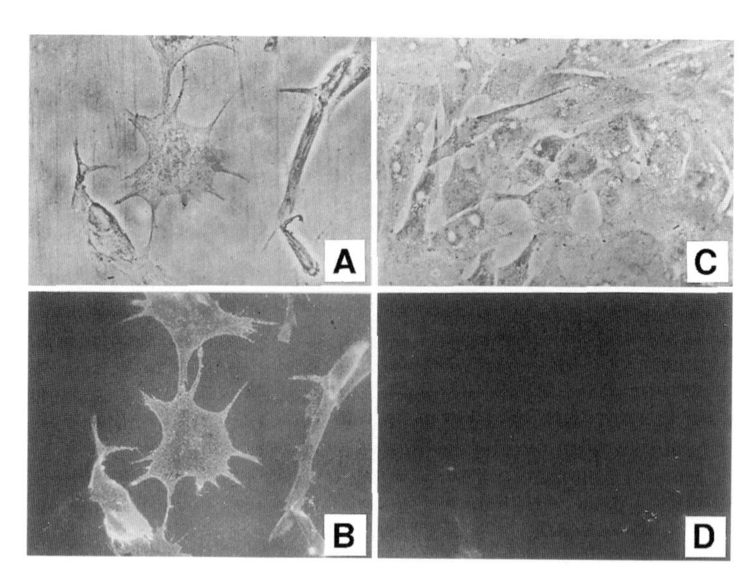

Fig. 4. Identification of pericytes by immunofluorescence with monoclonal antibody 3G5. **(A)** Phase-contrast photomicrograph of bovine retinal microvascular pericyte culture; **(B)** Same field as **A** showing 3G5 immunofluorescence; **(C)** phase-contrast photomicrograph of bovine microvascular ECs; and **(D)** same field as **C** showing lack of 3G5 immunofluorescence. (Reproduced from: Nayak R.C. et al, The J. Experiment. Med. (1988); 167, 1003–1015, *by copyright permission of the Rockefeller University Press*).

3.5.2. Immunophenotyping

In addition to morphological criteria, identification of pericytes and contaminating cells may be facilitated by immunological and biochemical markers ***(11,26–29)***. **Table 1** shows some markers that are commonly employed in the characterization of pericyte cultures.

For immunophenotyping studies cells must first be grown on multiple glass cover slips, followed by fixation and permeabilization as required. Antibodies to marker antigens are then bound and visualized by fluorescent secondary antibodies (indirect immunofluorescence). The marker reagents listed above are available commercially and should be used in accordance with the manufacturer's instructions. The following general protocol for immunofluorescence can be adapted for use with each reagent.

Place sterile cover slips (autoclaved) in the bottom of 12-well cluster plates and pipet 5,000 viable cells in growth medium into each well with a total medium volume of 1 mL. Exchange with fresh medium the next day. On the second day after plating, there should be sufficient cell numbers (check by phase-contrast microscopy) for immunostaining.

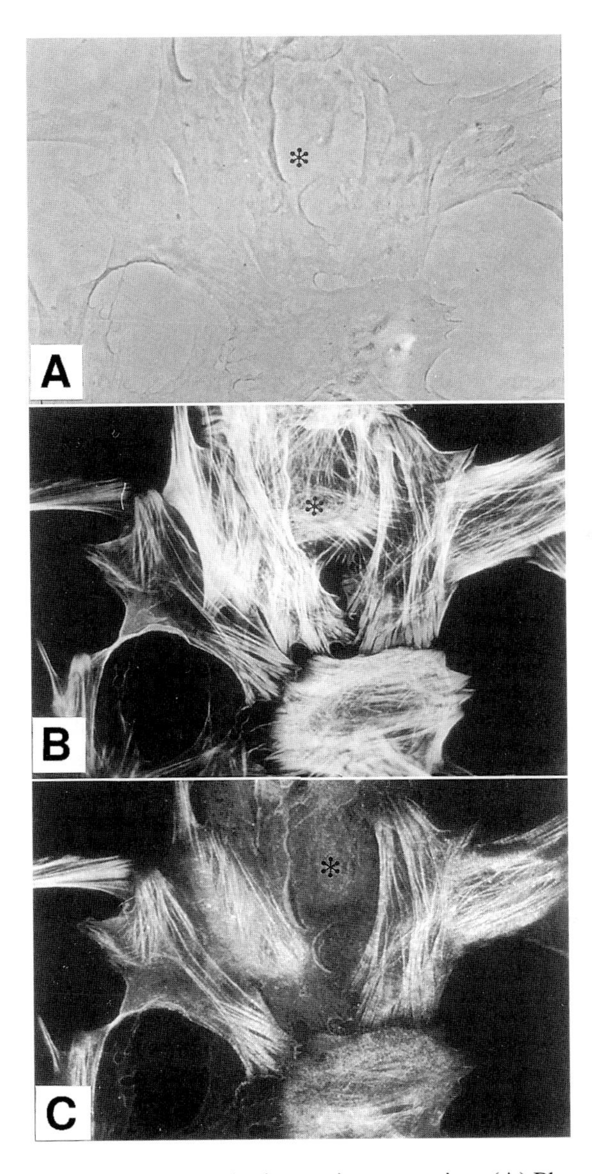

Fig. 5. Identification of pericytes by iso-actin expression. **(A)** Phase-contrast photomicrograph; **(B)** same field as in **A**. Stained for total F-actin (filamentous actin) with phalloidin; **(C)** same field as in **A** and **B** stained with antibody to vascular smooth muscle-actin; asterisk (*) identifies a cell which is not a pericyte (i.e., negative for vascular smooth muscle actin expression).

Table 1
Markers for Identification of Pericytes and Potential Cell Contaminants

	Pericytes	EC	SMC	Fibroblasts	RPEC	Astrocytes
Smooth muscle actin	+	–	+	–	–	–
Cortical nonmuscle actin	+	+	+↓	+	+	+
3G5	+	–	–	–	–	+
GFAP	–	–	–	–	–	+
Cytokeratin 18	–	–	–	–	+	–
FVIIIRA	–	+	–	–	–	–
Acetylated LDL uptake	–	+	–	–	–	–

EC, endothelial cells; SMC, smooth muscle cells; RPEC, retinal pigment epithelial cells; ↓, quantitatively decreased.

3.5.3. Fixation and Permeabilization ***(27)***

1. Wash cells once with DMEM (no serum) to remove serum components.
2. Add 1.0 mL of fixative to each well and incubate 5 minutes at room temperature.
3. Wash three times with 1.5 mL of PBS (containing 0.02% sodium azide), 5 min per wash; change the first wash immediately.
4. Add 1.0 mL of lysis buffer for cell permeabilization and incubate for 60–90 sec at room temperature.
5. Wash three times with 1.5 mL of PBS (containing 0.02% sodium azide), 5 min per wash; change the first wash immediately.

3.5.4. Immunostaining

1. Cover a small piece of perspex (plexiglass) with parafilm and drop 25 μL of diluted primary antibody per cover slip on it.
2. Place cover slips, cell side down, on the droplet of antibody solution.
3. Incubate in a humidified box for 45 min at room temperature.
4. Return the cover slips to the 12-well plate.
5. Wash three times with 1.5 mL of PBS (containing 0.02% sodium azide), 5 min per wash, change the first wash immediately.
6. Drop 25 μL of diluted secondary antibody per cover slip on a parafilm sheet, place cover slips, cell side down, on the droplet of antibody solution, and incubate in a humidified box for 45 min at room temperature.
7. Return the cover slips to the 12-well plate and wash three times with 1.5 mL of PBS (containing 0.02% sodium azide), 5 min per wash; change the first wash immediately.

3.5.5. Mounting

1. To labeled microscope slides add 1 drop of mounting fluid (9/1 glycerol–PBS/azide) with a pasteur pipet.
2. Rinse back of cover slips with water to remove salts and dry with a lint free paper tissue (e.g., Kimwipes) then mount cover slips cell side down on glycerol droplet.
3. Wick away excess glycerol with a Kimwipe and seal the cover slip edges onto the slides with nail polish.
4. When the nail polish has hardened the cells can be viewed with a fluorescence microscope.

3.5.6. Dil-acetylated LDL Staining **(28)**

Dil-acetylated LDL staining must be performed on viable cells, therefore the cells on cover slips should be fixed after the Dil-acetylated LDL staining procedure given below.

1. Dil-acetylated LDL at 10 μg/mL in serum supplemented culture medium is added to cells on cover slips (1.0 mL).
2. Incubate for 4 h at 37°C in a tissue culture incubator, then wash with PBS, and fix as described above.
3. Mount and view as described above.

3.6. Cryopreservation and Recovery of Pericytes

3.6.1. Cryopreservation of Pericytes

Pericytes can be cryopreserved in liquid nitrogen using the same methods that are in general use for other cell types. Pericytes can be recovered from liquid nitrogen at least 1–2 yr after freezing them.

1. Trypsinize and centrifuge pericyte cultures by the procedure in **Subheading 3.4.**
2. Resuspend pericytes in 10 mL of growth medium, remove 10 μL of the cell suspension and add it to 10 μL of trypan blue solution . After 5 min count the number of trypan blue-excluding (viable) cells in a hemocytometer. Adjust the pericyte suspension to 1×10^6 cells/mL and add 1mL/vial.
3. Add an equal volume of freezing medium and transfer the vials to a polystyrene foam tube rack and cover with a second rack (taping them together). Place the assembly in a –80°C freezer.
4. The next day transfer the frozen vials to a liquid nitrogen freezer.

3.6.2. Recovery of Cryopreserved Pericytes

1. Remove a vial of pericytes from the liquid nitrogen freezer and defrost rapidly in a 37°C water bath.
2. In the tissue culture hood, transfer the defrosted cell suspension to a sterile 10 mL tube and dilute slowly by dropwise addition of 9 mL of growth medium, shaking the tube gently after each drop.

3. Pellet the cells by centrifugation and aspirate off the supernatant. Resuspend the cells in growth medium and estimate the viable cell number by trypan blue exclusion as described in **step 2** of **Subheading 3.6.1.** Approx 250,000 viable cells are pipeted onto each sterile 10-cm Petri dish and the medium topped up to 8 mL total volume/plate.
4. The following day exchange with fresh medium.

4. Notes

1. The vitreous may have to be gently teased to squeeze it out of the eye-cup. Avoid disturbing the pigment epithelium beneath!
2. When removing the retina, start at an edge and gently stroke the retina in toward the optic nerve head.
3. The Swinnex filter holders can be attached to each other allowing sequential filtration over both meshes at once.
4. Do not leave out the "weeding" procedure, it is essential! The cultures will be overgrown with nonpericytes if they are not weeded.
5. The number of times the cells can be passaged depends on the structure/function that is to be studied and the conditions under which the pericytes are maintained ***(17)***. This must be determined empirically for the specific function that you wish to study.
6. The single best definable criterion for the identification of pericytes is their anatomical location within the microvascular basement membrane in vivo ***(18)***. Clearly such a definition is of no value in culture systems as the anatomical relationships are not preserved. We are therefore constrained in using biochemical and immunological means of identifying pericytes in vitro.
7. Iso-actin distribution (**Fig. 5**) is often used to distinguish pericytes from SMCs ***(29)***. Both pericytes and SMCs express smooth muscle and cortical (subplasmalemmal) nonmuscle actin. In SMCs, there is quantitatively less cortical nonmuscle actin than in pericytes. This may often be observed by fluorescent staining as a lack of cortical nonmuscle actin in smooth muscle cells ***(10)***. When staining total F-actin with phalloidin, extreme caution is necessary (phalloidin is very toxic). Phalloidin staining can be performed using the antibody staining procedure detailed above. Add 0.45 U of phalloidin stock (0.3 units/μL in methanol), i.e, 1.5 μL per cover slip; added to 25 μL of DMEM or PBS. The Sigma-Aldrich Chemical Company also sells monoclonal antibodies to α-vascular smooth muscle- and β-actin that are of different IgG isotypes and would be amenable to double staining using the appropriate fluorescently labeled secondary antibodies.
8. The 3G5 antibody stains pericytes (**Fig. 4**) well at a protein concentration of 5–10 μg/mL. It does react with retinal neurons, but this is not problematic because they do not survive in culture. Monoclonal antibody 3G5 also reacts with a subpopulation of GFAP positive neonatal rat brain astrocytes as well as a subpopulation of GFAP positive human astrocytoma cells (RCN, unpublished observations). The 3G5 monoclonal antibody will also react with a subpopulation of SMCs in subconfluent cultures but does not react with smooth muscle

cultures that have been maintained at confluence for several days (RCN, unpublished observations).

9. GFAP (glial fibrillary acidic protein) is the classic astrocyte marker although not expressed by all astrocytes. In vivo, nonastrocytic cells have been shown to express GFAP after tissue injury ***(19)***.
10. Cytokeratin 18 is a marker of simple epithelia (gastrointestinal, respiratory, urinary, liver, and glandular) but does not react with most stratified squamous epithelia (oesophageal, epidermal) ***(20)***. However, recent studies indicate that microvascular endothelium from the synovium of rheumatoid arthritis patients express cytokeratin 18 ***(21)*** while endothelia from other vascular beds do not ***(22)***.
11. Antifactor VIII-related antigen (von Willebrand's factor) immunohistochemistry and Di-I-acetylated LDL uptake are both extensively used as markers for vascular endothelial cells ***(23)***. Fluorescent staining with both reagents should show punctate fluorescence associated with intracellular vesicles, although there are organ and species specific exceptions ***(23)***. Care must be taken not to add too much reagent when performing acetylated LDL uptake studies as nonspecific uptake by pinocytosis will occur in both endothelial and nonendothelial cells.
12. Secondary antibodies are available from many commercial sources. When performing double indirect immunofluorescence studies it is crucial that the secondary reagents are absolutely specific for their respective primary antibodies. Any cross-reactions will invalidate the results. When the primary antibodies are from different species, the secondary antibodies should have been absorbed with serum from the other species. For example, if the primary antibodies are produced in rabbits and mice, then the antimouse secondary reagent should have been absorbed with rabbit serum and the antirabbit secondary reagent with mouse serum.
13. Shepro and colleagues have recently described a procedure for eliminating endothelial and epithelial cells from retinal pericyte cultures by selective killing with L-leucine-methyl ester ***(24)***. We have tried this technique twice and it appears to work as described. It is a much less tedious way of enriching pericytes than weeding, and may replace weeding in the standard protocol.

Acknowledgments

We thank Alice Lin, Michael Papetti, Albert Perdon and Gregory Sieczkiewicz for critical reading of the manuscript and insightful comments. We would also like to thank Patricia Griffiths for assistance with the preparation of the manuscript. Dr. Nayak is supported by NIH grants EY12607 and EY13054. Dr. Herman is supported by NIH grants GM55110 and EY09033.

References

1. Merimee, T. J. (1990) Diabetic Retinopathy: A Synthesis of Perspectives. *N. Engl. J. Med.* **322,** 978–983.
2. Folkman, J. (1995) Angiogenesis in cancer, vascular, rheumatoid and other disease. *Nature Med.* **1,** 27–31.

3. Saunders, K. B., Antonelli-Orlidge, A., Smith, S., and D'Amore, P. A. (1990) Cell communication and the control of endothelial cell growth: In: *Endothelial Cell Function in Diabetic Microangiopathy: Problems in Methodology and Clinical Aspects*. Frontiers in Diabetes, (Molinatti, G. M., Bar, R. S., Belfiore, F., and Porta, M., eds.) Basel Karger **9,**183–191.
4. Meezan, E., Klaus, B., and Carlson, E. C. (1974) Isolation of a purified preparation of metabolically active retinal blood vessels. *Nature* **251,** 65–67.
5. Klaus, B., Meezan, E., and Carlson, E. C. (1974) Isolated brain microvessels: A purified, metabolically active preparation from bovine cerebral cortex. *Science* **185,** 953–955.
6. Buzney, S. M., Frank, R. N., and Robison, W. G., Jr. (1975) Retinal capillaries: Proliferation of mural cells in vitro. *Science* **190,** 985–986.
7. Del Vecchio, P., Ryan, U. S., and Ryan, J. W. (1977) Isolation of capillary segments from rat adrenal gland. *J. Cell Biol.* **75,** 73a.
8. Folkman, J., Haudenschild, C. C., and Zetter, B. R. (1979) Long-term culture of capillary endothelial cells. *Proc. Natl. Acad. Sci. (USA)* **76,** 5217–5221.
9. Gitlin, J. D. and D'Amore, P. A. (1983) Culture of retinal capillary cells using selective growth media. *Microvasc. Res.* **26,** 74–80.
10. Herman, I. M. and D'Amore, P. A. (1985) Microvascular pericytes contain muscle and non-muscle actins. *J. Cell. Biol.* **101,** 43–52.
11. Nayak, R. C., Berman, A. B., George, K. L., Eisenbarth, G. S., and King, G. L. (1988) A monoclonal antibody (3G5) defined ganglioside antigen is expressed on the cell surface on microvascular pericytes. *J. Exp. Med.* **167,** 1003–1015.
12. Davies, P., Smith, B. T., Maddalo, F. B., Langleben, D., Tobias, D., Fujiwara, K., and Reid, L. (1987) Characteristics of lung pericytes in culture including their growth inhibition by endothelial substrate. *Microvasc. Res.* **33,** 300–314.
13. Carson, M. P. and Haudenschild, C. C. (1986) Microvascular endothelium and pericytes: High yield, low passage cultures. *In Vitro Cell Dev. Biol.* **22,** 344–354.
14. Challier, J. C., Kacemi, A., and Olive, G. (1995) Mixed culture of pericytes and endothelial cells from fetal microvessels of the human placenta. *Cell Mol. Biol.* **41,** 233–241.
15. Herman, I. M., Newcomb, P. M., Coughlin, J. E., and Jacobson, S. (1987) Characterization of microvascular cell cultures from normotensive and hypertensive rat brains: pericyte-endothelial cell interactions in vitro. *Tissue & Cell* **19,** 197–206.
16. Burke, J. M., Skumatz, C. M., Irving, P. E., and McKay, B. S. (1996) Phenotypic heterogeneity of retinal pigment epithelial cells in vitro and in situ. *Cur. Eye Res.* **62,** 63–73.
17. Newcomb, P. M. and Herman, I. M. (1993) Pericyte growth and contractile phenotype: Modulation by endothelial-synthesized matrix and comparison with aortic smooth muscle. *J. Cell Phys.* **155,** 385–393.
18. Shepro, D. and Morel, N. M. L. (1993) Pericyte physiology. *FASEB J.* **7,** 1031–1038.
19. Tassignon, M.-J., Stempels, N., Nguyen-Legros, J., De Wilde, F., and Brihaye, M. (1991) The effect of wavelength on glial fibrillary acidic protein immunoreac-

tivity in laser-induced lesions in rabbit retina. *Graefe's Arch. Clin. Exp. Ophthal.* **22,** 380–388.
20. Levy, R., Czernobilsky, B., and Geiger, B. (1988) Subtyping of epithelial cells of normal and metaplastic human uterine cervix, using polypeptide-specific cytokeratin antibodies. *Differentiation* **39,** 185–196.
21. Mattey, D. L., Nixon, N., Wynn-Jones, C., and Dawes, P. T. (1993) Demonstration of cytokeratin in endothelial cells of the synovial microvasculature in situ and in vitro. *Brit. J. Rheum.* **32,** 676–682.
22. Anderson, D. R. and Davis, E. B. (1996) Glaucoma, Capillaries and Pericytes: II. Identification and characterization of retinal pericytes in culture. *Ophthalmologia* **210,** 263–268.
23. Gorfien, S., Spector, A., DeLuca, D., and Weiss, S. (1993) Growth and physiological functions of vascular endothelial cells in a new serum-free medium (SFM). *Exp. Cell Res.* **206,** 291–301.
24. Lee, C. S., Patton, W. F., Chung-Welch, N., Chiang, E. T., Spofford, K. H., and Shepro, D. (1998) Selective propagation of retinal pericytes in mixed microvascular cell cultures using L-leucine-methyl ester. *Biotechniques* **25,** 482–494.
25. Helmbold, P., Nayak, R. C., Holzhausen, H. J., Marsch, W. C., and Herman, I. M. Isolation of human dermal microvascular pericytes and identification in culture. *Microvasc. Res.* (in press).
26. Helmbold, P., Wohlrab, J., Marsch W. C., and Nayak, R. C. Human dermal pericytes express 395 ganglioside: a new approach for microvessel histology. *J. Curtaneous Pathol.* (in press).
27. Herman, I. M. (1988) Developing probes and methods for morphological and biochemical analyses of cytoskeletal elements in vascular cells. *CRC Crit. Rev. Anatom. Sci.* **1,** 133–148.
28. Voyta, J. C., Via, D. P., Butterfield, C. E., Zettner, B. R. (1984) Identification and isolation of endothelial cells based on their increased uptake of acetylated-low density lipoprotein. *J. Cell Biol.* **99,** 2034–2040.
29. Herman, I. M. and D'Amore, P. A. (1985) Microvascular pericytes contain muscle and non-muscle actions. *J. Cell Biol.* **101,** 43–52.

Index

A

Actin, smooth muscle, 239, 242, 250, 257, 258
Adenovirus, transduction by, 227, 230
ADEPT, 19
Adhesion, 7, 38
Adipose tissue, microvessel endothelial cells from, 213
AGM1470, 10, 15, 16, 18
Alginate microbead release system, 53
Alkaline phosphatase, 30, 32
Angiogenesis, protocols
- 3D macromolecular matrices, 163
- alginate microbead, 53
- aortic ring assay, 185
- chick chorioallantoic membrane assay, 107
- collagen gel assays, 145
- corneal assay, 131
- disc assay, 59
- dorsal skinfold chamber preparation, 95
- hollow fiber assay, 87
- matrigel in vitro assay, 47
- matrigel in vivo, 205
- microscopic assessment of tumor angiogenesis, 29
- sponge implant assay, 77

Angiostatin, 3, 10, 11, 12, 13
Angiotensin-converting-enzyme (ACE), 8
Antibody-directed enzyme prodrug therapy, 19
Aortic ring assay, 185
Arthritis, 4
Autoradiography, 62

B

Basement membrane, 7, 67, 145, 205
Batimastat, 18

C

Calcium phosphate, transfection by, 229
Capillary, 8, 30, 68, 205
Carboxy-aminoimidazole (CAI), 18
Cartilage-derived inhibitor, 10, 11, 14
CD31 (PECAM-1), 30, 31, 34, 134, 214, 223, 224
CD34, 30
Cell
- proliferation assay, 60
- survival assay, 87

Chalkley counting, 31, 34, 35, 63, 156, 176
Chamber preparation, 95
Chemokinesis, 163
Chemotaxis, 163
Chemotherapeutic agents, 6, 13, 15, 20
Chorioallantoic membrane (CAM) assay, 107
CM101, 11, 15, 18, 20
Collagen
- gel, preparation of, 149, 150, 166, 187, 189
- native type I, preparation of, 148, 150, 165
- synthesis, assay of, 110, 113
- synthesis, inhibition of, 17
- type I, 147
- type IV, 13, 30, 123
- type VIII, 123
- type XVIII, 12, 13

Collagenase, 14, 16, 17
Corneal assay, 12
Cryptic peptides, 12, 146
Cyclooxygenase inhibitors, 68

D

DiI-acLDL, 250, 258, 259, 261
Disc assay, 59
DNA synthesis assay, 111, 116
Dorsal skinfold chamber preparation, 95

From: *Methods in Molecular Medicine, vol. 46, Angiogenesis Protocols*
Edited by: J. C. Murray © Humana Press Inc., Totowa, NJ

E

E-9, 36
ECGS, 48, 49, 50, 206
Eectroporation, 230, 232
Elvax, 60, 132, 136, 140
EN7/44, 36
Endoglin, 36
Endosialin, 36
Endostatin, 3, 10, 11, 12, 13
Endothelial cells, large vessel, 213
Endothelial cells, microvessel markers, 5, 29, 214
  characterization , 218, 219
  preparation of adipose, 213
  preparation of lung , 222
  preparation of stomach , 223
Englebreth-Holm-Swarm (EHS) tumor, 47, 205
Epidermal growth factor (EGF), 20, 60, 66, 67, 69
E-selectin (ELAM-1), 11, 20, 214, 220
ExGen500, 228, 231
Extracellular matrix, 7, 37

F

Factor VIII-related antigen (von Willebrand), 30, 47, 188, 202, 220, 258, 261

F

FGF receptor , 8, 19
Fibrin gel, preparation of, 187, 190
Fibroblast growth factor (FGF), 5, 8, 15, 19, 20, 47, 49, 50, 60, 62, 68, 69, 70, 123, 136, 153
Fibroblasts,
  in sponge assays, 66, 73, 84
  maintenance and growth, 166, 169
Fibronectin, 11, 12, 13
FLK-1 (VEGFR-2), 5, 15, 19, 214, 222
FLT-1 (VEGFR-1), 5, 214, 222
FLT4 (VEGFR-3), 214, 222
Fumagillin (AGM1470, TNP470), 10, 15, 16

G

Gene therapy, 19, 20, 233
Growth factors, 5, 10, 53, 54, 55, 56

H

Heparin, 11, 17, 67
Heparinases, 14, 16
Hepatocyte growth factor (HGF)5, 13, 20, 205
Hollow fiber assay, 87
Hot-spots, vascular, 5, 29, 30, 31, 33, 34, 36
HuMMEC, 218, 229
HUVEC27, 206, 208, 209, 228, 229
Hyaluronic acid, 16
Hyperthermia, 6, 68

I

IL-1, 17, 220
Il-12, 11, 16, 18
Image analysis, 36, 64, 91, 94, 197
Implant, sponge assay, 77
Inhibitors, angiogenesis, 3
Integrins, 7, 11, 15, 38
Interferon (IFN), 8, 11
Interleukins, 6, 11, 16, 17, 18, 220

K

Kaposi's sarcoma, 4, 5, 18
KDR(VEGFR-2), 5, 15, 37
Ketorolac, 69, 70

L

Laminin, 13, 16, 205, 206
Lectins, 30
Lung, microvessel endothelial cells from, 222

M

Magnetic beads, for isolating endothelial cells, 214
Mags (microscopic angiogenesis grading system), 29
Marimastat, 18
Matrigel, 47, 205
Matrigel assay, in vitro, 47-
Matrigel assay, in vivo, 205
Matrix, macromolecular, 7, 37, 146, 185
Metalloprotease, matrix (MMP), 11, 13, 18, 38, 186
Microvascular density (MVD), 20, 29, 31, 35, 37, 38

Migration assay, 47, 167, 172

P

p53, 10
Penicillamine, 11, 17
Pericytes, retinal
  characterization, 249, 254, 258
  cryopreservation 250, 259
  preparation, 248, 249, 251, 252
  weeding, 249, 253, 261
Plasminogen, 12
Plasminogen activator (PA), 17, 38
Plasminogen activator inhibitor (PAI) 15, 38
Platelet factor 4, 10, 11, 13, 18
Platelet-derived growth factor (PDGF), 5, 17
Pleiotrophin, 19
Point counting, 31, 34, 35, 63, 156, 176
Proliferation assay, tritiated thymidine, 60
Promoters, in endothelial gene transfer, 19, 233
Prostaglandins, 67, 69, 70
Protamine , 10, 17
Protease, 7, 10, 12, 14, 38
Psoriasis, 4, 9

R

Radiation, X-, 6, 68
Retinoblastoma tumor-suppressor gene, 12
Retinoids, 11, 17
Retinopathy, 7, 14
RGD peptides, 11, 13, 16, 186

S

Sandwich assay, three layer, 165, 171
Sandwich assay, two layer, 165, 170
Smooth muscle cells,
  characterization, 5, 237, 242
  cryopreservation, 243
  preparation by enzymatic dissociation, 239, 240
  preparation by explant culture, 241
Sparc, 13
Sponge
  assays, 59, 77
  materials, 60, 78
Sprouting, 145, 153
Steroids, 10, 11, 17, 68, 69
Stomach, microvessel endothelial cells from, 223
Suramin, 11, 14, 17

T

Taxanes, 11, 13, 15
Thalidomide, 18
Thrombospondin, 10, 14
Thymidine phosphorylase, 38
Thymidine, tritiated, 7, 60, 61, 111
Tie, receptors, 214, 222
Tie-2
  tissue inhibitor of matrix metalloproteinase (TIMP), 14
TNP470, 18
Transduction, adenoviral, 227, 230
Transfection,
  methods , 227
  efficiency, estimation of , 228, 232
Transforming growth factor (TGF), 11, 20, 71, 72, 158, 177
Tube formation, 13, 47
Tumor necrosis factor (TNF), 20, 70, 220

U, V, W

Ulex europaeus, 214, 223, 224
Vascular endothelial growth factor (VEGF), 5, 11, 12, 14, 15, 18, 20
VEGFR-1 (FLT-1), 5, 214, 222
VEGFR-2 (KDR/FLK-1), 5, 15, 19, 37, 214, 222
Vimentin, 30
Vitelline membrane, 108
Von Hippel-Lindau disease, 12
Washout, xenon, 77, 78, 81